CULTURE OF IMMORTALIZED CELLS

Culture of Specialized Cells

Series Editor

R. Ian Freshney

CULTURE OF EPITHELIAL CELLS
R. Ian Freshney, Editor

CULTURE OF HEMATOPOIETIC CELLS
R. Ian Freshney, Ian B. Pragnell and Mary G. Freshney, Editors

CULTURE OF IMMORTALIZED CELLS
R. Ian Freshney and Mary G. Freshney, Editors

CULTURE OF IMMORTALIZED CELLS

Editors

R. Ian Freshney
CRC Department of Medical Oncology
University of Glasgow
Glasgow, Scotland

Mary G. Freshney
Beatson Institute for
Cancer Research
Bearsden, Glasgow, Scotland

WILEY-LISS
A JOHN WILEY & SONS, INC., PUBLICATION
New York • Chichester • Brisbane • Toronto • Singapore

Cover Illustration

Rat O-2A optic nerve glial progenitor cells, immortalized by retroviral infection with the proto-oncogene c-*myc* (see Chapter 16). Stained by indirect immunofluorescence for the panneurectodermal marker A2B5 after differentiation induction. *Photograph by courtesy of Dr. Susan C. Barnett.*

Address All Inquiries to the Publisher
Wiley-Liss, Inc., 605 Third Avenue, New York, NY 10158-0012

Printed in the United States of America

The text of this book is printed on acid-free paper.

While the authors, editors, and publishers believe that drug selection and dosage and the specification and usage of equipment and devices, as set forth in this book, are in accord with current recommendations and practice at the time of publication, they accept no legal responsibility for any errors or omissions, and make no warranty, express or implied, with respect to material contained herein. In view of ongoing research, equipment modifications, changes in governmental regulations and the constant flow of information relating to drug therapy, drug reactions, and the use of equipment and devices, the reader is urged to review and evaluate the information provided in the package insert or instructions for each drug, piece of equipment, or device for, among other things, any changes in the instructions or indication of dosage or usage and for added warnings and precautions.

Library of Congress Cataloging in Publication Data

Culture of immortalized cells / editors, R. Ian Freshney, Mary G. Freshney.
p. cm. — (Culture of specialized cells)
Includes index.
ISBN 0-471-12134-7 (paper)
1. Continuous cell lines—Laboratory manuals. 2. Cell culture—Laboratory manuals.
3. Cell transformation—Laboratory manuals.
I. Freshney, R. Ian. II. Freshney, Mary G. III. Series.
[DNLM: 1. Cells, Cultured. 2. Cell Line, Transformed. QH 585.4 C968 1996]
QH585.45.C85 1996
574.87′072—dc20
DNLM/DLC
for Library of Congress 95-38556

10 9 8 7 6 5 4 3 2 1

Contents

Contributors

Adolphe, Monique, Laboratoire de Pharmacologie Cellulaire, Ecole Pratique des Hautes Etudes, 15 rue de l'Ecole de Médecine, Paris 75006, France

Barnett, Susan C., Departments of Neurology and Medical Oncology, University of Glasgow, Garscube Estate, Glasgow G61 1BD, Scotland

Benoit, Bénédicte, Laboratoire de Pharmacologie Cellulaire, Ecole Pratique des Hautes Etudes, 15 rue de l'Ecole de Médecine, Paris 75006, France

Bolton, Bryan J., European Collection of Animal Cell Cultures, Centre for Applied Microbiology and Research, Porton Down, Salisbury, SP4 OJG, England

Burke, John F., Biochemistry Group, School of Biological Sciences, University of Sussex, Falmer, Brighton, BN1 9QJ, England

Caputo, Jane L., American Type Culture Collection, 12301 Rockville Pike, Rockville, Maryland 20852

Clarke, J. B., Cell Resources Department, CAMR, Porton Down, Salisbury, Wiltshire, SP4 0JG, England

Cuthbert, Andrew P., Human Cancer Genetics Unit, Department of Biology and Biochemistry, Brunel University, Uxbridge UB8 3PH, England

De Silva, Ruwani, Children's Medical Research Institute, 214 Hawkesbury Road, Westmead, Sydney 2145, Australia

Duncan, Emma L., Children's Medical Research Institute, 214 Hawkesbury Road, Westmead, Sydney 2145, Australia

Harris, Curtis C., NCI, NIH, Bethesda, Maryland 20892

Macé, K., Nestlé Research Centre, P.O. Box 44, CH-1000, Lausanne, 26, Switzerland

Mayne, Lynne V., Trafford Centre for Medical Research, University of Sussex, Falmer, Brighton BN1 9RY, England

Moorwood, K., Trafford Centre for Medical Research, University of Sussex, Falmer, Brighton BN1 9RY, England

Moy, Elsa L., Children's Medical Research Institute, 214 Hawkesbury Road, Westmead, Sydney 2145, Australia

Newbold, Robert F., Human Cancer Genetics Unit, Department of Biology and Biochemistry, Brunel University, Uxbridge UB8 3PH, England

Noble, Mark, Huntsman Cancer Institute, Room 720, Biopolymers Building, University of Utah Health Sciences Center, Salt Lake City, Utah 84132

Parkinson, E. Ken, CRC Beatson Laboratories, Garscube Estate, Switchback Road, Bearsden, Glasgow G61 1BD, Scotland

Pfeifer, Andrea M. A., Nestlé Research Centre, P.O. Box 44, CH-1000, Lausanne 26, Switzerland

Price, T. N. C. Trafford Centre for Medical Research, University of Sussex, Falmer, Brighton BN1 9RY, England

Punchard, Neville A., Department of Applied Sciences, University of Luton, Luton LU1 3JU, England

Ravid, Katya, Department of Biochemistry, Boston University School of Medicine, Boston, Massachusetts 02118

Reddel, Roger R., Children's Medical Research Institute, 214 Hawkesbury Road, Westmead, Sydney 2145, Australia

Shaw, Mick, Zeneca Pharmaceuticals plc, Alderley Park, Macclesfield, SK10 4TG, England

Spurr, Nigel K., Human Genetic Resources, Imperial Cancer Research Fund, Clare Hall Laboratories, Blanche Lane, South Mimms, Potters Bar, Herts., EN6 3LD, England

Steinberg, Mark L., Biochemistry Division, The City College of New York, Convent Ave. and 138th Street, New York, New York 10031

Thenet-Gauci, Sophie, Laboratoire de Pharmacologie Cellulaire, Ecole Pratique des Hautes Etudes, 15 rue de l'Ecole de Médecine, Paris 75006

Thompson, Richard, Gastrointestinal Laboratory, The Rayne Institute, St. Thomas' Hospital, London SE1 7EH, England

Watson, Duncan, Gastrointestinal Laboratory, The Rayne Institute, St. Thomas' Hospital, London SE1 7EH, England

Wynford-Thomas, David, Cancer Research Campaign Thyroid Tumour Biology Research Group, Department of Pathology, University of Wales College of Medicine, Cardiff CF4 4XN, Wales

Preface

Cell line transformation has been an element of cell culture behavior that has been known and studied almost since the beginning of tissue culture and certainly since primary cell cultures were first passaged into cell lines. The association of cell line transformation with malignancy was established very early, but it is only in recent years that a distinction has been made between immortalization, the acquisition of an infinite lifespan, and the transformation of the growth characteristics of a cell line usually associated with the development of the malignant phenotype. It has long been recognized that continuous cell lines represent a very valuable resource, but it has taken many hours of debate, not yet concluded, to accept that an infinite lifespan may be achieved without a marked increase in malignant potential. This is important in terms of the study of the biology of growth and differentiation and senescence as well as our understanding of the aberrations in proliferative and positional control in malignancy.

The development of the biotechnology industry and the requirement for eukaryotic cells in the posttranslational processing of protein products of cells has placed a great deal of pressure on both the scientific and regulatory communities to provide continuous cell lines from specific cell lineages that are safe to use and with phenotypic properties relevant to their tissue of origin. While these objectives are far from completely realized, a large number of different cell lines are now available from different tissues, and the techniques for establishing them are becoming more commonplace. Full expression of the characteristic cell phenotype is not achieved in many cases, but the systems are in place for this to develop in future years either by regulation of the culture environment or by genetic manipulation of the cells. This book attempts to provide the basic technology for creating continuous cell lines from many different types of tissue, and, by analogy, to provide guidelines for extrapolation to other tissues not yet described.

It will be clear from a cursory glance at the chapters in this book that the majority of laboratories have opted for the SV40 large-T antigen gene, or whole SV40 viral genome, as the major genetic initiator of immortalization. This has been achieved by transfection of the LT gene, either constitutively expressed or modified for expression regulated by temperature or another faculative condition. This is emerging as an important concept as it may be necessary for continuous propagation to have the immortalizing gene expressed, but its expression should be repressed when the expression of the normal differentiated phenotype is required.

Although the emphasis is on SV40LT (Simian virus 40 large-T antigen) in these chapters, other genes, or viruses, have been implicated in immortalization. Epstein–Barr virus (EBV) is classically associated with B-lymphocyte immortalization (see Chapter 13) and adenovirus E1a, and papilloma viruses have also been used with epithelial cells (see Chapters 1 and 5). However, it has not been the purpose of this book to analyze the genetic events of immortalization, other than in the first review chapter, but rather to select those techniques in most common use, and to present them as established techniques.

The idea for this book arose from a meeting arranged under the auspices of the European Tissue Culture Society, U.K. Branch, organized by Jane Cole and Colin Arlett in the MRC Cellular Mutagenesis Unit at the University of Sussex, England, and it is to them that the editors owe a considerable debt of gratitude for assembling the core of (unsuspecting) authors that eventually led to this book. It is not, it must be said, a report of the meeting, as manuscripts were not presented at that time and the eventual publication arose from chapters written *de novo* for the purposes of this book, and not the meeting. However, the meeting provided an excellent opportunity to discuss the possibilities of this publication with a vital core of interested scientists and in an atmosphere capable of generating both positive and negative criticism of the concept of the book. I am grateful not only to those authors who did contribute but also to the views expressed by those who did not, as they helped to mold the ultimate style of the book.

This is the third book in the series Culture of Specialized Cells, and, like the others, is presented as a practical guide to established procedures, with enumerated protocols, lists of reagents, and an index of suppliers, so that the reader should be able to repeat these procedures without reference to the primary literature. As these techniques are at an advanced level, a basic knowledge of basic tissue culture and molecular biology procedures is assumed, but where further basic information is required, the reader is referred to standard texts like Freshney (1994), *Culture of Animal Cells,* Wiley-Liss, New York and Sambrook et al. (1989), *Molecular Cloning, a Laboratory Manual,* 2nd ed, Cold Spring Harbor Laboratory Press. Similarly, regular items of equipment normally found in a tissue culture laboratory, such as laminar flow biosafety cabinets, incubators, centrifuges, inverted microscopes, liquid nitrogen freezers, water baths, and a range of flasks and graduated pipettes, are not specified. Most reagents are specified before each protocol. Water is referred to as ultrapure water (UPW) on the assumption that many laboratories will prepare water in different ways but all will be of a very high purity, with at least four stages of processing, such as distillation (or reverse osmosis), deionization, carbon filtration, and micropore filtration, or an equivalent series of steps.

There is an extensive list of abbreviations and, rather than repeat these with each chapter, a comprehensive list is provided at the beginning of the book, following this preface. Suppliers are listed with each individual chapter, and their addresses listed at the end of the book. It is always difficult to keep these up to date, as companies merge, change their names, move, or go out of business with alarming frequency, but we have tried to provide the most up-to-date information possible.

Safety regulations are presented in individual sections in some chapters and reviewed in Chapter 2. It should be emphasized that these are suggestions provided by individual authors and should not be construed as legally binding regulations, as these will vary from country to country and even from one institution to another. It is essential that you consult your local safety officer and national guidelines to establish the legal position in your own laboratory. Chapter 2 has been prepared by Jane Caputo from the point of view of a typical U.S. laboratory, with an appendix by Nigel Spurr with relevance to a U.K. laboratory. Both are advisory only and do not replace local or national guidelines and regulations.

I am greatly indebted to my wife, Mary, for acting as coeditor of this volume. Her sound practical knowledge of cell culture and molecular biology techniques and her experience in the instruction of scientific and technical staff have proved of immense value.

R. Ian Freshney

List of Abbreviations

ACD	Acid-citrate dextran
ACDP	U.K. Advisory Committee on Dangerous Pathogens
ACE	Angiotensin-converting enzyme
ACGM	U.K. Advisory Committee on Genetic Manipulation
aFGF	Acidic fibroblast growth factor
AGMK	African green monkey kidney
AT	Adenine theophylline solution
ATCC	American Type Culture Collection
ATP	Adenosine triphosphate
B(a)P	Benzo(*a*)pyrene (3,4 benzpyrene)
BES	*N,N*-Bis(2-hydroxyethyl)-2-aminoethanesulfonic acid
bFGF	Basic fibroblast growth factor
BICR	Beatson Institute for Cancer Research
BM	Basal medium
BSA	Bovine serum albumin
BSS	Balanced salts solution
CATCH	Citrate adenine theophylline solution
CB	Cytochalasin B
Cdc	Cell division cycle (as in cdc genes or proteins)
CDK	Cell division kinase
Cfu	Colony-forming units
CJA	Creutzfeldt–Jacob agent
CNBr	Cyanogen bromide
CNS	Central nervous system
COSHH	Control of substances hazardous to health
CPE	Cytopathic effect
CYP450	Cytochrome P450
DAB	Diaminobenzidine
DABA	Diaminobutyric acid
DAPI	4,6-Diamidino-2-phenylindole
DHCB	Dihydrocytochalasin B
DMEM	Dulbecco's modification of Eagle's medium
DMEM-BS	Bottenstein and Sato's modification of DMEM
DMEM-PF	DMEM-BS with bFGF and PDGF
DMSO	Dimethylsulfoxide

DNAse	Deoxyribonuclease
EBV	Epstein–Barr virus
ECACC	European Collection of Animal Cell Cultures
ECGS	Endothelial cell growth supplement
ECM	Extracellular matrix
EDTA	Ethylenediamine tetracetic acid
EGF	Epidermal growth factor
EGTA	Ethylene glycol-bis(β-aminoethyl ether)*N,N,N′,N′*-tetraacetic acid
ELISA	Enzyme-linked immunosorbent assay
EPV	Epstein–Barr virus
F_1	First filial generation
FACS	Fluorescence-activated cell sorter
FBS	Fetal bovine serum (= fetal calf serum in this context)
FCS	Fetal calf serum (= fetal bovine serum in this context)
FGF	Fibroblast growth factor
FITC	Fluorescein isothiocyanate
G-11	Giemsa staining at pH11
GABA	γ-Aminobutyric acid
Gal-c	Galactocerebroside
β-GAL	β-Galactosidase gene
GFAP	Glial fibrillary acidic protein
GM-CSF	Granulocyte–macrophase colony-stimulating factor
GST	Glutathione S-transferase
HAT	Hypoxanthine aminopterin thymidine
HBS	HEPES-buffered saline
HBS	HEPES-buffered PBS (Macé et al.)
HBSS	Hanks' balanced salt solution
HBV	Hepatitis B virus
HDF	Human dermal fibroblast
HEPA	High-efficiency particulate air
HEPES	*N*-2-Hydroxyethylpiperazine-*N′*-2-ethanesulfonic acid
HGH	Human growth hormone
HGPRT	Hypoxanthine–guanosine phosphoribosyltransfersase
HIV	Human immunodeficiency virus
HMW	High molecular weight
HPV	Human papilloma virus
HSE	Health and Safety Executive (U.K.)
HTLV	Human T-cell leukemia virus
HUVEC	Human umbilical vein endothelial cells
Hyg	Hygromycin B
ICRF	Imperial Cancer Research Fund
IgE	Immunoglobulin E
IFN	Interferon

IL	Interleukin
IMDM	Iscove's modification of Dulbecco's medium
LAV	Lymphoadenopathy associated virus
LCM	Liver cell medium
LOH	Loss of heterozygosity
LT	Gene for "large T" antigen of SV40
LTAg	Large-T antigen of SV40
LTR	Long terminal repeat
M199Ir	Medium 199 with Earle's salts, supplemented with 10% FCS and 10% M1 serum-free concentrate
M199S	Medium 199 supplemented with 10% FCS and 10% NBCS
M199SF	Serum-free Medium 199
M199SG	Medium 199 supplemented with 10% FCS, 10% NBCS, and 15 μg/ml endothelial cell growth supplement (ECGS)
MCDB	Molecular, Cellular and Developmental Biology (Department of University of Colorado at Boulder)
ME	2-mercaptoethanol
MEM	Eagle's minimal essential medium
MMCT	Micronucleus monochromosome transfer
MMWR	Morbidity and Mortality Weekly Report
MOI	Multiplicity of infection
NBCS	Newborn calf serum
NIH	National Institutes of Health
NHBE	Normal human bronchial epithelium
NP40	Nonidet P40
NT-3	Neurotrophin-3
O-2A	Oligodendrocyte–type 2 astrocyte
OSHA	Occupational Safety and Health Administration
P	Passage
PBS	Dulbecco's phosphate buffered saline
PBSA	Dulbecco's phosphate buffered saline without Ca^{2+} and Mg^{2+}
PBSA+B	Dulbecco's phosphate buffered saline with Ca^{2+} and Ca^{2+}
PCR	Polymerase chain reaction
PD	Population doubling
PDGF	Platelet-derived growth factor
PDL	Poly-D-lysine
P/E	Phosphoethanolamine/ethanolamine
PEG	Ployethylene glycol
PET	PVP/EDTA/trypsin
PF4	Platelet Factor 4

PFU	Plaque-forming unit
PGI_2	Prostaglandin-I_2
PHA	Phytohemaglutinin
PLL	Poly-L-lysine
PVP	Polyvinylpyrolidone
RIA	Radio immunoassay
RPMI	Rosewell Park Memorial Institute
SBTI	Soyabean trypsin inhibitor
SCC	Squamous cell carcinoma
SCC-HN	Squamous cell carinoma of head and neck
SDI	Senescence-derived inhibitor
SDS-PAGE:	Sodium dodecyl sulfate–polyacrylamide gel electrophoresis
SFM	Serum-free medium
SHD	Syrian hamster dermal
SIV	Simian immunodeficiency virus
SOP	Standard operating procedure
SSM	Serum-supplemented medium
STF	Subchromosomal transferable fragment
STS	Sequence-tagged site
SV40	Simian virus 40
SV40 T	Simian virus 40 large-T antigen
T_3	Triiodothyronine
*ts*LT	Temperature-sensitive LT
TBE	Tris-buffered EDTA
TE:	Tris-EDTA
TGF	Transforming growth factor
THLE	Transformed human liver cells
TNF	Tumor necrosis factor
tPA	Tissue-type plasminogen activator
TRA	Telomeric repeat array
TSC	Trisodium citrate
UEA	*Ulex europaeus* aggluttinin
UPW	Ultrapure water
VCM	Virus-conditioned medium
vWF	Von Willebrand factor
W/V	Weight per volume
YAC	Yeast artificial chromosome

CULTURE OF IMMORTALIZED CELLS

I

Human Keratinocyte Immortalization: Genetic Basis and Role in Squamous Cell Carcinoma Development

E. K. PARKINSON

CRC Beatson Laboratories, Garscube Estate, Switchback Road, Bearsden, Glasgow G61 1BD, Scotland

1. INTRODUCTION

It has been known for some time that normal human somatic cells [Hayflick, 1965], including keratinocytes [Rheinwald and Green,

Culture of Immortalized Cells, Edited by R. Ian Freshney and Mary G. Freshney.
ISBN 0-471-12134-7

1975], exhibit a finite proliferative lifespan culminating in senescence. In contrast, most common carcinomas [Paraskeva et al., 1989; Edington et al., in press] and melanoma [Mancianti and Herlyn, 1989] contain variant cells that escape senescence and are immortal. Furthermore, using phenotypically stable rodent fibroblasts, it has been shown that cellular immortality is a prerequisite for oncogenic transformation by chemicals [Newbold et al., 1982] and oncogenes [Newbold and Overell, 1983].

The immortal phenotype is recessive to that of senescence [Bunn and Tarrant, 1980] and is usually a late event in human tumor progression [Paraskeva et al., 1989, Mancianti and Herlyn, 1989, Edington et al., in press] indicating that there may be several mechanisms of suppression of the immortal phenotype. Indeed, many senescence genes may be tumor suppressor genes [Sager, 1988], and p53 [Linzer and Levine, 1979; Lane and Crawford, 1979] and CDKN2 [Nobori et al., 1994; Kamb et al., 1994] are examples of these.

Since a high fraction of end-stage, clinically lethal, squamous cell carcinomas are immortal [Edington et al., in press], and end-stage ovarian cancers express markers of cellular immortality *in vivo* [Counter et al., 1994a; Kim et al., 1994], it is possible that immortality of the tumor population is important for clinical recurrence. Certainly cellular immortality would be expected to facilitate multiple clonal selections leading to tumor population diversity, which in turn is the basis for rapid tumor progression and therapeutic resistance.

Two of the events that are associated with cellular immortality, p53 alteration [Bischoff et al., 1990; Harvey et al., 1993], and telomerase activation [Morin, 1989; Counter et al., 1992, 1994a; Kim et al., 1994] would be good targets for chemotherapeutic drug design, since the reversal of the conformation of mutant p53 into a normal one [Milner, 1994] or the inhibition of the telomerase enzyme [Greider, 1990; Harley et al., 1990] would be predicted to be lethal to immortal end-stage cancer cells, but not to their normal counterparts.

Thus the need to understand human cellular immortality is increasingly important. In this review I will summarize some of the current ideas and how they fit in to the process of keratinocyte immortalization during the development of human squamous cell carcinoma (SCC).

2. SENESCENCE AND THE CELL CYCLE

When cultured human fibroblasts senesce, they do not undergo a population doubling for at least 4 weeks [Bunn and Tarrant, 1980], ^{3}H-thymidine incorporation is low [Cristofalo and Sharf, 1973] and the cells enlarge, giving a low nucleus:cytoplasm ratio [Hayflick,

1965]. Senescent fibroblasts are thought to be blocked at late G1 of the cell cycle since they contain the 2N complement of DNA [Hart and Setlow, 1974]. Senescent human keratinocytes also appear to accumulate in G1 as determined by FACS analysis [Wille et al., 1984]. Many studies have shown that virtually all early to mid-G1-associated genes are expressed in senescent fibroblasts but that the majority of late G1 genes are not (summarized in Fig. 1.1). Furthermore, phosphorylation of the Rb-1 protein, pRb, is absent in senescent fibroblasts [Stein et al., 1990], and, because this protein is phosphorylated by the cyclin-cdk kinase complexes, much recent attention has focused on their role in senescence.

Cyclins and cyclin-dependent kinases (Cdks) drive the cell cycle, possibly in response to growth factors acting at G1 (reviewed in Sherr [1993]) and the cell cycle is shown in Fig. 1.2. Cyclins D and E seem to act at G1/S and A and B cyclins at S and M phases. Cdks 2 and 4 are strong candidates for phosphorylating pRb and thus allowing exit

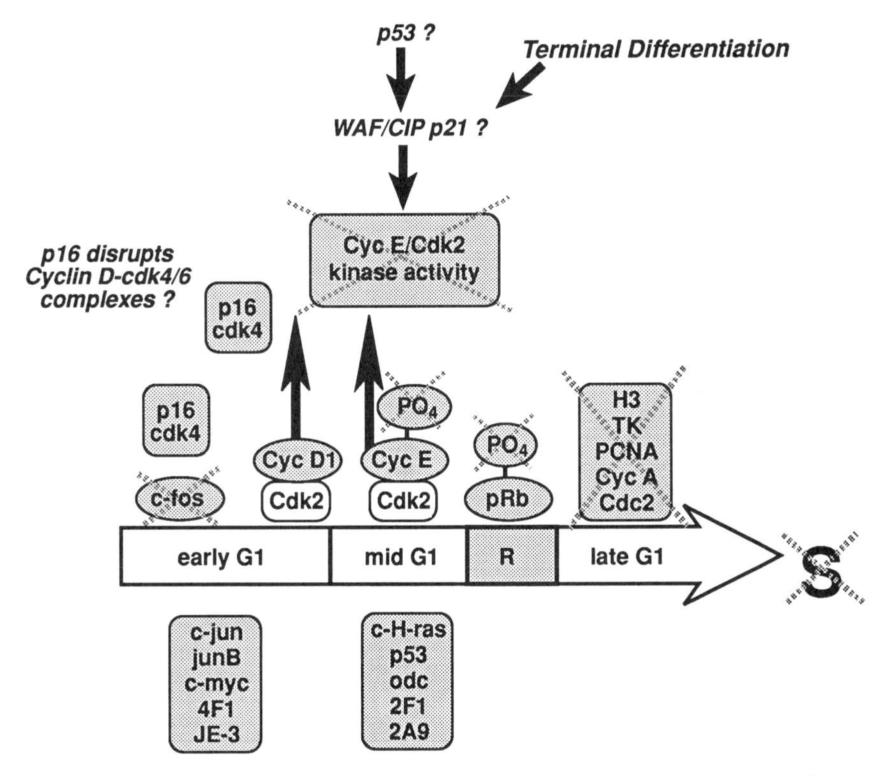

Fig. 1.1. A summary of serum-inducible events in senescent and early-passage quiescent cells. Senescent cells are deficient in the events crossed out with dashed lines and have an overabundance of cyclin D1 and E complexes as indicated by the arrows. Placement of pRb phosphorylation at R is not proven. (H3 = histone H3, TK = thymidine kinase, PCNA = proliferating cell nuclear antigen, odc = ornithine decarboxylase.) Adapted from Dulic et al. [(1993a,b)].

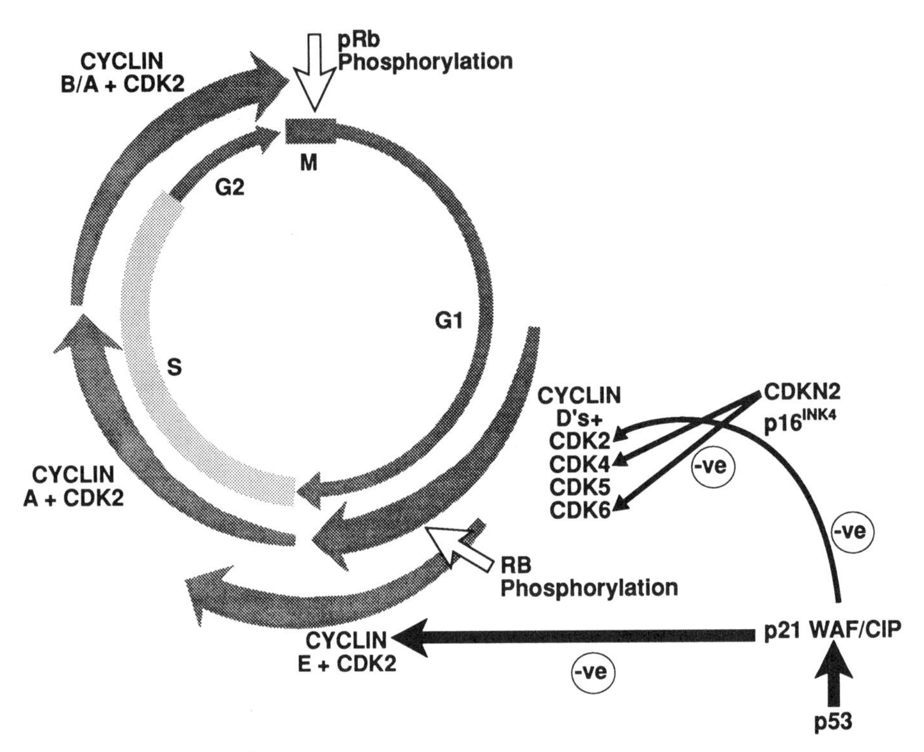

Fig. 1.2. The interaction of cyclins and cyclin-dependent kinases during the cell cycle. Adapted from Sherr [1993].

from G1, possibly targeted by the D and E cyclins [Akiyana et al., 1992; Kato et al., 1993; Sherr, 1993]. It has been found that both cyclins A and B and their associated kinase Cdk1 (p34cdc2) are down regulated in senescent fibroblasts and are not expressed on serum stimulation [Richter et al., 1991; Stein et al., 1991]. Cdk2 and Cdk4 are also suppressed in senescent fibroblasts, but surprisingly, cyclins D1 and E are elevated [Lucibello et al., 1993; Dulic et al., 1993a]. Dulic et al. [1993a] have also found that cyclin E is unphosphorylated and inactive in senescent fibroblasts, Cdk2 is unphosphorylated and inactive, and cyclin D1-Cdk2 complexes contained exclusively unphosphorylated Cdk2. They suggested that the failure to activate cyclin E-Cdk2 kinase activity in senescent cells accounts for the inability of the cells to phosphorylate pRb in late G1, which then blocks expression of late G1 genes such as cyclin A (Fig. 1.1). Even more recent interest has focused on the cyclin inhibitors p21 CIP/WAF and CDKN2 p16. The former is particularly interesting since it is found in cyclin-D-Cdk2 and cyclin-E-Cdk2 complexes and renders them inactive. p21 CIP/WAF is also induced by p53 in response to genetic damage [El Deiry et al., 1993] and accumulates in senescent human fibroblasts [Noda et al., 1994]. Therefore the inactive state of the cyclin E-Cdk2 complexes in senescent cells may well be due to the accumulation of p21 WAF/CIP.

Another cyclin-Cdk inhibitor recently identified is the $p16^{INK4}$, which appears to work by disrupting the cyclinD-Cdk4 complex by associating with Cdk4 [Serrano et al., 1993]. The inhibitor is encoded by the MTS-1 [Nobori et al., 1994; Kamb et al., 1994] gene, which is a candidate suppressor of both tumorigenesis and immortality [Loughran et al., 1994; Cuthbert et al., in press]. This gene is now termed CDKN2 (Hugo Nomenclature Committee; see Cairns et al.1994). The CDKN2 p16 is proposed to function both upstream and downstream of pRb protein and to act as a negative feedback signal for CyclinD-Cdk4 activity [Serrano et al., 1993]. It is suggested that when pRb becomes phosphorylated by cyclinD-Cdk4 it ceases to downregulate CDKN2 p16, and increased expression of CDKN2 p16 inhibits the cyclinD-Cdk4 activity [Serrano et al., 1993]. This model is supported by several observations showing that when pRb is missing or inactivated, the CDKN2 p16 is strongly overexpressed [Serrano et al., 1993; Malliri and Ozanne, unpublished data]. In this case, however, the negative feedback signal is proposed to be futile since the function of pRb is permanently inactivated [Serrano et al., 1993]. At present there is no direct evidence to implicate CDKN2 in the process of cellular senescence, and the activity of cyclinD-Cdk4 complexes in senescent cells has not been tested. However, alterations of the CDKN2 gene seem to be associated with immortalization [Nobori et al., 1994; Kamb et al., 1994; Loughran et al., 1994; Cuthbert et al., in press], suggesting that this cyclin inhibitor may play at least an indirect role in senescence.

3. TELOMERIC SHORTENING: A UNIFYING THEORY OF CELLULAR SENESCENCE?

There are at least three major theories of how human cells undergo cellular senescence, all of which have recently been discussed in various review articles [Goldstein, 1990; Harley, 1990; Kirkwood, 1991]. Orgel [1963] theorised that errors in newly synthesized proteins, especially those involved in replication, transcription, and translation, could feed back and cause additional errors culminating in a lethal error catastrophe [Orgel, 1970]. This theory is supported by the observation that gene mutations dramatically increase in cultured human fibroblasts as they approach senescence [Fulder and Holliday, 1975] and the observations that humans possessing DNA repair defects often age prematurely; examples are Cockayne syndrome and ataxia telangiectasia [Martin, 1978]. Also cells from the premature aging syndromes Werner's syndrome and Hutchinson's progeria both show increased rates of oxidized protein accumulation [Stadtman, 1992].

A gradual increase in the rate of DNA damage could result in the signaling of a G1 arrest by the ability of p53 to increase the expression of the p21 cyclin-dependent kinase inhibitor, CIP-1/WAF-1 [El-Deiry et al., 1993] and inhibit the cyclinD1-Cdk2 and cyclinE-Cdk2 [Dulic et al., 1993b] from phosphorylating the retinoblastoma protein [Stein et al., 1990], thus achieving a G1 arrest [Hart and Setlow, 1974; Cristofalo, 1973, Seshadri and Campisi, 1990]. Consistent with this theory, CIP/WAF was independently noted to be greatly overexpressed in cultures of senescent human fibroblasts, where it was termed *senescence-derived inhibitor* (Sdi) [Noda et al., 1994].

A second, more recent theory of cellular senescence, is based on the observation that the repeat sequences found at the ends of human chromosomes, the telomeric repeat arrays (TRAs), decrease in size in normal somatic cells with age both *in vivo* [Hastie et al., 1990] and *in vitro* [Harley et al., 1990]. This observation is also true for hemopoietic stem cells [Vaziri et al., 1994]. The loss of TRAs from human chromosomes is connected with the observation that during DNA synthesis the lagging strand must have an RNA primer that is removed at the end of DNA replication, thus creating a gap at the 5′ ends of newly synthesised strands [Greider, 1990] (see Fig. 1.3), which is at least the length of the primer [Walmsley, 1987]. The initial length of the telomere in donor cells strongly correlates with the proliferative lifespan

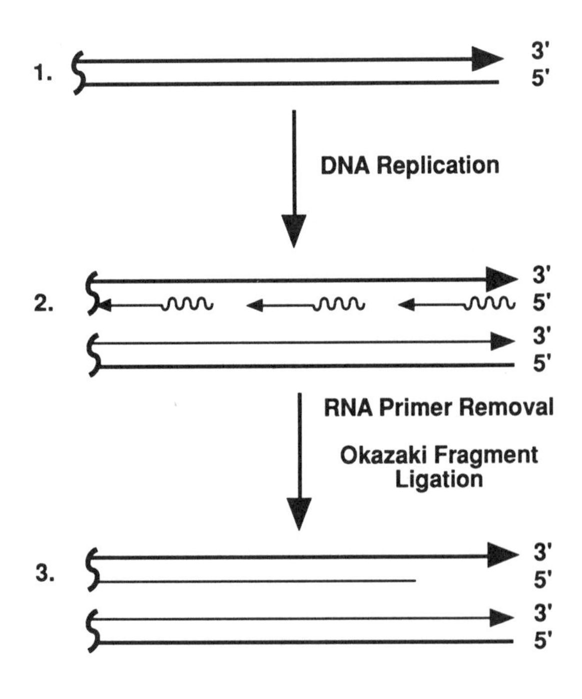

Fig. 1.3. The incomplete relication problem: (1) the molecular end of a DNA molecule is shown; (2) leading strand relication proceeds to the end of the DNA molecule, while lagging strand replication utilizes RNA primers and Okazaki fragment synthesis; (3) removal of the RNA primers and Okazaki fragment ligation leaves a region at one end of each daughter molecule unreplicated. If there is no mechanism to fill the gap, the chromosome will get shorter with each round of replication. Taken from Greider [1990].

that the fibroblasts undergo [Allsop et al., 1992]. Furthermore, progeria fibroblast telomeres are already very much shorter than their age-matched counterparts at birth, perhaps explaining their short *in vitro* lifespan and premature aging *in vivo.* Thus the telomere length may well be the basis of a somatic cell division clock that is sensitive to a variety of influences.

The shortening of the telomeres would clearly be disasterous if it occurred in the germ line of multicellular animals and in unicellular animals since it would ultimately result in the extinction of the species. However, at least one enzyme has been discovered in *Tetrahynena* [Greider and Blackburn, 1985], *Euplotes* [Shippen-Lentz and Blackburn, 1989], and *Oxytricha* [Zahler and Prescott, 1988], which is capable of preventing telomere loss by *de novo* addition. This enzyme has now been detected in *Xenopus* oocytes [Mantell and Greider, 1994] and mouse testes [Prowse and Greider, 1994], and its function is suggested in the human germ line since the telomeres of human sperm do not decrease in length with age [Allsop et al., 1992] and some germ line chromosomal truncations are healed by telomeric addition [Wilkie et al., 1990]. The presence of telomerase in yeast is also suggested by the ever shorter telomeres (*est*-1) mutation [Lundblad and Szostak, 1989], which is thought to encode a protein component of an essential yeast telomerase [Lundblad and Blackburn, 1990]. Very interestingly yeast *est* mutants eventually senesce and this correlates with a progressive decrease in telomere length as the cells divide [Lundblad and Szostak, 1989]. Thus a direct link between telomerase malfunction and cellular senescence is suggested.

The telomere shortening theory of cellular senescence might not be so very different from aspects of Orgel's hypothesis, since it has been suggested [Dulic et al., 1993a,b] that telomeric attrition might at some point be sensed by the cell as genetic damage and could achieve a G1 arrest through the p53/WAF pathway (see above and Fig. 1.4). Intriguingly, when a portion of yeast telomere is removed from a nonessential chromosome, growth arrest does not occur in G1 but occurs in G2 via a RAD9–RAD52-dependent mechanism [Sandell and Zakian, 1993] and the lack of p53 in yeast [Nigro et al., 1992] may explain this result. It should be stressed that there is as yet no direct evidence to implicate p53 in the induction of WAF/CIP during cellular senescence, and p53 is not stabilized in senescent cells as it appears to be in response to genetic damage [Kastan et al., 1991], although other mechanisms of activating p53 are possible. Also p21 WAF/CIP is induced when cells terminally differentiate by a p53 independent pathway [Jiang et al., 1994; Steinman et al., 1994], and terminal differentiation is a consequence of both fibroblast [Bayreuther et al., 1991] and keratinocyte [Rheinwald and Green, 1975] senescence. However,

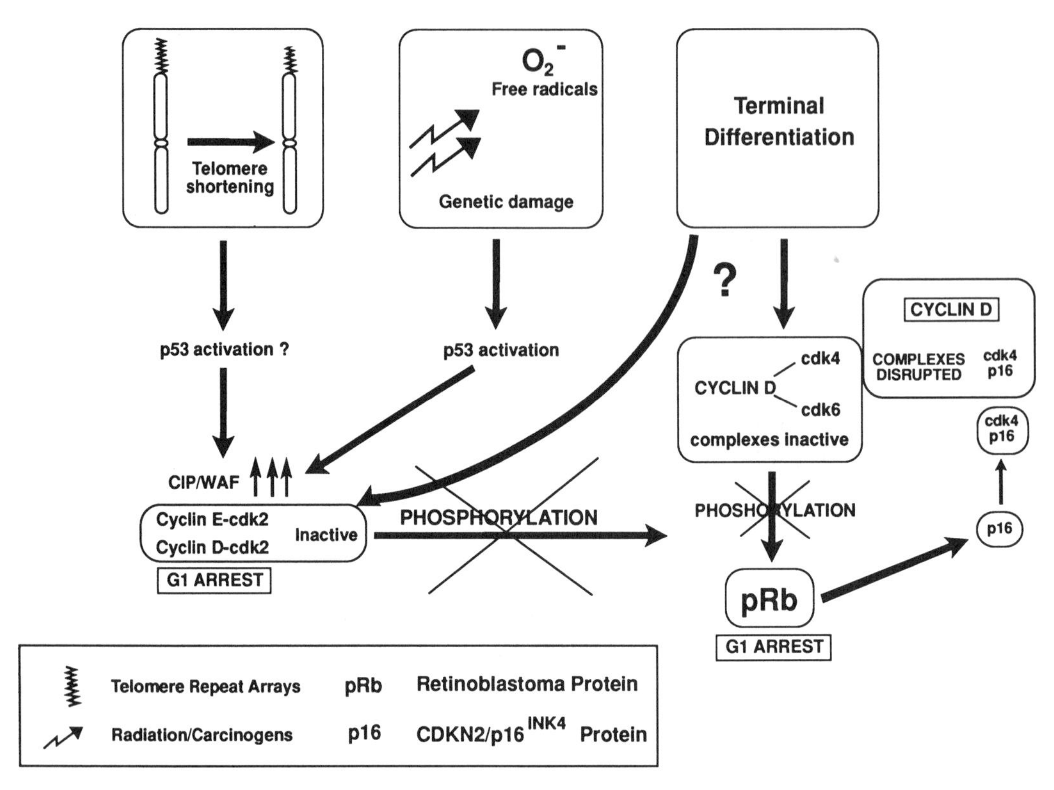

Fig. 1.4. G1 checkpoints in cellular senescence.

since yeast posess the ability to respond to mammalian p53 by arresting in G1 [Nigro et al., 1992, Bischoff et al., 1992], the idea could be tested.

One major difficulty with the first two theories is that it has been known for some time that senescence is stochastic, not synchronous, and that clonogenic cells with a very short proliferative lifespan can be identified in primary cultures of human fibroblasts [Smith and Whitney, 1980] and keratinocytes [Barrandon and Green, 1987]. This has lead some authors to hypothesise that cells committed to senescence [Holliday et al., 1977, 1981] or terminal differentiation [Bell et al., 1978] compete out the more slowly dividing stem cells of greatest proliferative power. Alternatively, the stem cells may terminally differentiate when deprived of a special microenvironment or niche [Schofield, 1978]. Another observation supporting the terminal differentiation theory is that proliferative lifespan can be extended and manipulated by the culture conditions [Poiley et al., 1978, Rheinwald and Green, 1977; Green, 1978; Loo et al., 1987].

There is also evidence that clonogenic cells in culture represent not only stem cells and committed cells but also maturing cells recruited back into the clonogenic pool [Wilke et al., 1988]. Furthermore, the culture conditions affect the extent to which the maturation process can be reversed [Barrandon and Green, 1985], and this correlates with proliferative lifespan [Rheinwald and Green, 1977]. A further problem with the telomere theory is that laboratory strains of mice have very long telomeres [Kipling and Cooke, 1990; Starling et al., 1990] but have a short *in vitro* and *in vivo* lifespan. However, laboratory mouse cells could contain one or more chromosomes with very short telomeres, or their telomeres could be interrupted by nontelomeric DNA near the tips [de Lange, 1994]. It also remains possible that telomere loss as a means to suppress tumor formation is not a property of all mammalian species, and this may explain the high spontaneous immortalization frequency of mouse cells *in vitro* [de Lange, 1994]. Indeed, one way to explain all the observations would be to suggest that telomere shortening acts as the senescence clock, which ultimately results in a cell cycle checkpoint in G1 (Fig. 1.5) but that the clock can be short-circuited if the cells accumulate too many errors in their DNA repair machinery [Orgel, 1963, 1970] or if the cells find themselves in a microenvironment where terminal differentiation signals induce the cells to arrest irreversibly in G1 [Schofield, 1978; Bell et al.,1978]. Interestingly, it has been shown by two groups that p21 WAF/CIP is induced when HL60 cells terminally differentiate in the absence of functional p53 [Jiang et al., 1994; Steinman et al., 1994].

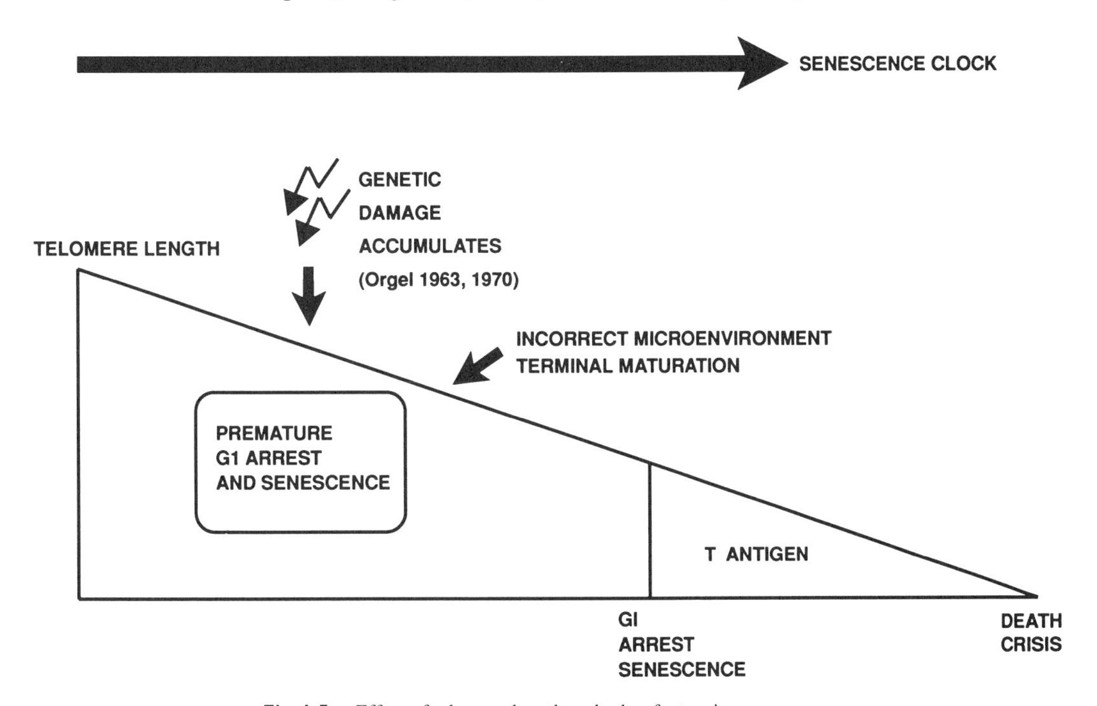

Fig. 1.5. Effect of telomere length and other factors in senescence.

Unraveling the mechanism of telomere replication and maintenance may initially depend on the use of yeast models to identify the components of the telomerase complex and discover how the enzyme is regulated. The repressor–activator protein 1 (RAP1p) binds to telomere sequences [Buchmann et al., 1988] and appears to be important for telomere length regulation since mutations in its carboxy terminus [Conrad et al., 1990] or in the associated RAP1-interacting factor, RIF-1 [Hardy et al., 1992] result in telomere elongation. Petite integration frequency-1, PIF-1 [Lahaye et al., 1991] is also essential for the correct regulation of telomerase, and its loss results not only in telomere elongation but in *de novo* telomere addition to broken chromosomes [Schulz and Zakian, 1994]. It has been suggested that PIF-1 could reduce telomere replication by acting as a telomeric DNA-telomerase RNA helicase [Schulz and Zakian, 1994]. According to their model, PIF-1p keeps telomeres short and prevents inappropriate telomere addition onto broken ends by dissociating telomerase from its substrate. In the absence of PIF-1p, telomerase would add telomeric repeats onto sequences that have little or no telomere homology, as is found in the germ line of humans [Wilkie et al., 1990] and during the macromolecular development of *Tetrahhymena* [Yu and Blackburn, 1991]. Thus PIF-1 could be a developmentally regulated specificity factor for telomerase. RIF-1 and PIF-1 may also be candidates for suppressors of the immortal phenotype by suppressing telomerase activity.

4. EXPERIMENTAL MODELS OF HUMAN CELLULAR IMMORTALIZATION

When human fibroblasts are transfected with Simian virus 40 (SV40) or its large–T antigen (SV40LT), cell cultures with an extended proliferative lifespan are generated and there is persuasive evidence that one of the functions of the SV40LT antigen is to abrogate both the p53 and the Rb tumor suppressor gene pathways (Fig. 1.6). This evidence comes largely from experiments showing that the LT antigen function can be supplanted by other DNA virus genes that collectively, like LT antigen, bind both p53 and pRb [Shay et al., 1991]. However, other proteins are bound by these DNA-virus gene products such as p107, p130, and p300, and the elimination of the function of these proteins may also play a part in the extension of the human fibroblast lifespan. Indeed, elimination by mutation of the pRb-binding domain of HPV E7 does not remove the ability of HPV to immortalize primary human keratinocytes [Jewers et al., 1992]. This work suggests that HPV at least encodes information that renders the binding of pRb redundant in the absence of terminal maturation signals but does not

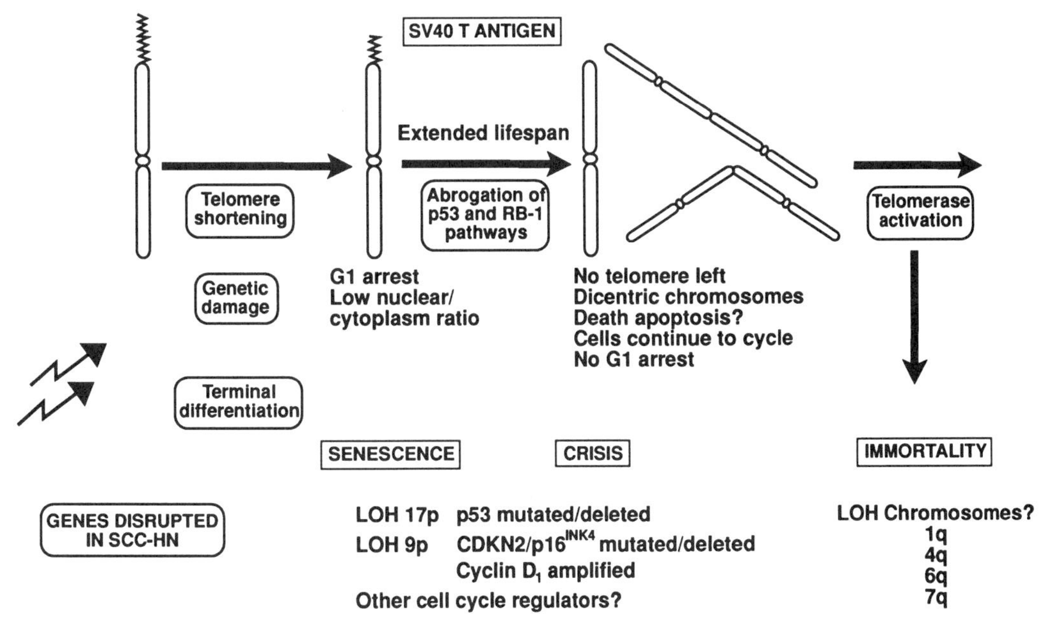

Fig. 1.6. SV40 T-antigen-induced fibroblast immortality.

exclude the possibility that abrogation of the Rb-1-directed pathway makes a contribution in other situations.

In the SV40 human fibroblast model (Fig. 1.6), it was shown independently by two groups [Wright et al., 1989; Radna et al., 1989] that SV40LT antigen was capable only of extending the lifespan of human fibroblasts and following a period of crisis a mutation in a cellular gene appeared to be responsible for the rare emergence of immortal variants. During crisis the SV40 transfected cells continue to traverse the cell cycle [Stein, 1985] but do not increase in number because of an excessive amount of cell death. Using the conditional expression of SV40LT antigen, these groups were able to show that, even after a cellular mutation had rendered the cells immortal, the expression of the LT antigen was still required for the maintenance of the immortal phenotype. Based on these observations Wright et al. [1989] suggested that there were two "mortality" mechanisms M1 and M2. M1 was suggested to be overridden by LT antigen and M2 function to be abolished by the mutation of a cellular gene. Wright et al., [1989] also pointed out that, in mouse fibroblasts, M2 appeared to be vestigial or absent, since SV40LT antigen would immortalise these cells without an obvious crisis.

When human cells are transfected with SV40 the telomeres continue to shorten beyond the point at which senesence would normally occur even though the proliferative lifespan is extended [Counter et al., 1992]. This presumably occurs because the T antigen has disrupted cell cycle control and desensitized it to the senescence "clock." However, when the telomeres reach a critically short length, vital sequences are lost and end to end chromosomal fusions result as a consequence of complete telomere repeat loss and the exposure of free ends. This results in dicentric chromosome formation and death [Counter et al., 1992]. When SV40 transfected cells pass through crisis the telomere shortening is arrested, telomere length and dicentric chromosome number are stabilized and the cells become immortal. When this happens, the cells also reactivate the enzyme telomerase, which can maintain telomere length [Counter et al., 1994a]. This enzyme is also reactivated in naturally occurring human cell lines [Morin, 1989; Counter et al., 1992] and human tumors *in vivo* [Counter et al., 1994a; Kim et al., 1994]. Thus the reactivation of the telomerase enzyme may be a vital step in the immortalization of human tumor cells also.

Telomere shortening has also been noted in lymphocytes infected with Epstein–Barr virus, and when these cells pass through crisis, the telomere shortening is arrested and telomerase is activated [Counter et al., 1994b]. The situation is slightly different, however, when human keratinocytes are transfected with HPV16, or HPV18 E6 and E7 genes together. As with the other two viruses, the HPV is not capable of arresting telomeric attrition, but when the cells pass through crisis, not only is the telomere shortening halted, but the telomeres begin to lengthen [Klingelhutz et al., 1994], suggesting a mechanism of immortalization different from that of the other two viral systems and most immortal human tumor lines. HeLa cells have long telomeres [de Lange et al., 1990], harbor HPV18 [Schneider-Gadicke and Schwarz, 1986] and have activated telomerase [Morin, 1989] but have presumably deregulated the enzyme in a different manner.

It is also notable that the telomeres of laboratory mice are extremely long to begin with, so that once the cell cycle response to the senescence clock is overridden, an obvious crisis would not be predicted, and this is precisely what is seen when cells are transfected with SV40 [Wright et al., 1989]. Furthermore, the unusually long telomere length of laboratory mouse chromosomes may partly explain the ability of some investigators to grow apparently normal mouse embryo cells for 200 population doublings without any sign of senescence [Loo et al., 1987].

5. THE RECESSIVE NATURE OF THE SV40-INDUCED IMMORTAL PHENOTYPE: INVOLVEMENT OF SUPPRESSOR GENES IN IMMORTALIZATION

Somatic cell genetics experiments have shown that when SV40-immortalized fibroblasts are fused with normal fibroblasts, senescent hybrids result [Muggleton-Harris and DeSimone, 1980; Pereira-Smith, and Smith, 1981]. Moreover, when the hybrids senesced, the SV40LT antigen was still shown to be expressed in the nucleus of the hybrid cells [Pereira-Smith and Smith, 1981], supporting the notion that M2 may have been inactivated by a recessive mutation of a cellular gene. Some of the possible negative regulators of cell proliferation inactivated by SV40LT antigen have been mentioned already and these include p53 [Lane and Crawford, 1979; Linzer and Levine, 1979], p21 CIP/WAF [El Deiry et al., 1993], pRb-1 [Friend et al., 1986], and p107 [Ewen et al., 1991], which have all been shown to be inhibitory to cellular proliferation [El-Deiry et al., 1993; Zhu et al., 1993; Baker et al., 1990; Huang et al., 1988].

The recessive mutation(s) that inactivate M2 are far less well understood, but some somatic cell genetics experiments suggest that there is more than one genetic mechanism of M2 inactivation [Pereira-Smith and Smith, 1983, 1988; Whitaker et al., 1992], and four complementation groups—A, B, C, and D—have been identified in immortal cells [Pereira-Smith and Smith, 1988]. It is also probable that the inactivation of M2 requires the reactivation or upregulation of telomerase [Counter et al., 1992, 1994a,b; Kim et al., 1994], although the question of cause and effect has not been established directly [Greider, 1990]. Several groups using the monochromosome transfer approach have shown that genes residing on chromosomes 1 [Sugawara et al., 1990], 4 [Ning et al., 1991], 6 [Hubbard Smith et al., 1992a,b], 7 [Ogata et al., 1993], 9 [Porterfield et al., 1992], 17 [Casey et al., 1993], and X [Wang et al., 1992] can all induce a phenomenon reminiscent of cellular senescence. The chromosomal regions responsible for the effect have been mapped to 1q25 and 1q42 [Cuthbert et al., in press], 6q13, 7q31–32 [Ogata et al., 1993], 9p21–22 [Porterfield et al., 1992], and Xp22-*ter* [Wang et al., 1992]. It has been suggested that these genes may encode transacting repressors of telomerase [Allsop et al., 1992]. These repressors may affect telomerase transcription, but it is also possible that they are regulators of telomerase activity such as the yeast genes RIF-1 [Hardy et al., 1992] and PIF-1 [Schulz and Zakian, 1994], since there is no formal evidence that the telomerase enzyme complex is actually absent from normal somatic cells. The unraveling of telomerase function and the cloning of the senescence genes described above will enhance the understanding of telomerase reactivation in normal somatic cells.

6. THE GENETIC BASIS OF IMMORTALIZATION OF HUMAN SQUAMOUS CELL CARCINOMA CELLS

Despite extensive investigations into the immortalization of human cells *in vitro*, until recently, very little evidence existed for a similar process in human tumors.

Our preliminary investigations into the immortal phenotype of neoplastic head and neck keratinocytes indicated that it was a late event in tumor progression [Edington et al., in press, Edington et al., in preparation] and that, like the SV40 transformation model [Muggleton-Harris and DeSimone, 1980; Pereira-Smith and Smith, 1981], immortality was genetically recessive to senescence [Berry et al., 1994]. These results suggested that human keratinocyte immortality required multiple genetic changes and that some of these may involve recessive mutations, perhaps of tumor suppressor genes. We therefore turned our attention to the genetic analysis of head and neck squamous cell carcinomas (SCC-HN) expressing the immortal SCC-HN phenotype using the loss of heterozygosity (LOH) approach as an indicator of loss of tumor suppressor gene function [Cavanee et al., 1983; Vogelstein et al., 1989]. To do this we established neoplastic cultures from different stages of SCC-HN progression and characterized their *in vitro* proliferative lifespan [Edington et al., in press]. In addition, we obtained normal fibroblasts or lymphocytes from the same patients so that LOH studies could be carried out [Edington et al., in press; Loughran et al., 1994].

Our results to date suggest that most, if not all, immortal SCC-HN have LOH at chromosome 17p13 [Loughran et al., unpublished data] and 9p21 [Loughran et al., 1994] where the p53 and CDKN2 genes reside, respectively. Accordingly, we and others have shown that nearly all immortal SCC-HN can be shown to harbour p53 mutations [Sakai and Tsuchida, 1992; Somers et al., 1992; Burns et al., 1993] and CDKN2 deletions [Loughran et al., 1994; Yeudall et al., 1994; Zhang et al., 1994] and mutations [Yeudall et al., 1994, Zhang et al., 1994]. However, CDKN2 alterations and LOH of chromosome 9p21 are never seen in neoplastic head and neck keratinocytes that senesce [Loughran et al., 1994] and in the same cultures p53 mutations are rarely seen [Burns et al., 1993, 1994a]. However, even when a homozygous p53 mutation [Burns et al., 1993] and a homozygous CDKN2 deletion [Loughran et al., 1994] are present in the same cell culture, BICR7, this is still an insufficient condition for immortality and the culture enters crisis after around 30–40 population doublings [Edington et al., in press]. Culture BICR7 is thus proposed to lack at least one genetic change necessary for the immortal phenotype, and this may include mutations affecting the control of telomerase.

It is possible that abrogation of the p53 and Rb-1 pathways may be achieved in SCC-HN keratinocytes by mutation of the p53 gene itself and deletion or mutation of the CDKN2 gene. Loss of p53 function would lead to a reduction of cell cycle control as mediated by p21 CIP/WAF, which could desensitize the response of the cell cycle checkpoints to the senescence clock [Dulic et al., 1993a,b] in addition to rendering the cells genetically unstable [Lane, 1992; Yin et al.,

1992, Livingstone et al., 1992]. The loss of CDKN2, which encodes the p16^{INK4} cyclin D-Cdk4 inhibitor, would result in an enhancement of the cyclin D-Cdk4 kinase activity. This, in turn, would be expected to override the inhibitory effects of pRb on cell cycle progression [Dowdy et al., 1993] and perhaps increase the rate of entry into S phase.

It is as yet unclear whether CDKN2/p16^{INK4} is part of the senescence clock, but any disruption of cell cycle control might be expected to desensitize the response of the cell cycle to the clock. Mutations of the p53 gene can be detected in SCC-HN *in vivo,* including the original tumor from which the immortal cells were derived [Sakai and Tsuchida, 1992; Burns et al., 1993], and the same p53 mutation is sustained throughout tumor progression [Burns et al., 1994b], indicating that loss of p53 function confers a selective advantage *in vivo.* Similarly, CDKN2 gene mutations have been detected in SCC-HN *in vivo,* [Cairns et al., 1994; Zhang et al., 1994], and the same deletions have been reported in cell lines derived from primary tumors as the corresponding lines derived from metastases of the same tumors [Yeudall et al., 1994]. These results indicate that the CDKN2 mutations are present in the SCC-HN *in vivo* and that, like p53 mutations, they confer a selective advantage to the keratinocytes *in vivo.* It is still unclear whether other genetic changes are required to ablate M1 in human keratinocytes, but our studies have revealed consistent LOH at other genetic loci [Loughran et al., manuscript in preparation], and some of them may play a part in cellular immortalization. It has been proposed that genes involved in senescence are located on chromosomes 1, 4, and 7 are responsible for the complementation groups C [Hensler et al., 1994], B [Ning et al., 1991], and D [Ogata et al., 1993], respectively, but the senescence gene located on chromosome 6 [Hubbard-Smith et al., 1992a,b] is not complementation-group-specific [Pereira-Smith, personal communication]. However, the immortal phenotype of SCC-HN cells is not readily complemented by fusion with other immortal cells [Berry et al., 1994], and our more recent results indicate that many SCC-HN cell lines show LOH on more than one of the above chromosomes [Loughran et al., manuscript in preparation].

Other groups have also reported that some cell lines map to more than one complementation group for immortality [Duncan et al., 1993] or that they fail to complement altogether [Ryan et al., 1994]. Some of these results could be explained by the fact that many immortal variants from tumors, as opposed to *in vitro* immortalized cells, may have altered more than one of the senescence genes [Berry et al., 1994, Loughran et al., manuscript in preparation] or that in some studies the hybrids were not maintained under selection [Berry et al., 1994, Ryan et al., 1994]. The reason why the maintenance of selection should affect the result is not entirely clear, but senescence genes have been reported on chromosomes 17 [Casey et al., 1993] and X [Wang et al., 1992], which carry the thymidine kinase (TK) and hypoxanthine–guanosine phosphoribosyltransfersase (HGPRT) genes, respectively. Although these chromosomes have not been assigned complementation groups, their loss may affect the detection of senescence since their retention would be ensured during continuous selection in HAT medium. It is not as yet clear whether immortal SCC-HN keratinocytes harbor short telomeres or whether they activate telomerase but telomerase has been detected in other human tumor cells that lack DNA tumor viruses [Counter et al., 1992; Prowse et al., 1994; Kim et al., 1994]. Furthermore, telomerase activity can always be detected in late-stage carcinoma *in vivo* [Counter et al., 1994a; Kim et al., 1994].

7. IMMORTAL VARIANTS AND CANCER TREATMENT: PROSPECTS

Not all primary tumors contain immortal variants, but early indications at least suggest that patients with tumors harboring immortal variants have a poor prognosis [Clark and Parkinson, unpublished data]. There are also reports that early-stage SCC-HN harboring p53 mutations, a genetic marker of the immortal phenotype [Bischoff et al., 1990; Harvey et al., 1993], have a much poorer prognosis [Gluckman et al., 1994] and suffer from early metastasis to the lymph nodes [Clark and Parkinson, unpublished data]. Also all recurrent late stage SCC-HN are immortal [Edington et al., in press], and all end-stage ovarian carcinomas *in vivo* have activated telomerase [Counter et al., 1994a].

Telomerase has been suggested as a target for chemotherapeutic drugs [Harley et al., 1990; Greider, 1990] since the immortal cancer cells with short telomeres would be dependent on telomerase for their continued proliferation and perhaps survival. Also, p53 mutants that can be reverted to normal [Milner, 1994] may also prove to be a fruitful therapeutic target. As the mutations that give rise to the immortal phenotype are discovered and the molecular pathways to cellular immortality in different tumors are established, any potential for developing diagnostic aids and new therapeutic strategies against cancer [Greider, 1990; Harley et al., 1990; Milner, 1994] and other diseases [Blackburn 1991] should be realized.

ACKNOWLEDGMENTS

I wish to thank my colleagues J. Burns, M. Baird, I. Berry, L. Clark, K. Edington, O. Loughran, R. MacFarlane, R. Mitchell, D. Soutar, and

G. Robertson for their contributions to some of the work described in this chapter. I would also like to thank John Wyke for critical reading of the manuscript and the Cancer Research Campaign for their support of our work.

REFERENCES

Akiyama T, Ohuhi T, Sumida, S, Matsumoto K, Toyoshima K (1992): Phosphorylation of the retinoblastoma protein by cdk2. Proc Natl Acad Sci USA 89: 7900–7904.

Allsop RC, Vaziri H, Patterson C, Goldstein S, Younglai EV, Futcher AB, Greider, CW, Harley CB (1992): Telomere length predicts replicative capacity of human fibroblasts. Proc Natl Acad Sci USA 89: 10114–10118.

Baker SJ, Markovitz S, Fearon ER, Willson JK, Vogelstein, B (1990): Suppression of human colorectal carcinoma cell growth by wild-type p53. Science 249: 912–915.

Barrandon Y, Green H (1985): Cell size as a determinant of the clone-forming ability of human keratinocytes. Proc Natl Acad Sci USA 82: 5390–5394.

Barrandon Y, Green H (1987): Three clonal types of keratinocyte with different capacities for multiplication. Proc Natl Acad Sci USA 84: 2302–2306.

Bayreuther K, Francz PI, Gogl J, Hapke C, Maier M, Heinrath HG (1991): Differentiation of primary and secondary fibroblasts in cell culture systems. Mut Res 256: 233–242.

Bell E, Marck LF, Levinstone DS, Merrill C, Sher S, Young IT, Eden M (1978): Loss of division potential *in vitro:* aging or differentiation? Science 202: 1158–1163.

Berry IJ, Burns JE, Parkinson EK (1994): Assignment of two epidermal squamous cell carcinoma cell lines to more than one complementation group for the immortal phenotype. Mol Carcinog 9: 134–142.

Bischoff JR, Casso D, Beach D (1992): Human p53 inhibits growth in *Schizosaccharomyces pombe.* Mol Cell Biol 12: 1405–1411.

Bischoff FZ, Yim SO, Pathak S, Grant G, Siciliano MJ, Giovanella BC, Strong LC, Tainsky MA (1990): Spontaneous abnormalities in normal fibroblasts from patients with Li Fraumeni cancer syndrome: aneuploidy and immortalization. Cancer Res 50: 7979–7984.

Blackburn EH (1991): The structure and function of telomeres. Nature 350: 569–573.

Buchmann AR, Lue NF, Kornberg RD (1988): Connections between transcriptional activators, silencers, and telomeres as revealed by functional analysis of a yeast DNA-binding protein. Mol Cell Biol 8: 5086–5099.

Bunn CK, Tarrant GM (1980): Limited lifespan in somatic cell hybrids and cybrids. Exp Cell Res 127: 385–396.

Burns JE, Baird MC, Clark LJ, Burns PA, Edington KG, Chapman C, Mitchell R, Robertson G, Soutar D, Parkinson EK (1993): Gene mutations and increased levels of p53 protein in human squamous cell carcinomas and their cell lines. Br J Cancer 67: 1274–1284.

Burns JE, Clark LJ, Yeudall WA, Mitchell WA, Mitchell WA, Mitchell R, MacKenzie K, Chang SE, Parkinson EK (1994a), The p53 status of cultured premalignant oral keratinocytes. Br J Cancer 70: 591–595.

Burns JE, McFarlane R, Clark LJ, Mitchell R, Robertson G, Soutar D, Parkinson EK (1994b): Maintenance of identical p53 mutations throughout progression of squamous cell carcinomas of the tongue. Oral Oncol Eur J Cancer 30B: 335–337.

Cairns P, Mao L, Merlo A, Lee DJ, Schwab D, Eby Y, Tokino K, van der Riet P, Blaugrund JE, Sidransky D (1994): Rates of p16 (MTS-1) mutations in primary tumors with 9p loss. Science 265: 415–416.

Casey G, Plummer S, Hoetge G, Scanlon D, Faschling C, Stanbridge EJ (1993): Functional evidence for a breast cancer growth suppressor gene on chromosome 17. Hum Mol Genet 2: 1921–1927.

Cavanee WK, Dryja T, Philips RA, Benedict WF, Godbout R, Gallie BL, Murphee AL, Strong LC, White RL (1983): Expression of recessive alleles by chromosomal mechanisms in retinoblastoma. Nature 305: 779–784.
Conrad MN, Wright JH, Wolf AJ, Zakian VA (1990): RAP-1 protein interacts with yeast telomeres *in vivo:* overproduction alters telomere structure and decreases chromosome stability. Cell 63: 739–750.
Counter CM, Avilion AA, Le Feuvre CE, Stewart NG, Greider CW, Harley CB, Bacchetti S (1992): Telomere shortening associated with chromosome instability in arrested in immortal cells which express telomerase activity. EMBO J 11: 1921–1929.
Counter CM, Hirte HW, Bacchetti S, Harley CB (1994a): Telomerase activity in human ovarian carcinoma. Proc Natl Acad Sci USA 91: 2900–2904.
Counter CM, Botelho FM, Wang P, Harley CB, Bacchetti S (1994b): Stabilization of short telomeres and telomerase activity accompany immortalization of Epstein-Barr virus-transformed human B lymphocytes. J Virol 68: 3410–3414.
Cristofalo VJ (1973): Cellular senescence: factors modulating cell proliferation *in vitro*. INSERM 27: 65–92.
Cristofalo VJ, Sharf BB (1973): Cellular senescence and DNA synthesis. Thymidine incorporation as a measure of population age in human diploid cells. Exp Cell Res 76: 419–427.
Cuthbert AP, Trott DA, England NL, Jezzard S, Ekong, R, Todd CM, Themis M, Newbold RF (in press): Mapping human antiproliferative genes by microcell-mediated chromosome transfer: evidence for cellular senescence genes on chromosomes 1 and 9. in press.
de Lange T (1994): Activation of telomerase in a human tumor. Proc Natl Acad Sci USA 91: 2882–2885.
de Lange T, Shiue L, Myers R, Cox DR, Naylor SL, Killary AM, Varmus HE (1990): Structure and variability of human chromosome ends. Mol Cell Biol 10: 518–527.
Dowdy SF, Hinds PW, Louie K, Reed SI, Arnold A, Weinberg RA (1993): Physical interaction of the retinoblastoma protein with human D cyclins. Cell 73: 499–511.
Dulic V, Drullinger, LF, Lees E, Reed SI, Stein GH (1993a): Altered regulation of G1 cyclins in senescent human diploid fibroblasts: accumulation of inactive cyclin E-cdk2 and cyclinD1-cdk2 complexes. Proc Natl Acad Sci USA 90: 11034–11038.
Dulic V, Kaufmann WK, Wilson SJ, Tlsty T, Lees E, Harper JW, Elledge SJ, Reed SI (1993b): p53-dependent inhibition of cyclin-dependent kinase activities in human fibroblasts during radiation-induced G1 arrest. Cell 76: 1013–1023.
Duncan EL, Whitaker NJ, May EL, Reddel RR (1993): Assignment of SV40-immortalized cells to more than one complementation group. Exp Cell Res 205: 337–344.
Edington KG, Berry IJ, O'Prey M, Burns JE, Clark LJ, Mitchell R, Robertson G, Soutar D, Coggins LW, Parkinson EK (in press): *In vitro* analysis of multistage squamous cell carcinoma: defective terminal maturation is an early and ubiquitous event. In Freshney R I (ed): "Culture of Tumor Cells, Culture of Specialised Cells," Vol.4. New York: Wiley-Liss.
El-Deiry WS, Tokino T, Velculescu VE, Levy DB, Parsons R, Trent JM, Lin D, Mercer E, Kinzler KW, Vogelstein B (1993): WAF1, a potential mediator of p53 tumor suppression. Cell 75: 817–825.
Ewen ME, Xing YG, Lawrence JB, Livingston DM (1991): Molecular cloning, chromosomal mapping, and expression of the cDNA for p107, a retinoblastoma gene product-related protein. Cell 66: 1155–1164.
Friend SH, Bernards R, Rogelj S, Weinberg RA, Rapaport JM, Albert DM, Dryja TP (1986): A human DNA segment with properties of the gene that predisposes to retinoblastoma and osteosarcoma. Nature 323: 643–646.
Fulder SJ, Holliday R (1975): A rapid rise in cell variants during the senescence of populations of human fibroblasts. Cell 6: 67–73.
Gluckman JL, Stanbrook PJ, Pavelic ZP (1994): Prognostic significance of p53 protein accumulation in early stage T1 oral cavity cancer. Oral Oncol Eur J Cancer 30B, p281.

Goldstein S (1990): Replicative senescence: the human fibroblast comes of age. Science 249: 1129–1133.
Green H (1978): Cyclic AMP in relation to proliferation of the epidermal cell: a new view. Cell 15: 801–811.
Greider CW (1990): Telomeres, telomerase and senescence. Bioessays 12: 363–369.
Greider CW, Blackburn EH (1985): Identification of a specific telomere terminal transferase activity in *Tetrahymena* extracts. Cell 43: 405–413.
Hardy CF, Sussel L, Shore S (1992): A RAP1-interacting protein involved in transcriptional silencing and telomere length regulation. Genes Devel 6: 801–814.
Harley CB (1990): Telomere loss: mitotic clock or genetic time bomb? Mutation Res 256: 271–282.
Harley CB, Futcher AB, Greider CW (1990): Telomeres shorten during aging of human fibroblasts. Nature 345: 458–460.
Hart CW, Setlow RB (1974): Correlation between DNA excision repair and lifespan in a number of mammalian species. Proc Natl Acad Sci USA 71: 2169–2173.
Harvey M, Sands AT, Weiss RS, Hegi ME, Wiseman RW, Pantazis P, Giovanella BC, Tainsky MA, Bradley A, Donehower LA (1993): *In vitro* growth characteristics of embryo fibroblasts isolated from p53-deficient mice. Oncogene 8: 2457–2467.
Hastie ND, Dempster M, Dunlop MG, Thompson AM, Green DK, Allshire RC (1990): Telomere reduction in human colorectal carcinoma and with aging. Nature 346: 866–868.
Hayflick L (1965): The limited *in vitro* lifetime of human diploid cell strains. Exp Cell Res 37: 614–636.
Hensler PJ, Annab LA, Barrett JC, Pereira-Smith OM (1994): A gene involved in control of human cellular senescence on human chromosome 1q. Mol Cell Biol 14: 2291–2297.
Holliday R, Huschscha LJ, Kirkwood TBL (1981): Further evidence for the commitment theory of cellular aging. Science 214: 1505–1508.
Holliday R, Huschscha LI, Tarrant GM, Kirkwood TBL (1977): Testing the commitment theory of cellular aging. Science 198: 366–372.
Huang H-JS, Yee J-K, Shew J-K, Chen C-L, Bookstein R, Friedmann T, Lee EY-HP, Lee W-H (1988): Suppression of the neoplastic phenotype by replacement of the RB gene in human cancer cells. Science 242: 1563–1566.
Hubbard-Smith K, Patsalis P, Pardinas JR, Jha KK, Henderson AS, Ozer HL (1992a): Altered chromosome 6 in immortal human fibroblasts. Mol Cell Biol 12: 2273–2281.
Hubbard-Smith K, Sandu AK, Kaur GP, Jha KK, Athwal RS, Ozer HL (1992b): Induction of senescence of immortal human fibroblasts by chromosome 6. Cytogenet Cell Genet 62: p85 (abstract).
Jewers RJ, Hildebrandt P, Ludlow JW, Kell B, McCance DJ (1992): Regions of human papillomavirus type 16 E7 oncoprotein required for immortalisation of human keratinocytes. J Virol 66: 1329–1335.
Jiang H, Lin J, Su Z-Z, Collart FR, Huberman E, Fisher PB (1994): Induction of differentiation in human promyelocytic HL-60 leukemia cells activates p21, WAF1/CIP1, expression in the absence of p53. Oncogene 9: 3397–3406.
Kamb A, Gruis NA, Weaver-Feldhaus J, Liu Q, Harshman K, Tavtigian SV, Stockert E, Day RS III, Johnson BE, Scolnick MHA (1994): A cell cycle regulator potentially involved in genesis of many tumor types. Science 264: 436–440.
Kastan MB, Onyekwere O, Sidransky D, Vogelstein B, Craig RW (1991): Participation of p53 protein in the cellular response to DNA damage. Cancer Res 51: 6304–6311.
Kato J-Y Matsushime H, Hiebert SW, Ewen ME, Sherr CJ (1993): Direct binding of cyclin D to the retinoblastoma gene product (pRb) and pRb phosphorylation by the cyclin dependent kinase cdk4. Genes Devel 7: 331–342.
Kim NW, Piatyszak MA, Prowse KR, Harley CB, West MD, Ho PL, Coviello GM, Wright WE, Weinrich SL, Shay JW (1994): Specific association of human telomerase activity with immortal cells and cancer. Science 266: 2011–2014.

Kipling D, Cooke HJ (1990): Hypervariable ultra-long telomeres in mice. Nature 347: 400–402.
Kirkwood TBL (1991): Genetic basis of limited proliferation. Mutation Res 256: 323–328.
Klingelhutz AJ, Barber SA, Smith PP, Dyer K, MacDougall JK (1994): Restoration of telomeres in human papillomavirus-immortalized human anogenital epithelial cells. Mol Cell Biol 14: 961–969.
Lahaye A, Stahl H, Thines-Sempoux D, Foury F (1991): PIF1: a DNA helicase in yeast mitochondria. EMBO J 10: 997–1007.
Lane DP (1992): *p53,* guardian of the genome. Nature 358: 15–16.
Lane DP, Crawford LV (1979): T antigen is bound to a host protein in SV40-transformed cells. Nature 278: 261–263.
Linzer DI, Levine AJ (1979): Characterization of a 54K Dalton cellular SV40 tumor antigen present in SV40-transformed cells and uninfected embryonal carcinoma cells. Cell 17: 43–52.
Livingston LR, White A, Sprouse J, Levaros E, Jacks T, Tlsty TD (1992): Altered cell cycle arrest and gene amplification potential accompany loss of wild type *p53*. Cell 70: 923–935.
Loo DT, Fuquay JI, Rawson CL, Barnes DW (1987): Extended culture of mouse embryo cells without senescence: inhibition by serum. Science 236: 200–202.
Loughran O, Edington KG, Berry IJ, Clark LJ, Parkinson EK (1994): Loss of heterozygosity of chromosome 9p21 is associated with the immortal phenotype of neoplastic human head and neck keratinocytes. Cancer Res 54:5054–5049.
Lucibello FC, Sewing A, Brusselbach S, Burger C, Muller R (1993): Deregulation of cyclins D1 and E and suppression of cdk2 and cdk4 in senescent human fibroblasts. J Cell Sci 105: 123–133.
Lundblad V, Blackburn EH (1990): RNA-dependent polymerase motifs in EST-1: tentative identification of a protein component of an essential yeast telomerase. Cell 60: 529–530.
Lundblad V, Szostak JW (1989): A mutant with a defect in telomerase elongation leads to senescence in yeast. Cell 57: 633–643.
Mancianti M-L, Herlyn M (1989): Tumor progession in melanoma: the biology of epidermal melanocytes *in vitro*. In Conti CJ, Slaga TJ, Klein-Szanto AJP, (eds): "Carcinogenesis, a Comprehensive Survey," Vol 11: Skin Tumors Experimental and Clinical Aspects." New York: Raven Press, pp 369–386.
Mantell, L K, Greider, C W (1994). Telomerase activity in germline and embryonic cells of *Xenopus*. EMBO J 13: 3211–3217.
Martin GM (1978): "Genetic Effects on Aging." New York: Liss, 1978, pp 5–39.
Milner J (1994): Forms and functions of p53. Sem Cancer Biol 5: 211–219.
Morin GB (1989): The human telomere terminal transferase enzyme is a ribonucleoprotein that synthesises TTAGGG repeats. Cell 59: 521–529.
Muggleton-Harris AL, DeSimone DW (1980): Replicative potentials of various fusion products between W138 and SV40-transformed W1-38 cells and their components. Somat Cell Genet 6: 689–698.
Newbold RF, Overell RW (1983): Fibroblast immortality is a prerequisite for transformation by EJ-HA-RAS oncogene. Nature 304: 648–651.
Newbold RF, Overell RW, Connell JR (1982): Induction of immortality is an early stage event in malignant transformation of mammalian cells by carcinogens. Nature 299: 633–635.
Nigro JM, Sikorski R, Reed SI, Vogelstein B (1992): Human p53 and CDC2Hs genes combine to inhibit the proliferation of *Saccharomyces cerevisiae*. Mol Cell Biol 12: 1357–1365.
Ning Y, Weber JL, Killary AM, Ledbetter DH, Smith JR, Pereira-Smith OM (1991): Genetic analysis of indefinite division in human cells: evidence for a cell senescence-related gene(s) on human chromosome 4. Proc Natl Acad Sci USA 88: 5635–5639.
Nobori T, Miura K, Wu, DJ, Lois A, Takabayaski K, Carson DA (1994): Deletions of

the cyclin-dependent kinase-4 inhibitor gene in multiple human cancers. Nature 368: 753–755.
Noda A, Ning Y, Venable SF, Pereira-Smith OM, Smith JR (1994): Cloning of senescent cell-derived inhibitors of DNA synthesis using an expression screen. Exp Cell Res 211: 90–98.
Ogata T, Ayusawa D, Namba M, Takahashi E, Oshimura M, Oishi M (1993): Chromosome 7 suppresses indefinite division of non-tumorigenic immortalized human fibroblast cell lines KMST-6 and SUSM-1. Mol Cell Biol 13: 6036-6043.
Orgel LE (1963): The maintenance of the accuracy of protein synthesis and its relevence to aging. Proc Natl Acad Sci USA 49: 517–521.
Orgel LE (1970): The maintenance of the accuracy of protein synthesis and its relevance to aging: a correction. Proc Natl Acad Sci USA 67: 1476.
Paraskeva C, Harvey A, Finerty S, Powell S (1989): Possible involvement of chromosome 1 in *in vitro* immortalisation: evidence from progression of a human adenoma-derived cell line *in vitro*. Int J Cancer 43: 743–746.
Pereira-Smith OM, Smith JR (1981): Expression of SV40 T antigen in finite lifespan hybrids of normal and SV40 transformed fibroblasts. Somat Cell Genet 7: 411–421.
Pereira-Smith OM, Smith JR (1983): Evidence for the recessive nature of cellular immortality. Science 221: 964–966.
Pereira-Smith OM, Smith JR (1988): Genetic analysis of indefinite division in human cells: identification of four complementation groups. Proc Natl Acad Sci USA 85: 6042–6046.
Poiley JA, Shuman RF, Pienta RJ (1978): Characterization of normal human embyro cells grown to over 100 population doublings. In Vitro 14: 405-412.
Porterfield BW, Diaz MO, Rowley JD, Olopade OI (1992): Induction of senescence in two human neoplastic cell lines by microcell transfer of human chromosome 9. Proc Am Assoc Cancer Res 33: 73 (abstract).
Prowse KW, Greider CW (1994): Cited in Mantell LL, Greider C W (1994): Telomerase activity in germline and embryonic cells of *Xenopus*. EMBO J 13: 3211–3217.
Radna RL, Caton Y, Jha KK, Kaplan P, Li G, Traganos F, Ozer HL (1989): Growth of immortal Simian Virus 40 tsA-transformed human fibroblasts is temperature dependent. Mol Cell Biol 9: 3093–3096.
Rheinwald JG, Green H (1975): Serial cultivation of strains of human epidermal keratinocytes: the formation of keratinizing colonies from single cells. Cell 6: 331–344.
Rheinwald JG, Green H (1977): Epidermal growth factor and the multiplication of cultured human epidermal keratinocytes. Nature 265: 421–424.
Richter KH, Afshari CA, Annab LA, Burkhart BA, Owen RD, Boyd J (1991): Downregulation of cdc2 in senescent human and hamster cells. Cancer Res 51: 6010–6013.
Ryan PA, Maher VM, McCormick JJ (1994): Failure of indefinite lifespan human cells from different immortality complementation groups to yield finite lifespan hybrids. J Cell Physiol 159: 151–158.
Sager R (1988): Tumor suppressor genes: the puzzle and the promise. Science 246: 1406–1412.
Sakai E, Tsuchida N (1992): Most human squamous cell carcinomas in the oral cavity contain mutated *p53* tumor suppressor genes. Oncogene 7: 927–933.
Sandell LL, Zakian VA (1993): Loss of a yeast telomere: arrest, recovery and chromosome loss. Cell 75: 729–739.
Schneider–Gadicke A, Schwarz E (1986): Different human cervical carcinoma cell lines show similar transcription patterns of human papillomavirus type 18 early genes. EMBO J 5: 285–291.
Schofield R (1978): The relationship between the spleen colony-forming cell and the haemopoietic stem cell. Blood Cells 4: 7–25.

Schulz VP, Zakian VA (1994): The *Saccharomyces* PIF1 DNA helicase inhibits telomere elongation and *de novo* telomere formation. Cell 76: 145–155.
Serrano M, Hannon GJ, Beach D (1993): A new regulatory motif in cell cycle control causing specific inhibition of cyclinD/cdk4. Nature 366: 704–707.
Seshadri T, Campisi J (1990): Repression of C-FOS transcription and an altered genetic programme in senescent human fibroblasts. Science 247: 205–209.
Shay JW, Pereira-Smith OM, Wright WE (1991): A role for both RB and p53 in the regulation of human cellular senescence. Exp Cell Res 196: 33–39.
Sherr CJ (1993): Mammalian G1 cyclins. Cell 73: 1059–1065.
Shippen-Lentz D, Blackburn EH (1989): Telomere terminal transferase activity in the hypotrichous ciliate *Euplotes crassus*. Mol Cell Biol 9: 2761–2764.
Smith JR, Whitney RG (1980): Intraclonal variation in proliferative potential of human diploid fibroblasts: stochastic mechanism for cellular aging. Science 207: 82–84.
Somers KD, Merrick MA, Lopez ME, Incognito LS, Schechter GL, Casey G (1992): Frequent p53 mutations in head and neck cancer. Cancer Res 52: 5997–6000.
Stadtman ER (1992): Protein oxidation and aging. Science 257: 1220–1224.
Starling JA, Maule J, Hastie ND, Allshire RC (1990): Extensive telomere repeat arrays in mouse are hypervariable. Nucl Acids Res 18: 6881–6888.
Stein GH (1985): SV40-transformed human fibroblasts: Evidence for cellular aging in precrisis cells. J Cell Physiol 125: 36–44.
Stein GH, Beeson M, Gordon L (1990): Failure to phosphorylate the retinoblastoma gene product in senescent fibroblasts. Science 249: 666–669.
Stein GH, Drullinger LF, Robetorye RS, Pereira-Smith OM, Smith JR (1991): Senescent cells fail to express cdc2, cyclin A and cyclin B in response to mitogen stimulation. Proc Natl Acad Sci USA 88: 11012–11016.
Steinman RA, Hoffman B, Iro A, Guillouf C, Lieberman DA, El-Houseini ME (1994): Induction of p21 (WAF-1/CIP1) during differentiation. Oncogene 9: 3389–3396.
Sugawara O, Oshimura M, Koi M, Annab LA, Barrett JC (1990): Induction of cellular senescence in immortalized cells by human chromosome 1. Science 247: 707–710.
Vaziri H, Dragowska W, Allsop RC, Thomas TE, Harley CB, Lansdorp PM (1994): Evidence for a mitotic clock in human hematopoietic stem cells: Loss of telomeric DNA with age. Proc Natl Acad Sci USA 91: 9857–9860.
Vogelstein B, Fearon ER, Kern SE, Hamilton SR, Preisinger AC, Nakamura Y, White R (1989): Allelotype of colorectal carcinomas. Science 244: 207–211.
Walmsley RM (1987): Yeast telomeres: the end of the chromosome story? Yeast 3: 139–148.
Wang XW, Lin X, Klein CB, Bhamra RK, Lee Y-W, Costa M (1992): A conserved region in human and Chinese hamster X chromosomes can induce cellular senescence of mickel-transformed Chinese hamster cell lines. Carcinogenesis 15: 555–561.
Whitaker NJ, Kidston CL, Reddel RR (1992): Finite lifespan of hybrids formed by fusion of different Simian Virus 40-immortilised cells. J Virol 66: 1202-1206.
Wilke MS, Edens M, Scott RE (1988): Ability of normal human keratinocytes that grow in culture in serum-free medium to be derived from suprabasal cells. J Natl Cancer Inst 80: 1299–1304.
Wilkie AOM, Lamb J, Harris PC, Finney RD, Higgs DR (1990): A truncated human chromosome 16 associated with a thalassaemia is stabilized by addition of telomeric repeat (TTAGGG)n. Nature 346: 868–871.
Wille JJ Jr, Pittelkow MR, Shipley GD, Scott RE (1984): Integrated control of growth and differentiation of normal human prokeratinocytes cultured in serum-free medium: clonal analyses, growth kinetics and cell cycle studies. J Cell Physiol 121: 31–44.
Wright WE, Pereira-Smith OM, Shay JW (1989): Reversible cellular senescence: implications of immortalization of normal human diploid fibroblasts. Mol Cell Biol 9: 3088–3092.

Yeudall WA, Crawford RY, Ensley JF, Robbins KC (1994): MTS1/CKD4I is altered in cell lines derived from primary and metastatic oral squamous cell carcinomas. Carcinogenesis 15: 2683–2686.

Yin Y, Tainsky MA, Bischoff FZ, Stroma LC, Wahl GM (1992): Wild type p53 restores cell cycle control and inhibits gene amplification in cells with mutant p53 alleles. Cell 70 : 937–948.

Yu G-L, Blackburn EH (1991): Developmentally programmed healing of chromosomes by telomerase in *Tetrahymena.* Cell 67: 823–832.

Zahler AM, Prescott DM (1988): Telomere terminal transferase activity in the hypotrichous ciliate *Oxytricha nova* and a model for replication of the ends of linear DNA molecules. Nucl Acids Res 16: 6953–6972.

Zhang S-Y, Klein-Szanto AJP, Souter ER, Shafarenko M, Mitsunaga S, Nobori T, Carson DA, Ridge JA, Goodrow TL (1994): Higher frequency of alterations in the p16/CDKN2 gene in squamous cell carcinoma cell lines than in primary tumours of the head and neck. Cancer Res 54:5050–5053.

Zhu L, van der Heuvel S, Helin K, Fattaey A, Ewen M, Livingston D, Dyson N, Harlow E (1993): Inhibition of cell proliferation by p107, a relative of the retinoblastoma protein. Genes Devel 7: 1111–1125.

2

Safety Procedures

Jane L. Caputo

American Type Culture Collection, 12301 Rockville Pike, Rockville, Maryland 20852

Culture of Immortalized Cells, Edited by R. Ian Freshney and Mary G. Freshney.
ISBN 0-471-12134-7

1. INTRODUCTION

Safety procedures for the establishment and replenishment of cell cultures have evolved over many years. While the historical concern emphasized protection of the cell cultures from contamination with viruses, mycoplasma, bacteria, and fungi, current procedures also stress the protection of the scientific personnel. Human and other primate tissue samples may be infected with any number of infectious viruses, rickettsiae, bacteria, fungi and parasites. Hepatitis B virus (HBV), human immunodeficiency virus (HIV) and *Mycobacterium tuberculosis* are the agents of principal concern to the developers of immortalized human cell lines ["Biosafety in the Laboratory," 1989]. Clinical materials may also be infected with Creutzfeldt–Jacob agent (CJA), the human herpesviruses, including Herpes simplex viruses 1 and 2, varicella, cytomegalovirus, and Epstein–Barr virus. Human and rodent tissue may be infected with Hantaan virus (Korean hemorrhagic fever), while primate tissue may be infected with Simian immunodeficiency virus (SIV). Immortalized cells, however established, may contain a transforming virus or transfectant.

While various national guidelines have been published, it is ultimately the responsibility of each individual laboratory to access the risk potential of the material in use and to determine the level of Biosafety necessary for appropriate protection. This chapter may serve as a guide to aid in that determination. Each institution is advised to establish an institutional biosafety committee and customize an institutional biosafety manual and/or human bloodborne exposure plan. The nature of the research or production will determine the content of these documents. Standard operating procedures (SOPs) should be in place to document exactly how each procedure should be performed in a specific institution.

Research workers should be informed of the risks known or potential for the material under investigation. They should have access to literature and training programs that reinforce safety procedures. New laboratory workers should be trained using material known to be of little risk and should not be allowed to work with potentially biohazardous material until they are proficient in all techniques required for a particular application [Centers for Disease Control, 1988b]. Strict adherence to these guidelines will contribute to a safer environment in research laboratories and production facilities.

2. ENGINEERING CONTROLS

Engineering controls include biosafety cabinets, puncture resistant sharps containers, and mechanical pipetting devices. They remove the hazard or help isolate the worker from exposure.

2.1. Primary

2.1.1. Biological safety cabinets

The primary method of containment is the use of a vertical laminar flow cabinet certified as either Class II or Class III. These cabinets provide a double benefit: the researcher is protected from biological hazards, and the work area is protected from contamination. All manipulation with primate tissue, oncogenic viruses, and transforming viruses should ideally take place in this type of cabinet. A particulate-free environment is provided for the tissue cultures, and the researcher is protected from airborne contamination. Aerosols are transported to high-efficiency particulate air (HEPA) filters, where they are removed from the airstream before the air is exhausted from the cabinet enclosure.

Class I cabinets, which function like chemical fume hoods, do not provide protection to the research material. Class II, Type A and Type B, cabinets provide personnel and product protection. Class II, Type A cabinets are suitable for work with low to moderate risk biological agents in the absence of volatile toxic chemicals and volatile radionuclides. They require a minimum air inflow of 75 ft/min (0.4 m/s), and they recirculate approximately 70% of the total air volume through the supply HEPA filter and exhaust approximately 30% through the exhaust (HEPA) filter. Class II, Type B1 cabinets are connected to an exhaust duct and maintain a minimum air intake velocity of 100 ft/min (0.5 m/s); they recirculate approximately 70% of the total air volume. Class II, Type B2 do not recirculate and are totally exhausted through a HEPA filter and exhaust duct. No air is recirculated into the room. Class III cabinets are totally enclosed gastight cabinets with a plate glass view screen and rubber gloves. Horizontal laminar airflow clean benches in which air blows from the side facing the researcher are not biological safety cabinets and should never be used for handling toxic, infectious, radioactive, or sensitizing materials. They provide no personnel or environmental protection and can expose the worker to aerosol or proteinaceous materials [U.S. Department of Health and Human Services, 1993].

Although Class II, Type A cabinets exhaust into the laboratory, they may be used with all Class I, II , or III etiological agents as classified by the U.S. Department of Health, Education, and Welfare, Center for Disease Control and with oncogenic viruses specified as low to moderate risk by the National Cancer Institute. They are not suitable for use with Class IV biological agents or oncogenic viruses specified as high risk. Since the HEPA filters are completely porous to gases and vapors, they have little ability to collect volatile chemicals. They should not be used with agents that have a significant amount of hazardous volatile components. Vapors that are hazardous from a toxic, radioactive, or flammability standpoint should not be used in cabinets which recirculate part of their air ["Biosafety in the Laboratory," 1989].

In the United States these cabinets must be inspected before initial use, at yearly intervals and whenever moved. This inspection for containment should be performed by professional laminar flow technicians according to the National Sanitation Foundation Standard 49. Tests II, IX, and X are performed. Test II determines the integrity of the HEPA filter. Test IX measures the velocity of the air moving through the cabinet workspace, and Test X measures the airflow at the work access opening (face velocity). Filters are replaced as needed by professional laminar flow technicians [National Sanitation Foundation Standard 49, 1983].

Aerosols The use of biological safety cabinets will reduce the risk of infection from aerosols. All procedures that might produce infectious aerosols should be performed in these cabinets, including, but not limited to, centrifuging without the use of safety cups, grinding, blending, vigorous mixing or pipetting and sonication. Centrifuge tubes should be opened in the cabinet as well as all specimen containers. Gauze or absorbent paper may be wrapped around the opening of the specimen containers to contain aerosols that may be produced and to reduce the risk from splashing. While the risk of transmission of HBV, HIV, or SIV by aerosols is thought to be rare, splashes or droplets of contaminated fluids could cause infection thorough exposure through the mucous membranes of the eyes, nose or mouth or through direct contact with abraded skin.

2.1.2. Pipetting devices

Although the safety cabinet is the principal containment device, its usefulness is limited to airborne contamination. Injection, ingestion and direct contact with potentially hazardous materials must also be avoided. The use of hand-held pipetting-devices is mandatory. *Under no circumstances should mouth-pipetting be allowed.* A wide variety of hand-held pipetting devices is available for use.

2.1.3. Centrifuge cups (or buckets)

The safety centrifuge cup, which is an enclosed container designed to prevent aerosols from being released during centrifugation, should be used to centrifuge all primate tissues and all cell lines that could be potentially infected. These materials can be centrifuged in the open laboratory, provided the safety cups are opened only in a biosafety cabinet that is rated as Class II or Class III. The centrifuge should never be opened while it is running; it should be completely stopped.

2.1.4. Puncture-resistant sharps containers

Rigid plastic and cardboard puncture-resistant containers are available for the disposal of needles, syringes, pipettes, and broken glass. These containers are leakproof on the sides and bottom, closable, and properly labeled. They can be capped when full and disposed of as a unit, and can subsequently be autoclaved or incinerated and disposed of without contact with the sharps. They provide a higher level of safety than biohazard bags.

2.2. Secondary

2.2.1. Facility design

Facilities should be designed to provide maximum protection to the environment while affording protection to the personnel within the laboratory. Laboratories should be easily cleaned, with benchtops impervious to water and resistant to acids, alkalies, and organic solvents. Autoclaves should be available for use within each laboratory or otherwise easily accessible. Hand-washing facilities must be located in each laboratory at, or close to, the exit.

3. WORK PRACTICE CONTROLS

Work practice controls such as hand-washing, exposure policies, and proper waste disposal techniques reduce the likelihood of exposure through the alteration of the manner in which the task is performed. The greatest risk of transmission of HBV and HIV and other bloodborne pathogens is by direct entry into the bloodstream. This risk is best minimized by meticulous attention to work practice controls that eliminate wherever possible the use of sharps and prevent contamination of open wounds or abrasions on the skin.

3.1. Technique and Hygiene

3.1.1. Routine exposure

Eating, drinking, gum chewing, the use of tobacco products, handling contact lenses, and applying cosmetics or lip balm are not permitted in any laboratory area.

Food or beverages must not be stored in refrigerators, freezers, shelves, or cabinets or on countertops or benchtops in any laboratory area; cabinets or refrigerators designated for the storage of food and beverages only must be used, and these should be located outside the laboratory area.

3.1.2. Hand-washing

Hand-washing facilities should be readily available, preferably in the same work area as the laminar flow cabinet. Hands are washed before and after handling cells, tissue samples, clinical material, and all blood or blood products. Hand-washing is required immediately after removing gloves and other personal protective devices. The use of gloves is not a substitute for frequent hand-washing. Contaminated gloves should be removed immediately and hands washed before putting on another pair of gloves. Hands or other exposed skin should be washed as soon as feasible following contact of such body areas with human blood or other potentially infectious material. Telephones, doorknobs, laboratory notebooks, or laboratory equipment should not be handled before contaminated gloves are removed and hands washed.

3.1.3. Sharp implements, needles, and ampules

The use of needles, syringes, and scalpel blades should be avoided whenever an alternative method is available. Needles should never be recapped, removed from the syringe, bent, sheared, or broken. All sharp implements should be discarded in a puncture-resistant container and properly decontaminated, preferably by autoclaving or incineration.

3.1.4. Broken glass, pasteur pipettes, and ampules

All broken glassware, pasteur pipettes, slides, capillary tubes, and ampules should be discarded in puncture-resistant containers that can be closed and either autoclaved or incinerated. Never handle broken glassware directly by hand; use mechanical means such as a brush and dustpan or forceps. Pasteur pipettes must never be discarded with reusable glassware or placed in plastic autoclave bags. As a general rule, anything that might puncture the discard bag must be put in a puncture-resistant container. Anything that might cause cleanup personnel injury such as pasteur pipettes or broken ampules must not be placed in discard pans with reusable glassware. Whenever possible, unbreakable plasticware should be substituted for glassware.

3.1.5. Contact Lenses

Contact lenses should not be worn around chemicals, fumes and other hazardous materials, and dust particles. When contact lenses are worn, eye protection, such as tight-fitting goggles or a face shield should be worn [U.S. Department of Health and Human Services, 1993].

Contact lenses do not provide protection to the eyes. In fact, the use of contact lenses may increase the risk of eye damage. Foreign material present on the surface of the eye may become trapped in the capillary space between the contact lenses and the cornea. Inert, but sharp particles, caustic chemicals, irritating vapors, and infectious agents in this space cannot be washed off the surface of the cornea. If the material that gets into the eye is painful, it becomes extremely difficult to remove the contact lens because of the muscle spasms that may develop [Office of Research Safety, 1978]. In some cases, the lens will actually adhere to the eye and not let the eye wash cleanse the eye. Some chemicals are solvents for plastic lenses and will cause the lens to melt to the eyeball surface.

3.1.6. Standard techniques

- The biological safety cabinet should be turned on 15 min before use to allow complete exhaustion of dirty air; the surface should be decontaminated with 70% ethyl alcohol or other suitable disinfectant to remove surface dust.
- The work operation should be preplanned. Everything needed for a complete procedure should ideally be placed in the cabinet before starting so that nothing passes in or out thorough the air barrier. Separate clean and dirty materials on opposite sides of the operator if possible. Equipment should be placed at least 4 in. (20 cm) from the front opening, and manipulations should be performed as deeply in the cabinet as possible. Substantial leakage from the cabinet can occur when work is performed within 4 in. (20 cm) of the opening. If the unit has a swing-up view screen, make certain that it is securely latched in the closed position. If the unit has a sliding-sash view screen, secure it at the 8 in (40 cm) opening.
- The work surface may be covered with plastic-backed absorbent lab toweling to capture and facilitate spill cleanup. Replace toweling frequently; decontaminate used toweling by incineration or autoclaving.
- After all the materials are placed in the cabinet, wait approximately 3 min before beginning work to allow purging of airborne contamination from the work area. Do not block the intake grilles; all material must be confined to the solid work surface, not the perforated.
- Cabinet surfaces should be decontaminated before another cell line or tissue sample is placed on the work surface. An autoclavable discard pan filled with water or with sodium hypochlorite (household bleach), depending on the decontamination method employed, is placed in the rear of the cabinet. All pipettes and other equipment used for handling cell cultures are immersed in this pan.

- The pan is covered before removal from the cabinet and preferably autoclaved before opening. Large plasticware (e.g. flasks) may be discarded into large biohazard bags that are autoclaved before disposal. The length of time needed for decontamination will depend on the size of the pans, bottles, and bags used and must be determined at each facility. In some cases, an autoclave time of >2 h will be required. Individual laboratories using chemical decontamination such as sodium hypochlorite must determine the safety of autoclaving these materials. Autoclave control indicators such as Kilit Ampules and Kilit Sporestrips 1 are commercially available and will facilitate this determination.
- Alternatively, some laboratories may prefer to decontaminate using chemical methods. Immersion in 2500 ppm available chlorine (1:20 dilution of 5.25% sodium hypochlorite, such as domestic bleach, e.g., Chlorox) for 2 h is generally sufficient. Discard pans containing hypochlorite should not need to be autoclaved. However, individual laboratories using chemical decontamination and subsequently autoclaving must determine the safety of autoclaving these materials. It is possible that, in some cases, poisonous chlorine gas will be released into the laboratory cleanup area; this might well prohibit this method of decontamination.
- The building vacuum system should be protected from biohazards by the use of a HEPA or hydrophobic filter between the collection vessel and the vacuum shutoff valve in the cabinet.

3.1.7 Accidents and spills

- All accidents and spills should be reported to the institutional biosafety officer and/or biosafety committee in accordance with institutional guidelines. An SOP should be in place to define proper procedures to be followed in the event a potentially infectious agent is spilled. A record of these incidents should be maintained.
- All spills should be attended to immediately. The institutional SOP should have specific instructions for handing large and small spills. Each worker must be familiar with the procedures outlined. Spills of liquids require the absorption of the liquid to minimize the spread of the contaminant. Spill kits containing absorbent materials capable of soaking up spills of various sizes should be available in every laboratory area.
- Accidental spills of human tissue or blood may be decontaminated with sodium hypochlorite solution (1:10 dilution of household bleach). This solution will inactivate most infectious agents. This solution should be freshly prepared and kept in the diluted form for no longer than one week.

- Procedures to follow after an accidental contamination of an exposed area of the skin should be detailed in the institutional SOP. Contaminated clothing should be removed and autoclaved; skin should be washed with a good liquid detergent and water and then rinsed copiously with water. Do not apply strong disinfectants to the skin and avoid abrading the skin by vigorous scrubbing.

3.2. Personnel Protective Equipment

3.2.1. Laboratory coats

- Protective laboratory coats, gowns, smocks, or uniforms are worn when handling all cell cultures, human blood, and other potentially infectious material. The exact type and characteristics of the covering will depend on the task and the exposure anticipated. Back, or side, fastening gowns may be preferable.
- They must be removed before leaving the laboratory. They are not permitted in office areas, lunchrooms, or libraries or outside the building.
- If a laboratory coat becomes contaminated, it must be replaced as soon as possible.
- All protective clothing is either disposed of in the laboratory or laundered by the institution; laboratory clothing should never be taken home for any purpose.

3.2.2. Gloves

The use of gloves will protect the operator from contamination and will also minimize the shedding of skin flora into the work area.

- Gloves are worn when skin contact with the material is unavoidable.
- Gloves should be worn, particularly when the worker has cuts, scratches, or other breaks in the skin.
- Gloves should be worn when handling human biopsy material and human blood.
- *Never reuse or wash disposable gloves.*
- The use of gloves is not a substitute for frequent hand washing.
- Hand-washing is required immediately after removing gloves.
- Double-gloving may be recommended when working with material known to be contain HIV, HBV, or other human pathogens.
- Contaminated, torn, or punctured gloves must be replaced as soon as possible and hands washed before putting on another pair of gloves.

The type of glove used will depend on the application and degree of dexterity required. Particular attention must be given to allergic reactions, rashes, and hand irritations, whether to the latex or to the powder used to facilitate use of the gloves. There are many types of vinyl, polyethylene, and powder-free gloves commercially available. The use of sleeves to protect the arm between the top of the glove and the bottom of the sleeve of the lab coat should be considered.

3.2.3. Face masks, eye protection, and face shields

Wear a mask and eye covering or a full-face shield when there is a potential for splashing or spraying or the creation of aerosols from any biological material especially blood or body fluids.

- A full-face mask must be worn when working with Class 2 agents or with bloodborne pathogens outside a biosafety cabinet.
- Wear a full-face mask when retrieving ampules from liquid nitrogen and when thawing ampules that have been stored in liquid nitrogen.
- Research workers who wear contact lenses must wear eye goggles or face shields.

3.3. Freezing and Thawing Ampules

3.3.1. DMSO

The chemical dimethylsulfoxide (DMSO) is a powerful solvent that can penetrate many synthetic and natural membranes, including skin and rubber gloves. Therefore, potentially harmful substances, such as carcinogens, could be carried into the body through the skin or even through rubber gloves [Freshney, 1994]. DMSO is used as the protective freezing additive in preserving almost all cell lines and must always be handled with caution.

Special safety precautions should be taken when retrieving glass or plastic ampules from liquid nitrogen storage; the potential exists that liquid nitrogen has entered the ampule and rapid expansion of the liquid nitrogen could cause an explosion creating flying fragments and disseminating the contents of the ampule.

- Protective gloves and clothing should be used when thawing ampules, and a face mask or safety goggles must be worn. Storing ampules in the liquid nitrogen vapor instead of submerging them in the liquid should eliminate this hazard.
- Ampules should be opened in a biological safety cabinet in a manner to prevent dispersion of the ampules contents.
- Wrapping the neck of the ampule between several folds of sterile

gauze will protect the operator from cuts and also will contain aerosols and droplets [Caputo, 1988].

3.4. Waste Disposal

3.4.1. Sharp implements

- Needles should be discarded in rigid plastic puncture-resistant containers that are clearly labeled or color-coded. The needle and syringe should fit into the container as a complete unit.
- Never separate the needle from the syringe.
- Never bend, shear or recap needles.
- Sharp-implement containers should be autoclaved or incinerated.
- Separate containers clearly labeled for broken-glass disposal should be made available for the disposal of broken glass, pasteur pipettes, and other sharp implements. These containers can be capped when full and disposed of as a unit.
- Pasteur pipettes must not be discarded in plastic bags or put into autoclave pans with reusable glassware.
- Nondisposable sharp implements, if used, must be placed in a hard-walled container for transport to the processing area for decontamination, preferably by autoclaving.

3.4.2. Biomedical waste

All institutions should establish a waste management plan for the collection, segregation, containment, treatment, and disposal of infectious and potentially infectious wastes. Biomedical waste must be disposed of following local and national regulations.

- All forms of blood and/or infectious materials should be placed in appropriate containers that are properly labeled. The labels should be orange or red-orange with the biohazard symbol in a contrasting color.
- Bags and pans to be transported to an autoclave should be placed in leakproof containers.
- After autoclaving, each institution should have in place a procedure for disposal; for example, bags are placed in specially marked or color-coded containers for pickup as biohazardous waste.

3.4.3. Contaminated laundry and housekeeping

Contaminated laundry should be placed in properly labeled or color-coded bags or containers and transported to the laundry facility. Laboratories must be scheduled for periodic cleaning and appropriate disinfection to ensure that the workplace is clean and sanitary. An insect and rodent control program should be in effect.

4. BIOSAFETY LEVELS, UNIVERSAL PRECAUTIONS, AND BLOODBORNE PATHOGENS

In the United States all etiologic agents have been classified by the Public Health Service on the basis of hazard as Class 1, 2, 3, or 4. [U.S. Department of Health, Education and Welfare, Public Health Service, 1974b]. The National Cancer Institute has classified oncogenic viruses as low, moderate, or high risk [U.S. Department of Health, Education and Welfare, 1974a].

The class I classification consists of etiologic agents of no recognized hazard when used under ordinary conditions of handling. Class 2 agents are those considered to be of ordinary potential harm. Class 3 consists of pathogens involving special hazard, and Class 4 consists of agents of potential danger to the public health or animal health, or both, or of extreme danger to laboratory personnel. A high risk oncogenic virus is one known and proven to cause cancer in humans. No virus is in that category at the present time. A virus not classified as moderate risk is considered to be low risk. The criteria for moderate risk are not absolute but can include the following:

(1) The virus is a suspected oncogenic virus isolated from humans.

(2) It transforms human cells *in vivo* as evidenced by a morphologic or functional alteration, or both, that is transferred genetically.

(3) It produces cancer without the aid of experimental modification, whether in a subhuman primate of any age or across other mammalian species barrier in juvenile or adult animals.

(4) It is a genetic recombinant between an animal oncogenic virus and microorganism infectious for humans.

(5) It is concentrated oncogenic virus or transforming viral nucleic acid.

The Centers for Disease Control and Prevention and The National Institutes of Health have described four biosafety levels that are a combination of laboratory practices and techniques, safety equipment, and laboratory facilities [U.S. Department of Health and Human Services, 1993]. These biosafety levels provide increasing levels of protection to laboratory personnel and to the environment.

When a cell line or tissue sample is known to contain an etiologic agent or an oncogenic virus, the cell line can be classified at the same level as that recommended for the agent. Therefore, following this guideline a cell immortalized with Epstein–Barr virus (EBV) or Simian virus 40 (SV40) could be handled at Biosafety Level 2. However, it is suggested that the biosafety level be raised to the next highest level when the workload increases from the research level to production quantities. Most cell lines currently available in the United States and the United Kingdom have not been thoroughly tested for the presence of viruses. In most cases tissue samples from humans cannot be certified as noninfectious. The precautions for Biosafety Level 2 containment should be used to handle all primate cell lines derived from lymphoid or tumor tissue, all cells exposed to or transformed by a primate oncogenic virus, all human tissues and fluids received from surgical resection, autopsy, or clinical studies. These precautions must be followed regardless of whether the samples have been shown to harbor a Class 2 agent. The possibility exists that human tissue samples may contain hepatitis viruses, human immunodeficiency viruses, tuberculosis, or other pathogens [U.S. Department of Health and Human Services, 1993].

Routine diagnostic work with HIV can be done in a Biosafety Level 2 facility using the practices and procedures recommended for Biosafety Level 2. Research work (including cocultivation, virus replication studies, or manipulations involving concentrated virus) can be done in a Biosafety Level 2 facility using Biosafety Level 3 practices and procedures. Virus production activities, including virus concentrations, require a Biosafety Level 3 facility with Biosafety Level 3 practices and precautions [U.S. Department of Health and Human Services, 1993]. All decisions regarding biosafety levels ultimately must be made by the institutional laboratory director after a careful risk assessment of the work to be performed with the goal of minimizing the potential for laboratory acquired infections.

Biosafety Level 2 is suitable for work involving agents that represent a moderate hazard to personnel and the environment. Personnel involved in handling potentially biohazardous materials must have specific training in handling biohazardous materials and they must be supervised by competent scientists. Unlike Biosafety Level 1, certain procedures must be conducted in biological safety cabinets. The following precautions are generally recognized as those to be followed for Biosafety Level 2 containment:

(1) Access to the laboratory is limited or restricted by the laboratory director.

(2) Persons with an increased risk of infection should not be allowed in the laboratory.

(3) Work surfaces are decontaminated before and after each procedure.

(4) Mechanical pipetting devices are used for all procedures; mouth-pipetting is never permitted.

(5) Eating, drinking, gum chewing, the use of tobacco products, handling contact lenses, and applying cosmetics or lip balm are not permitted in the laboratory; food is not stored in laboratory areas. Persons who wear contact lenses in laboratories should consider the use of goggles or a face shield.

(6) Laboratory coats are worn in and are removed before leaving the laboratory.

(7) Hands are washed before and after handling cells lines, tissue samples, and blood or blood products.

(8) A Class II or Class III vertical laminar flow biological safety cabinet is used for all cell manipulations that may create aerosols or splashes, regardless of whether the procedure requires sterility. These procedures include, but are not limited to, centrifuging without a safety centrifuge cup, grinding, blending, vigorous shaking, mixing or pipetting, and sonication. All initial specimens must be opened in a biosafety cabinet to contain aerosol dispersion and to minimize the risk of accidental exposure to mucous membranes of the eyes, nose, and mouth and to the skin. Benchtops may be used for manipulations that do not produce splashes or aerosols such as cell counting using a hemocytometer. Contaminated materials are placed in leakproof pans and decontaminated before disposal by an approved decontamination method such as autoclaving. These pans are closed before removal from the cabinet. All applicable local, state, and national regulations must be followed.

(9) Samples for cell counting should be put on the hemocytometer inside the biosafety cabinet and taken to the microscope on the bench for the cell count. The hemocytometer should then be returned to the biosafety cabinet, where it should be immersed in a 1:10 dilution of household bleach for at least 10 min.

(10) Large plastic ware (e.g., roller bottles, and tissue culture flasks, centrifuge tubes), which do not fit in the discard pan, and which can be closed, are discarded into autoclave bags.

(11) Contaminated materials are autoclaved before disposable materials are discarded and before reusable glassware is washed.

(12) Disposable gloves are worn when skin contact with infectious materials is unavoidable, or at the discretion of the worker. Individual laboratories may mandate that all workers wear gloves when manipulating potentially biohazardous materials because the possibility always exists that skin contact can occur if there is a splash or a leak in a container. All breaks in the skin of the hand and lower forearm must be covered with a dressing impervious to liquids. Arms should be covered above the glove; the use of sleeve protectors with elastic ends will protect the area of the arm between the top of the glove and the bottom of the sleeve of the lab coat.

(13) Laboratory personnel should receive appropriate immunizations or tests for the agents handled or potentially present in the laboratory.

(14) Baseline serum samples may be collected and stored before initial laboratory exposure and at yearly intervals.

(15) An institutional biosafety manual is prepared and available to all personnel.

(16) Appropriate personnel training is provided before handling any potentially biohazardous material and on a yearly basis thereafter.

(17) All specimens of human tissue and fluids must be transported in containers that cannot leak; double packaging may be advisable.

The guidelines listed for Biosafety Level 2 Containment are sufficient for working with all cell lines and tissue samples provided that the quantities in use are small and the amount of virus is low. Procedures involving research quantities of HIV or HIV in high concentration can be conducted in a Biosafety Level 2 facility with the additional practice and containment equipment recommended for Biosafety Level 3 [Centers for Disease Control, 1988a]. This recommendation should also apply to cell lines known to harbor other human viruses such as HBV and EBV and primate viruses such as SIV and SV40. Cell lines transformed with EBV and SV40 can be also be handled at Biosafety Level 2. As these cells lines may continue to produce virus, large-scale use would best be handled using the practices and equipment listed as Biosafety Level 3. Activities involving industrial-scale production or high concentrations would be conducted in Biosafety Level 3 facility. Animals including mice inoculated with HIV, HBV, or SIV may be chronically infected, and all tissue and blood samples and all cell lines developed from such animals should be handled with the above precautions [Locardi et al., 1992]. Hybridomas raised against human material which may be potentially infectious must be considered as potentially harboring those agents.

In addition to all the precautions listed above, modified Biosafety Level 3 containment would require

(1) Laboratory doors are kept closed when tissues or cells are been manipulated.

(2) Solid front or wraparound laboratory coats are worn; they are sterilized before being laundered.

(3) Vacuum lines are protected with HEPA filters and disinfectant traps.

(4) Baseline serum samples must be collected, both at time of employment and annually.

(5) Biological safety cabinets (Class I, II, or III) must be used for

all activities involving potentially infectious material whether or not aerosols are created.

(6) Gloves must be worn.

In addition, a Biosafety Level 3 facility would contain

(7) A laboratory physically separated from other areas by passage though two sets of doors. Access doors are self-closing.

(8) Penetration to laboratory walls, floors, ceilings, and windows are sealed.

(9) A steam autoclave, preferably in the laboratory.

(10) Laboratory air is not recirculated to other areas of the building.

(11) HEPA-filtered exhaust from the biological safety cabinet may be discharged in the laboratory only if the cabinet is certified annually; otherwise the exhaust must be discharged outside.

Bloodborne pathogens refers to pathogenic microorganisms that are present in human blood and can cause disease in humans, including, but not limited to, hepatitis B virus and human immunodeficiency virus. The Bloodborne Pathogens Regulation [U.S. Department of Labor, Occupational Safety and Health Administration, 1991] applies to all persons who may reasonably anticipate contact with blood or other potentially infectious material. The regulation requires the creation of a written *exposure control plan* that describes how all employers will protect their employees from exposure.

Universal precautions are an approach to infection control according to which all human blood and all human body fluids are treated as if known to be infectious for HIV, HBV, and other bloodborne pathogens. It is not possible to test all samples of human blood, and even if tested, the possibility could exist that a latent infection was present or that false-negative results were obtained. It is seldom possible to determine retrospectively whether a tissue source was infected with a bloodborne pathogen [Grizzle, and Polt, 1988].

5. INSTITUTIONAL SAFETY PROGRAM

5.1. Biosafety Manual

All institutions should prepare a *manual of safety* specific for their laboratories with safety and operational procedures carefully outlined. SOPs should be in place with specific step-by-step instructions for all laboratory procedures.

5.2. Human Bloodborne Pathogen Exposure Plan

In the United States each employer in institutions where blood or blood products are handled shall establish a written *exposure control plan* designed to eliminate or minimize employee exposure [U.S. Department of Labor, Occupational Safety and Health Administration, 1991]. This plan contains the following elements:

(1) *A list of employees whose job duties might include exposure to bloodborne pathogens.* Both primary researchers and those involved in disposal and cleanup such as glassware assistants are included. Duties can include but are limited to

(2) The separation, transformation, transfection, expansion, and characterization of cell lines from whole blood from human or other primates.

(3) Handling clinical material or other potentially infectious materials from surgical resection or autopsy.

- (a) Preparation of media from whole blood.
- (b) The extraction of DNA from whole blood or from cell lines not previously determined to be free of bloodborne pathogens.
- (c) Handling HIV or HBV containing cell cultures or viral cultures; handling molecular clones of HIV or HBV.
- (d) Handling all primate cell lines unless previously screened for HIV, HBV, and/or SIV.
- (e) Handling all laboratory animals inoculated with HIV, HBV, or SIV and all tissue and blood products obtained from those animals and all cell lines developed from these animals, including hybridoma cell lines.
- (f) Handling transgenic mice or cell lines carrying either in tact or partial HIV, HBV, or SIV genomes.

(4) *A schedule for hepatitis B vaccination.* This should be made available to anyone whose job duties might include exposure, including clean-up personnel.

(5) *A schedule for training.* Training should be made available to all employees whose job duties might include exposure to bloodborne pathogens during working hours and at no cost to the employee.

(6) *An exposure surveillance plan.* A plan must be in place for postexposure evaluation and follow-up. The exposure route and circumstances surrounding the incident must be documented. Appropriate medical and psychological support must be made available.

(7) *Methods of recordkeeping.* Each institution shall maintain in confidence medical and training records for each employee who may

be exposed or who has been exposed to a bloodborne pathogen. A signed form for each training session should be on file as well as dates and results of HBV inoculations and follow-up. All exposure incidents should be recorded along with follow-up medical care received.

The *human bloodborne pathogen plan* shall be updated at least annually; it must be modified whenever necessary to reflect new circumstances, either tasks or procedures, or new or revised employee positions.

5.3. Baseline Blood Samples

Baseline serum samples should be collected from all laboratory personnel and stored at −70°C at the time of employment and annually thereafter. These samples will be available for evaluation if exposure to an adventitious agent should occur.

5.4. Hepatitis B Immunization

In the United States all employees who may reasonably anticipate contact with blood or other potentially infectious materials must be offered the opportunity to receive hepatitis B immunization. Employees whose job duties include exposure to blood and who refuse this vaccination must sign a *hepatitis B vaccine declination.* Individual laboratories can at their discretion make hepatitis B vaccination mandatory. Hepatitis B virus may be found in all body fluids and secretions of infected individuals. It may be transmitted by blood transfusion, sexual contact, mucous membrane exposure, or puncturing of the skin by contaminated instruments. Many infections can occur without symptoms, or there may be mild symptoms. However, the symptoms may be very severe and can result in chronic hepatitis, permanent liver damage, or even death.

The hepatitis B vaccine is produced by recombinant DNA engineering; the vaccine contains a viral protein of the hepatitis B virus. The vaccine is not made from blood or blood products and is not infectious. Three doses of the vaccine are given over a 6-month period, and the vaccine is approximately 90% effective. The safety of this vaccine has been demonstrated by extensive studies. All vaccine recipients must have post-vaccination testing of their blood to determine that they are immune to HBV.

5.5. Annual Safety Training

Employees must receive annual training to ensure that they understand the hazards of working with bloodborne pathogens, the modes of transmission of bloodborne infectious agent, the risks involved and the methods to minimize those risks, the *exposure control plan,* the bloodborne regulations, and the use and limits of engineering control, work practices, and personal protection equipment. Emergency procedures, postexposure evaluations, hepatitis B virus (HBV) vaccine, and the use of signs and labels should also be discussed. New employees should receive training before exposure to bloodborne pathogens or transforming viruses. Supplemental training should be provided anytime a modification in procedures may affect the potential for exposure occurring.

6. NATIONAL GOVERNMENT AND LOCAL REGULATIONS

The first set of guidelines in the United States describing the four biosafety levels was published jointly by the Centers for Disease Control and the National Institutes of Health in 1984. This publication has been updated twice, and the current third edition of "Biosafety in Microbiological and Biomedical Laboratories" [U.S. Department of Health and Human Services, 1993] will serve as the most comprehensive guideline available in the United States for recommendations for biosafety levels for research and clinical work involving agents infectious or potentially infectious to humans. These biosafety levels are consistent with the general criteria used to assign agents to Classes 1–4 in "Classification of Etiological Agents on the Basis of Hazards" [U.S. Department of Health, Education and Welfare, Public Health Service, 1974b] and in the National Cancer Institutes *Safety* "Standard for Research Involving Oncogenic Viruses" [U.S. Department of Health, Education and Welfare, National Institutes of Health, 1974a]. The National Institutes of Health has published guidelines for research using recombinant DNA techniques; its recommendations are consistent with the descriptions of Biosafety Levels 1–4 previously described [Office of Research Safety, NCI, 1978]. In 1988, guidelines were promulgated for healthcare workers describing the concept of universal precautions for prevention of transmission of human immunodeficiency virus, hepatitis B virus, and other bloodborne pathogens in healthcare settings [Centers for Disease Control, 1988c].

Universal precautions and Biosafety Levels 1–4 were used as the basis of the most recent OSHA publication [U.S. Department of Labor, Occupational Safety and Health Administrations, 1991]. This document removes the guidelines for institutions working with bloodborne pathogens in the United States from voluntary to mandatory. This standard became effective on March 6, 1992.

The Centers for Disease Control provides updated information on bloodborne pathogens as well as other infectious agents. New developments, exposure incidents, and methods of infection control are published weekly in the "Morbidity and Mortality Weekly Report" (MMWR). New agents are discovered periodically, and descriptions are published as Agent Summary Statements in this report.

REFERENCES

"Biosafety in the Laboratory: Prudent Practices for the Handling and Disposal of Infectious Materials" (1989). National Research Council. Washington, DC: National Academy Press.

Caputo JL (1988): Biosafety procedures in cell culture. J Tiss Cult Meth 11: 223–227.

Centers for Disease Control (1988a): Agent summary statement for human immunodeficiency viruses (HIVs) including HTLV-III, LAV, HIV-1. and HIV-2. MMWR 37: S-4, 1–18.

Centers for Disease Control. (1988b): Occupationally acquired human immunodeficiency virus infections in laboratories producing virus concentrate in large quantities. MMWR 37: S-4, 19–22.

Centers for Disease Control (1988c): Update: universal precautions for prevention of transmission of human immunodeficiency virus, hepatitis B virus and other bloodborne pathogens in healthcare settings. MMWR 37: 377–382, 387, 388.

Freshney RI (1994): "Culture of Animal Cells. A Manual of Basic Technique," 3rd ed. New York: Wiley-Liss, 63.

Grizzle WE, Polt SH (1988): Guidelines to avoid personnel contamination by infective agents in research laboratories that use human tissue. J Tiss Cult Meth 11: 191–200.

Locardi C, Puddu P, Ferrantini M, Parlanti E, Sestili P, Varano F, Belardelli F (1992): Persistent infection of normal mice with human immunodeficiency virus. J Virol 66: 1649–1654.

National Sanitation Foundation Standard 49 (1983): "Class II (Laminar Flow) Biohazard Cabinetry," Ann Arbor, Michigan.

Office of Research Safety, National Cancer Institute, and the Special Committee of Safety and Health Experts. 1978. "Laboratory Safety Monograph: A Supplement to the NIH Guidelines for Recombinant DNA Research." Bethesda, MD.

US Department of Health, Education and Welfare, National Institutes of Health (1974a): "National Cancer Institute Safety Standard for Research Involving Oncogenic Viruses."

US Department of Health, Education and Welfare, Public Health Service, Center for Disease Control, Office of Biosafety (1974b). "Classification of Etiologic Agents on the Basis of Hazard," 4th ed.

US Department of Health and Human Services (1993): "Biosafety in Microbiological and Biomedical Laboratories," 3rd ed, publication (CDC) 93-8395, Centers for Disease Control, US Government Printing Office, Washington, DC.

US Department of Labor, Occupational Safety and Health Administration (1991): "Occupational Exposure to Bloodborne Pathogens," Final Rule, Fed Register 56: 64175–64182.

APPENDIX A: MATERIALS AND SUPPLIERS

Materials	Suppliers
Bags, labels, and tags (biohazard)	Lab Safety Supply
Biological safety cabinets	The Baker Company
Biotransport carrier	Nalge Company
Face mask	VWR Scientific
Filters	Gelman Sciences
Instrument or pipette sterilizing pan	Nalge Company

APPENDIX (*Continued*)

Materials	Suppliers
Kilit Ampules, BBL, and Kilit Sporestrips 1, BBL	VWR Scientific
Needle and sharp-implement discard containers	Lab Safety Supply
Pipette aids	VWR Scientific
Protective clothing (laboratory coats; sleeves; gloves)	Lab Safety Supply
Sodium hypochlorite (Chlorox)	Lab Safety Supply

APPENDIX B: PROCEDURES FOR HANDLING OF "LOW RISK" BLOOD SAMPLES AND EBV TRANSFORMED CELL LINES PER U.K. REGULATIONS*

Introduction

The main text of this chapter has dealt with safety procedures from a general standpoint as relevant to a laboratory in the United States. Safety regulations in the United Kingdom and the rest of Europe are similar, but it was thought that it might be useful to include the following chapter written from the point of view of a laboratory in the United Kingdom.

U.K. safety regulations

In the United Kingdom safety is managed by the Health and Safety Executive (HSE). They are responsible for all aspects of safety at work and are empowered to inspect sites and enforce government regulations. Their powers range from the serving of improvement orders on a organization to closure until appropriate safety standards have been met.

The genetic modification of organisms and their manipulation in contained laboratory facilities are part of U.K. and European Community law. These are "Genetically Modified Organisms (contained use) Regulations" [1992]. They have been in force in the United Kingdom since February 1, 1993. They cover all experiments using genetically modified microorganisms in contained laboratory use. These regulations replaced the "Genetic Manipulation Regulations" [1989].

The regulations define genetic modification of an organism as "the altering of the genetic material in that organism by a way that does not occur naturally by mating or natural recombination or both."

The main duties of persons wishing to carry out work under these regulations are

* This appendix (Appendix B) was contributed entirely by Nigel K. Spurr of Human Genetic Resources, Imperial Cancer Research Fund (see Contributors list for complete affiliation).

- To carry out an assessment of the risks to human health and the environment and to keep records.
- To establish a local genetic modification safety committee to advise on risk assessments.
- To classify all experiments according to guidelines.
- To notify the HSE of the use of premises and to obtain permission for experiments in higher containment categories.
- To adopt appropriate containment measures.

Anyone contemplating the use of genetically modified organisms *must* realize that these regulations are not just advisory but have the force of law. A laboratory found to be contravening the regulations can be closed down. A properly constituted, representative, and informed safety committee must consider all proposals to work with genetically modified organisms, assess the risk, and allocate the work to the appropriate level of containment. Work coming under Category 2 containment will require a 60-day notice to HSE. Information on procedures for risk assessment and categorisation can be obtained in the following publications from HSE.

Relevant publications

The classification and other guidelines are laid out in a series of guidance notes published by the HSE from their Advisory Committee on Genetic Modification (ACGM).

These guidance notes cover the following areas:

ACGM/HSE Note 1: guidance on construction of recombinants containing potentially oncogenic nucleic acid sequences.

ACGM/HSE Note 4: guidelines for the health surveillance of those involved in genetic manipulation at laboratory and large scale.

ACGM/HSE Note 5: guidance on the use of eukaryotic viral vectors in genetic manipulation.

ACGM/HSE Note 6: guidelines for the large-scale use of genetically manipulated organisms.

ACGM/HSE Note 7: guidelines for the risk assessment of operations involving the contained use of genetically modified microorganisms.

ACGM/HSE Note 8: laboratory containment facilities for genetic manipulation.

ACGM/HSE Note 9: Guidelines on work with transgenic animals.

ACGM/HSE Note 10: Guidance on work involving the genetic manipulation of plants and plant pests.

ACGM/HSE Note 11: Genetic manipulation safety committees.

Copies of these guidance notes can be obtained from Health Policy Division A3 (address listed below) and should be consulted before work commences.

The release of modified microorganisms into the environment and commercial sale is covered by a separate set of regulations, the "Genetically Modified Organisms (Deliberate Release) Regulations" [1992], which were made under the Environmental Protection Act of 1990. These regulations in the United Kingdom parallel similar EC Directives on the release of genetically modified microorganisms. These also came into force on February 1, 1993 and replaced existing Genetic Manipulation Regulations, 1989.

In addition to the above regulations all work is also covered by the management of Health and Safety at Work Regulations (1992) and the Control of Substances Hazardous to Health Regulations (1988). The latter mainly covers chemicals. However, these regulations are being altered to cover all microorganisms.

Health Hazards

The principal hazards to health from work with human blood and tissues are of infection with microorganisms, particularly hepatitis B virus (HBV) and human immunodeficiency virus (HIV). The greatest hazard of transmission of both viruses is by direct entry into the bloodstream. Under laboratory conditions this is most often by contamination of open wounds with blood or accidental injury by glassware or sharp implements. Transmission by aerosol is thought to be rare.

Acute HBV infection has a significant mortality and may also result in chronic hepatitis leading to long-term liver damage and death. However, an effective recombinant vaccine is available providing good protection against infection. In the case of HIV, only a minority of injuries involving infected blood result in infection. However once infected, most individuals eventually progress to AIDS and the majority die within 5–15 years after infection. Treatment may prolong life but is not curative.

Control of Exposure

These guidelines work only with blood and tissue samples from individuals with a relatively low chance of infection with HBV or HIV. They specifically avoid the use of blood from known high risk groups, including homosexual or bisexual males, injecting drug abusers, haemophiliac patients who have received untreated blood products, and persons from high-prevalence areas or their sexual partners. However, all human blood has a risk of containing these pathogens and for ethical reasons it is not possible to test blood routinely.

Samples

The blood samples that can be accepted for laboratory experiments fall into two categories.

(1) Samples tested for HBV and HIV prior to entry into the laboratory (from blood banks). These can be considered to be "minimal risk" samples and that handling under containment Level 2 is appropriate even if the samples are cultured for >100 h.

(2) All untested samples. (e.g., laboratory volunteers and patients not listed in the "high risk" groups detailed above). These samples are considered to be "low risk" and may be handled under Containment Level 2 unless cultured. For long term culture (>100 h) modified level 2 should be used. This is justified for the following reasons

> (a) In London the incidence of HIV seropositivity in women attending antenatal clinics is low (0.2%), and even among those attending genitourinary medicine clinics, which may be considered a high risk group, it is only 0.35% (excluding those known to be HIV-antibody-positive and known intravenous drug users).
>
> (b) Some experiments involve the isolation of T-cell clones and EBV transformed B-cell lines from donors. The total volume of such cultures rarely reaches 500 ml. Because of the lytic effects of HIV it is likely to be extremely difficult to generate HIV-infected T-cell clones. Infected B cells or B-cell lines have not been generated from patients although one EBV-transformed B lymphoblastoid line has been infected *in vitro.*

Procedures

(1) No staff should be allowed to handle any of the abovementioned materials without suitable training and supervision and should be able to demonstrate proficiency in handling such materials.

(2) All staff engaged on such work should be registered and advice should be available on immunisation.

(3) The local biological safety officer should be notified of all laboratories handling blood and tissue samples.

(4) All materials should be handled in Class II safety cabinets with double HEPA filters or cabinets exhausted to atmosphere.

(5) Cabinets must be monitored under COSHH Regulation 9. In addition containment testing should be carried out annually or whenever a cabinet is recommissioned. Cabinets must be made safe from any biological, radiochemical, or chemical hazard before engineers test the equipment.

(6) No remedial work is permitted within the cabinets without appropriate formaldehyde fumigation.

(7) Only designated cabinets may be used and access is permitted only for authorized staff.

(8) The door to the laboratory must be kept closed whilst work is in progress.

(9) Any tissue/cell disruption procedures likely to result in aerosol formation must be confined to a Class I biological safety cabinet.

Laboratory staff

(1) Laboratory coats must be worn at all times.

(2) Gloves must be worn when handling EBV transformed cells or blood samples.

(3) Appropriate training will be given to staff handling these materials.

(4) The use of sharps should be avoided.

(5) Mouth pipetting must not take place.

(6) Eating, drinking, smoking, and applying cosmetics must not take place in the laboratory.

(7) Food must not be stored in the laboratory, or in associated cupboards, refrigerators, and freezers.

Waste disposal

(1) Any glassware involved in cell culture activities must be soaked in 2.5% Chloros (hypochlorite disinfectant, similar to but distinct from Chlorox) for at least 12 h. It may then be drained and autoclaved if necessary.

(2) All disposable items must be double-bagged and transferred to the autoclaves in leakproof containers.

(3) All liquid waste must be placed into 10% Chloros (or equivalent) and left for at least 12 h prior to disposal. Autoclaving should not be necessary, but if it is required by local rules, adequate ventilation must be provided to remove the chlorine released from the Chloros.

Spills

(1) Small volumes: cover area with undiluted Chloros to give a final concentration of 10% or greater.

(2) Large spills: cover area with Klorsept granules sufficient to soak up all the liquid.

(3) Sweep up granules and wipe the area with 10% Chloros.

(4) In the event of a spillage inside a shaker or centrifuge, follow the following procedure (*do not open the shaker or centrifuge*):

(a) Wait at least 30 min for aerosols to settle

(b) The incident must be reported to the local biological safety officer immediately.

(c) Disinfect with 2% Cidex solution.

(d) This should be made up freshly from the concentrate supplied.

Cidex is a glutaraldehyde based disinfecting agent (there are reports of dermatitis occurring on repeated exposure). Care should be taken when handling Cidex (always wear gloves). *Always handle cidex concentrate in a fume hood.*

REFERENCES

"Genetically Modified Organisms (Contained Use) Regulations" (1992): SI 1992/3217, HMSO, ISBN 0-11-025332-9, £4.00.

"Genetically Modified Organisms (Deliberate Release) Regulations" (1992): SI 1992/3280, HMSO, ISBN 0-11-025216-0, £3.55.

"A Guide to the Genetically Modified Organisms (Contained Use) Regulations (1992): ISBN 0-7176-0473-X, price £5.00.

COSHH regulation number 9.

Useful Addresses

Copies of notification forms, details of the approved methods of risk assessment, copies of ACGM guidance notes, and advice on general procedures can be obtained from:

The Health and Safety Executive, Health Policy Division A3, Baynards House, 1 Chepstow Place, London W2 4TF
Tel. 0171 243 6000

Advice on how to comply with the regulations in particular circumstances can be obtained from HSE's Specialist Inspectorate at the following address:

The Health and Safety Executive, Technology Division 4B, Magdalen House, Stanley Precinct, Bootle, Merseyside L20 3QZ
Tel. 0151 951 4831

For copies of the ACGM guidance notes, or advice on any subject covered by the notes, contact:

ACGM Secretariat, Health and Safety Executive, Health Policy Division, Room 536, Baynards House, 1 Chepstow Place, London W2 4TF.
Tel. 0171 243 6149; fax 0171 243 6293

For information on the genetically "Modified Organisms (Deliberate Release) Regulations," contact:

Department of the Environment, Biotechnology Unit, Room A324, Romney House, 43 Marsham Street, London SW1P 3PY
Tel. 0171 276 8187; fax 0171 276 8333

Sources of Materials

Materials	Suppliers
Cidex	Southern Syringe Services, Ltd.
Chloros	Hays Chemical Distribution
Class II safety cabinets	Medical Air Technology (MAT)
	Hepaire
Klorsept	Jencons (Scientific) Ltd.

3

Mapping Human Senescence Genes Using Interspecific Monochromosome Transfer

Robert F. Newbold and Andrew P. Cuthbert

Human Cancer Genetics Unit, Department of Biology and Biochemistry, Brunel University, Uxbridge UB8 3PH, England

Culture of Immortalized Cells, Edited by R. Ian Freshney and Mary G. Freshney.
ISBN 0-471-12134-7

1. BACKGROUND

All adult mammalian cells derived from tissues capable of proliferation *in vivo* display a limited proliferative capacity *in vitro*. The resulting loss of division potential is usually termed *replicative senescence* [Goldstein, 1990]. This, apparently preprogrammed, limitation to cellular growth presents a formidable barrier to clonal evolution and malignant transformation in culture and may constitute an important tumor suppressor mechanism [Newbold, et al., 1982; Newbold and Overell, 1983; Newbold 1985; Trott et al., 1995a]. How mammalian cells control their mitotic potential so precisely is not known, although a very attractive mechanism to account for cell senescence and immortalization has recently been proposed based on telomere shortening and telomerase reactivation, respectively [Counter, et al., 1992, 1994; Grieder, 1990; Chapter 1, this volume]. A better understanding of the cell and molecular biology of these two alternative cellular proliferation phenotypes should contribute to our understanding of fundamental cancer mechanisms. In addition, improvements in knowledge in this area may also suggest means by which specialized mammalian cell types, particularly of human origin, might be efficiently immortalized, while at the same time maintaining their differentiated functions. Such an advance would, in turn, be likely to facilitate studies of the biology of mammalian cell differentiation and would provide cellular systems for the commercial production of differentiated cell proteins (eg growth factors and hormones) for medical and other uses. It is to the latter interest groups that the contents of this chapter are primarily directed.

Microcell-mediated monochromosome transfer is a powerful new technique in modern somatic cell genetic analysis [Saxon and

Stanbridge, 1987; Cuthbert et al., 1995]. The method permits a single chromosome (for example, a normal human chromosome) to be introduced into almost any homologous or heterologous mammalian cell background, and maintained therein (by selection) as an intact functional and structural entity. Using this technique, individual chromosomes can be screened for the presence of active genes specifying a particular cellular phenotype (e.g., those controlling the replicative senescence programme) on the basis of genetic complementation.

Heterospecific human:rodent microcell hybrids offer a major technical advantage over homospecific combinations in genetic complementation studies, in the sense that it is a relatively straightforward task to define the structural integrity of the foreign chromosome by PCR analysis of sequence-tagged-site (STS) genetic markers and chromosome painting. This facilitates deletion mapping of the introduced chromosome in rare candidate segregants that do not express the phenotype of interest. Such heterospecific transfers, involving the introduction of normal human chromosomes singly into newly immortalized rodent cells, have recently led to the identification of genes that may be important in controlling replicative senescence and their assignment to distinct human subchromosomal regions [Hensler et al., 1994; Cuthbert et al., 1995]. The following account describes in detail the procedures used in such analyses and summarizes additional technical developments (Section 7.3) that are currently being employed to isolate the genes responsible for senescence induction.

2. DERIVATION OF NEWLY IMMORTALIZED SYRIAN HAMSTER CELL LINES FOR USE AS RECIPIENTS IN MICROCELL TRANSFER EXPERIMENTS

2.1. Initiation of Primary Cultures from Syrian Hamster Dermis

In general, normal rodent cells in culture readily generate immortal variants, unlike their human counterparts, which are completely resistant to spontaneous immortalization and can be induced only to generate immortal variants (by carcinogens or DNA tumor virus early genes) with great difficulty [Bai et al., 1993]. Rodent cells from different species, however, differ greatly in their propensity to immortalize. For example, Syrian hamster cells (e.g., fibroblasts) are markedly more resistant to spontaneous immortalization than counterparts obtained from rat or mouse [Newbold et al., 1982; Trott et al., 1995a]. Moreover, in contrast to human cells, immortal variants of Syrian hamster cells can be readily induced against a zero spontaneous background by treating cultures with chemical carcinogens or ionizing radiation (x-rays, γ-rays, or fast neutrons). In addition to providing a suitable system for quantifying the immortalizing efficiencies of carcinogens, an immortalization assay based on Syrian hamster fibroblasts permits, with modest effort, the induction and isolation of large numbers of newly immortalized cell lines suitable for studying the genetics of mammalian senescence control.

Protocol 2.1

Reagents and Materials

- DMEM15: Dulbecco's modified MEM (DMEM) supplemented with 15% prescreened fetal calf serum (Gibco)
- Pronase, 0.01% in PBSA
- Trypsin solution, 0.05% in PBSA, supplemented with 1 mg/ml collagenase 1A

Protocol

(a) Kill newborn hamsters by neck dislocation and store on ice. Decapitate, eviscerate, and remove skins with fine scissors. Float skins overnight in 0.01% pronase at 4°C to facilitate separation of the epidermis from the dermis.

(b) Disaggregate isolated dermis by treatment at 37°C with 75 ml collagenase-trypsin in 100-ml spinner flasks (slow speed). Harvest supernatant every 20 min and spin out cells.

(c) Seed cell suspensions at a density of 1.5×10^7/175-cm^2 culture flask in 40 ml DMEM15.

(d) When almost confluent (i.e., still in log phase), cryopreserve stocks in liquid nitrogen using standard procedures. The cloning efficiency of recovered cells should be >15% (no feeder layer); otherwise discard the stock. Check samples for the presence of mycoplasma using Hoechst 33258 [Chen, 1977] and/or PCR-based methods (Stratagene).

2.2. Irradiations and Carcinogen Treatments

A single exposure to ionizing radiation or a chemical carcinogen is sufficient to induce immortal variants of SHD cells [Newbold et al., 1982]. The most potent immortalizing agents studied to date are fast neutrons and nickel chloride (Ni^{2+}). X-rays and γ-rays have also proved effective [Trott et al., 1995a].

Protocol 2.2

Reagents and Materials

- 75-cm^2, 175-cm^2 flasks
- 50-ml or 125-ml spinner flasks
- DMEM15 (see above)
- 0.25mM $NiCl_2$ (Sigma)
- ^{60}Cobalt source or Van der Graaf accelerator

Protocol

(a) Thaw ampules of cells and return to 175-cm^2 flasks; subculture (1:5 split ratio) after 1–2 days.
(b) For x- and γ-irradiations, trypsinize log phase cultures of passage 3 SHD cells and transfer aliquots of 10^7 cells to 50-ml spinner flasks in 25 ml DMEM15 (X-rays) or 125-ml flasks in 50-ml medium (γ-rays) and irradiate. Doses reducing cloning efficiency to around 50–60% of the control value (2.5–3 Gy 250–300 rads) are the most effective at immortalizing SHD cells. With neutrons (2.3-MeV neutron beam produced by a Van der Graaf accelerator; dose rate 0.71 Gy/min) irradiate cells as exponentially dividing monolayers in 75-cm^2 flasks.
(c) If nickel is to be used as the immortalizing agent, expose exponentially dividing monolayers to 0.25mM $NiCl_2$ for 18 h in complete medium (50–60% survival).

2.3. Isolation of Immortal Clones

Protocol 2.3

Reagents and Materials

- 9-cm petri dishes
- Medium: DMEM15
- Materials for clonal isolation and freezing [Freshney, 1994]

Protocol

(a) After treatment, replate cells immediately at 10^6 cells/9-cm dish (10–20 replicates) and passage when just confluent at a fixed split ratio of 1:5 initially reducing to 1:3 as the cells begin to enter senescence. Include an equal number of untreated, or solvent-treated, control cultures.
(b) Replace medium regularly every 3 days.
(c) Immortal variants are seen initially as rare colonies of proliferating cells on a background of senescent counterparts (for photomicrographs depicting examples of emerging immortal SHD clones, see Trott, et al. [1995a]).
(d) Expand newly immortalized clones to 2×10^7 cells and cryopreserve (10 ampules).

3. CONSTRUCTION OF A HUMAN–RODENT MONOCHROMOSOME HYBRID "DONOR" PANEL

3.1. Tagging Chromosomes in Normal Human Cells Using a Retroviral Vector Incorporating a Mammalian Selectable Marker

In order to be able to transfer human chromosomes individually from normal human cells to newly immortalized hamster cells, and to maintain them in the recipient by drug selection, a panel of monochromosome "donor" hybrids is constructed each carrying a different human chromosome [Cuthbert et al., 1995]. The first step in this procedure involves tagging chromosomes in a normal human fibroblast cell strain (early passage) with a selectable marker. The use of retroviral vectors for this purpose offers many advantages over plasmid transfection, particularly in terms of the efficiency of marker transfer (which can approach 100% of the infected cells) and in the stability of the resulting tagged chromosomes. We have obtained a great deal of success using an amphotropic pseudotype of the replication-defective retrovirus vector tgLS(+)HyTK (Fig.3.1) [see also Lupton et al., 1991]. A producer cell line was constructed [Cuthbert et al., 1995; Trott et al., 1995b] using the packaging cell line PA317 [Miller and Buttimore, 1986], which routinely generates virus titres in excess of 3×10^6 CFU/ml. Producer cell lines should at all times be maintained in selection medium (for the retroviral drug resistance marker) and tested for replication-competent helper virus production by marker rescue assay [Miller and Rosman, 1989]. Only producer lines that routinely test negative for helper virus should be used to infect human cells.

Protocol 3.1

Reagents and Materials

- 9-cm petri dishes
- Trypsin, 0.25% in PBSA
- Medium: DMEM15
- 50-ml centrifuge tubes
- 0.45-μm sterilizing filters, 25 mm for <50 ml, 47 mm for >50 ml and <500 ml

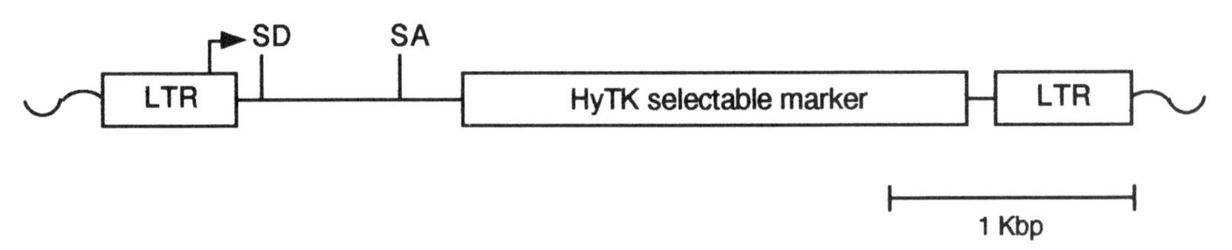

tgLS(+)HyTK provirus

Fig. 3.1. Structure of the tgLS(+)HyTK provirus used to tag human chromosomes with the selectable marker, *Hytk*. The *Hytk* is a fusion gene derived from the bacterial *hph* gene (conferring resistance to hygromycin B) and herpes simplex virus thymidine kinase (*TK*). Unlike the mammalian protein, HSV *TK* can utilize the acyclic nucleoside, Ganciclovir (Syntex Pharmaceuticals) as a substrate. The *Hytk* gene, therefore, may be used as an "in-out" selectable marker; selection for the marker ("in") with hygromycin B-supplemented medium, and selection against ("out") with Ganciclovir.

- Polybrene, 4-mg/ml stock, use at 4 μg/ml
- Hygromycin B, 100 U/ml, in DMEM15
- Freezing medium and materials [Freshney, 1994]

Protocol

(a) Replate virus producing cells in selection medium (9-cm dishes) at a density such that the cultures are 80% confluent after overnight incubation, then replace the medium with 10 ml drug-free medium.

(b) Replate semiconfluent cultures of early-passage adult human dermal fibroblasts (HDF) at 1×10^6 cells/9-cm dish.

(c) After 24 h, collect the virus-conditioned medium (VCM) from the producer cell cultures, centrifuge at 3500g (5 min), and filter (0.45 μm) to remove cells. Adjust the final concentration of FCS to 15% and add polybrene (final concentration 4 μg/ml) to the VCM.

(d) Aspirate the medium from the HDF cultures, replace with VCM. Expose the HDF to virus for 24 h before reverting to normal growth medium.

(e) After a 24 h recovery period replate virally infected HDF (1×10^6/9-cm dish) in medium containing 100 U/ml hygromycin B.

(f) Pool drug-resistant cells (i.e., those with *Hytk*-tagged chromosomes) before they reach confluence and cryopreserve for future use in the construction of monochromosomal human:rodent hybrid "donor" panels by direct microcell transfer.

3.2. Optimization of Micronucleus Formation

Exposure of mammalian cells to colcemid for extended periods results in the formation of micronuclei containing the genetic material of a small subset of the chromosome complement (sometimes a single chromosome). The efficiency and size of micronuclei produced using a single set of conditions, varies dramatically between cell types. For microcell-mediated *monochromosome transfer* (MMCT), the aim is to maximize the number of micronuclei derived from a single chromosome. Mouse A9 cells respond particularly well to colcemid, producing large numbers of small micronuclei. For this reason A9 cells represent a good choice for constructing human:rodent monochromosome hybrids suitable for use as chromosome donors. In contrast, the response of HDF to colcemid-induced micronucleus formation is relatively poor, and the choice of a suitable cell strain will depend on its response to colcemid, which will need to be determined empirically. A careful analysis of the response of selected HDF cell strains at early passage is therefore strongly recommended. Optimal colcemid doses vary between 0.1 and 4.0 μg/ml, with cells being exposed for 48 h in

20% FCS. Visualization and enumeration of HDF micronuclei is best achieved by staining fixed cells with Hoescht 33258 and observing micronuclei with a fluorescence microscope.

Protocol 3.2

Reagents and Materials

- 5-cm petri dishes
- DMEM with 20% FCS
- Colcemid, 100 μg/ml stock
- Ethanol, 100%
- Hoechst 33258 in Hanks' BSS

Protocol

(a) Replate early passage (P1–P10) quiescent HDF (previously maintained at confluence for 48 h) at 5×10^5–1×10^6 cells/5-cm dish in DMEM containing 20% FCS.

(b) After overnight incubation, add colcemid to partially synchronized HDF at a series of final concentrations ranging from 0.1–4.0 μg/ml, and incubate for 48 h.

(c) Fix cells with 100% ethanol and stain with Hoescht 33258. Enumerate and assess size of micronuclei under a fluorescence microscope.

3.3. Construction of Monochromosomal Human:Rodent Hybrid "Donor" Panel, by Direct Microcell Transfer

Protocol 3.3 (see Fig. 3.2)

Reagents and Materials

- A9 mouse fibroblasts
- 25-cm² flasks (special for centrifugation, Nunclon Cort. No. 1-52094A see Suppliers list at end of book)
- 9-cm petri dishes
- DMEM10: DMEM with 10% FCS
- DMEM20: DMEM with 20% FCS
- SFM: serum-free DMEM
- SFM/CB: serum-free DMEM with 5 μg/ml cytochalasin B
- SFM/PHA-P: SFM containing 100 μg/ml phytohaemagglutinin (PHA-P)
- DMEM/Hyg: DMEM10 with 400U/ml Hygromycin B
- Colchicine, 10^{-6}M in SFM
- 8-, 5-, and 3-μm polycarbonate filters
- 15-ml centrifuge tubes
- SFM/PEG: SFM containing 42.5% PEG (1000 MW) and 8.5% DMSO

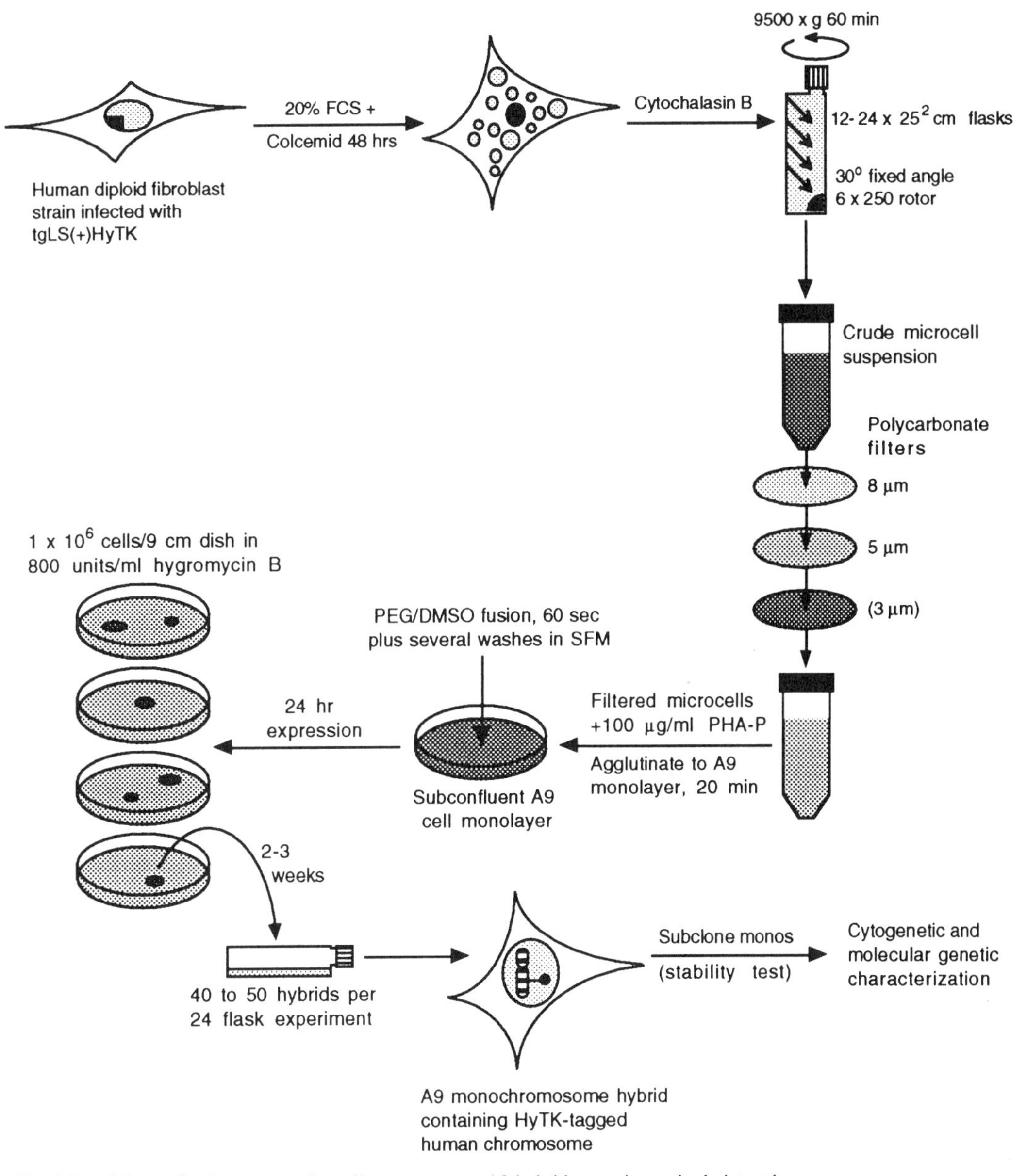

Fig. 3.2. Scheme for the construction of human:mouse-A9 hybrids carrying a single intact human chromosome tagged with the *Hytk* marker.

- Accurate top-pan balance
- Centrifuge with programmable acceleration and deceleration
- 30° fixed-angle 6×250-ml rotor

(a) (Day 1) Replate quiescent retrovirus-infected HDF cultures into 12 or 24, 25-cm^2 flasks in DMEM20.

(b) Add colcemid to final concentration of 4 μg/ml (for 1BR.2 strain) and incubate for 48 h.
(c) (Day 2) Replate rodent recipients (usually A9 cells) at 5×10^6 cells/9-cm dish 18–24 h before performing microcell fusions. By day 3 the monolayers should be 80–90% confluent.
(d) (Day 3) Aspirate the colcemid-containing medium from HDF cultures and replace with 30 ml warm SFM/CB.
(e) Balance pairs of flasks to within 0.02g prior to centrifugation.
(f) Carefully align the balanced flasks in a 30° fixed-angle 6×250-ml rotor (prewarmed to 37°C overnight) such that the growth surface of the flask is lowermost and the caps point toward the axis of the rotor.
(g) Centrifuge at 9500g for 1 h. (Centrifuges that have fully programmable acceleration and deceleration rates are the best machines for this purpose.) The acceleration phase (lasting 15 min) should be linear up to maximum speed, as should the deceleration phase, which should be of $\geq$10-min duration. The overall run time is approximately 85 min.
(h) Pool the crude pellets of microcells from the flasks, and filter in series through 8-, 5-, and (in experiments designed to maximise the yield of smaller chromosomes) 3-μm polycarbonate filters.
(i) Centrifuge the filtered microcells at 3500g for 5 min in 15-ml centrifuge tubes and resuspend in 3 ml warm SFM/PHA-P.
(j) Wash the A9 cell monolayers (including controls) with three changes of warm SFM before adding the microcell suspension.
(k) Allow the microcells to agglutinate to the 80–90% confluent monolayer for 20 min at 37°C.
(l) Fuse microcells to A9 cells by replacing medium with 3 ml prewarmed SFM/PEG for 1 min followed immediately by several thorough washes with warm SFM.
(m) Refeed with DMEM10 and incubate overnight.
(n) (Day 4) Replate fused cells (and mock-fused cells in control dishes) at 1×10^6 cells/9-cm dish into selection medium (normal growth medium supplemented with 800 units/ml hygromycin B).
(o) (Days 14–24) Microcell hybrid clones will be large enough for picking [see Chapter 7, this volume and Freshney 1994] after 10–20 days of incubation. Expand clones in DMEM/Hyg for cryopreservation of stocks and for preliminary cytogenetic analysis.

Note: It is essential that hybrid cultures are maintained in selective medium at all times and that they are not allowed to become overconfluent.

4. CHARACTERIZATION OF HYBRIDS

4.1. Preliminary Analysis by G-11 Staining

G-11 staining [Bobrow and Cross, 1974) is a differential chromosome staining procedure that allows one to discriminate between human (pale blue) and mouse A9 (magenta arms with pale blue centromeres) chromosomes in hybrids. Most human chromosomes display characteristic magenta-staining regions that can in some cases permit unequivocal identification, *per se*. However, the G-11 method will only permit identification of relatively large amounts of human genetic material and cannot be used reliably to assess the purity of candidate monochromosomal hybrids. Use of this valuable and inexpensive technique is therefore best confined to preliminary screening procedures aimed at eliminating hybrids with multiple human chromosomes.

Protocol 4.1

Reagents and Materials (Nonsterile)

- Colcemid-arrested cells
- 0.075M KCl
- 3:1 methanol:glacial acetic acid
- Microscope slides
- Coplin jars
- Double-distilled water
- Phosphate buffer: 50mM disodium phosphate adjusted to pH 11.6
- Staining solution (prepare fresh for each batch of slides): to a solution phosphate buffer (pH 11.6, 37°C) add azure B (Sigma) 0.084% (w/v), eosin Y (Sigma) 0.0024% (w/v), and Giemsa (BDH) 1:50 (v/v).

Protocol

(a) Prepare metaphase chromosome spreads using standard methods [Mandahl, 1992]. For hypotonic treatment, incubate colcemid-arrested cells (0.05 μg/ml for 50 min for A9 cells) in 0.075M KCl and then fix in 3:1 (v/v) methanol:glacial acetic acid (three changes).

(b) Drop fixed cells onto prewashed slides and air-dry.

(c) Age slides for 14 days at room temperature prior to G-11 staining.

(d) Preincubate aged slides for 90s at 37°C in a solution of 50mM sodium phosphate buffer, pH 11.6.

(e) Place in staining solution. Staining times are dependent on the density of the chromosome spreads but should range between 17 and 25 min. Low density spreads give the best results.

(f) Remove slides from stain (maximum 3 per slides per Coplin jar), quickly rinse in a large excess of double-distilled water.
(g) Mount well-stained preparations.

4.2. Characterization of Hybrids by Fluorescence In Situ Hybridization (Chromosome Painting)

A biotin-labeled or digoxigenin-labeled total human DNA probe (Oncor; follow instructions) is employed for conventional ("forward") chromosome painting of human chromosomes in human : mouse A9 microcell hybrids [Cuthbert et al., 1995]. This technique is used to visualize the human DNA content of hybrids, and to establish whether this is associated with a single chromosome or whether hybrids contain rearranged human genetic material not detectable by G-11 staining. The procedure may be repeated after extensive subculturing of selected hybrids and/or recloning to test for stability of the introduced chromosome. Reverse chromosome painting (chromosomal *in situ* suppression hybridization) to normal human metaphases, using probes prepared from *Alu*-PCR-amplified hybrid DNA, is performed to determine the integrity of the introduced human chromosome, and as a sensitive screen for the presence of chimaeric human chromosomes [Cuthbert et al., 1995; Cole et al., 1991; Lui et al., 1993].

4.2.1. Preparation of Alu-PCR probe for reverse painting

Protocol 4.2.1

Reagents and Materials

- (Kit available for this procedure from Boehringer, Cat. 1636 146)
- High-molecular weight (HMW) DNA: prepare from approximately 1×10^7 hybrid cells following standard protocols [Sambrook et al.,1989]
- Alu primer ALE 3
- Reaction mixture: 0.5 μg hybrid DNA, 1.2 μM ALE 3 primer, 5mM $MgCl_2$, 0.5mM dNTPs, and 4.5 U *Taq* polymerase (Promega) all in 1 × Promega Thermo DNA polymerase buffer.
- GeneClean
- Tris, 10mM/EDTA, 1mM, pH 8.0
- 2% agarose in tris/borate EDTA (TBE) buffer
- *2×SSC:* double-strength saline–sodium citrate buffer; 0.3M NaCl, 30mM sodium citrate
- SDS, 0.1% in 2×SSC
- Biotinylated probes purified through Sephadex-G50 columns containing 2×SSC/0.1% SDS
- Sonicated salmon sperm DNA (Sigma)
- Cot1 DNA (BRL)

Protocol

(a) Amplify inter-*Alu* (human) sequences in HMW hybrid DNA with the Alu primer ALE 3. (PCR amplification of sequences in mouse A9 DNA is not detected with this primer.) Perform reactions in a total volume of 100 μl reaction mixture.

(b) After denaturation for 5 min at 95°C followed by 3 min at 80°C (hot start) amplification is performed for 35 cycles of 94°C (1 min), 68°C (1 min) followed by extension at 72°C (5 min). Add *Taq* polymerase after the initial denaturation step. Monitor each PCR reaction by loading a 10-ml sample onto a 2% agarose gel in TBE buffer.

(c) Purify PCR products with GeneClean, resuspend in tris/EDTA, pH 8.0, and label by nick translation with biotin-14-dATP (BRL).

(d) Ethanol precipitate biotinylated probe in the presence of a 100-fold excess of sonicated salmon sperm DNA and a 20-fold excess of Cot1 DNA (BRL), dry, and resuspend in 16-μl hybridization solution [Carter et al., 1992].

4.2.2. *In situ hybridization (forward and reverse painting)*

Protocol 4.2.2

Reagents and Materials

- 2×SSC (see Reagents and Materials for Protocol 4.2.1)
- 20 μg/ml RNAse in 2×SSC
- 70% and 100% ethanol on ice
- Acetone
- 70% formamide in 2×SSC
- DNA probe
- Alu-PCR probe

Protocol

(a) Incubate aged slides in 20 μg/ml RNAse at 37°C for 60 min.

(b) Rinse in 2×SSC (65°C) for 15 min.

(c) Dehydrate through ice-cold 70% (twice) 90% (twice) and 100% ethanol.

(d) Fix in acetone for 10 min.

(e) Denature in a solution of 70% formamide in 2 x SSC (65°C, 2 min.)

(f) Quench in 70% ethanol.

(g) Dehydrate at room temperature by allowing the denatured slides to air-dry.

(h) Denature probe at 65°C for 10 min and store on ice prior to hybridization. In the case of reverse painting (*Alu*-PCR probe) preanneal probe to competitor DNA (37°C for 20 min) before hybridization.

(i) For hybridization, 20-μl total human DNA probe or 15-μl *Alu*-PCR probe is used for each denatured slide. Hybridization is performed under sealed coverslips at 37°C either overnight (total human probe) or for 7 days (*Alu*-PCR probe).

4.2.3. Detection and imaging (see also Fan, et al. [1990]

Protocol 4.2.3

Reagents and Materials

- 2× SSC
- 0.1× SSC
- 50% formamide in 2× SSC
- TNFM: 4× SSC containing 0.05% Tween-20 and 5% nonfat milk
- Fluorescein isothiocyanate (FITC)-avidin
- Biotinylated antiavidin antibody (oncor)
- 20 ng/ml propidium iodide, or 100 ng/ml DAPI, in Antifade-containing mounting solution (vector)

Protocol

(a) Wash slides twice in 50% formamide (in 2×SSC) followed by one 5/min. wash in 2×SSC and one 5-min wash in 0.1×SSC, all at 42°C.

(b) Block in TNFM for 30 min at 37°C.

(c) Detect hybridized biotinylated probe with FITC-avidin. Amplify signal by incubating slides for 20 min with biotinylated antiavidin antibody in TNFM at room temperature.

(d) Wash and incubate again with FITC-avidin. (Carry out one round of amplification with the total human DNA probe and two rounds with *Alu*-PCR probes.)

(e) Counterstain chromosomes with 20 ng/ml propidium iodide or 100 ng/ml DAPI in Antifade-containing mounting solution.

(f) Observe chromosome spreads and capture digitized images using a confocal laser scanning microscope. (We use a Bio-Rad MRC 600 scanner fitted to a Nikon Optiphot fluorescence microscope; for examples of painted monochromosome hybrids, see Fig. 3.3)

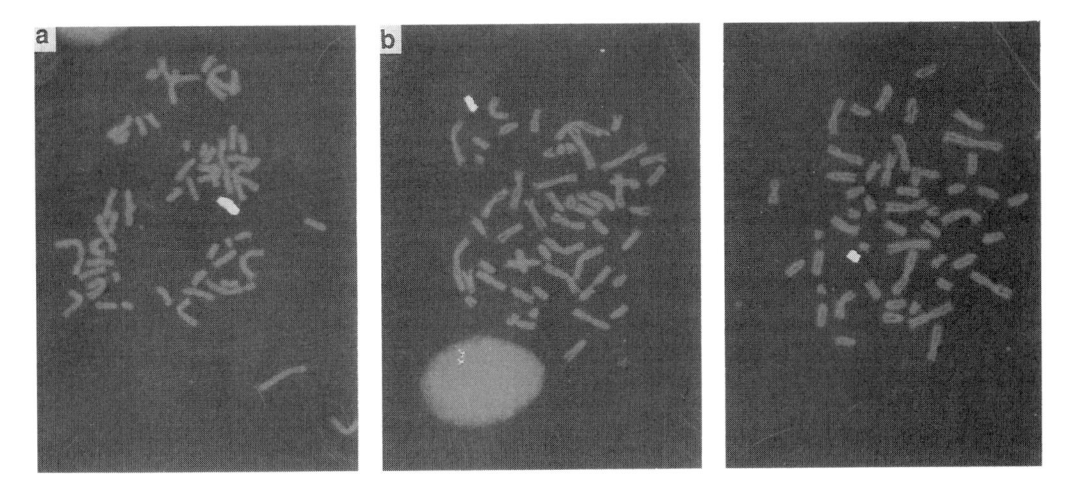

Fig. 3.3. Examples of monochromosomal human:mouse-A9 hybrid cell lines in which the single human chromosome has been revealed by fluorescence *in situ* hybridization (chromosome painting) using a total human DNA probe. Shown are A9 cell lines carrying human chromosomes X (a), 13 (b), and 19 (c).

5. SCREENING THE HUMAN CHROMOSOME COMPLEMENT FOR ANTIPROLIFERATIVE GENES

The human:rodent monochromosomal hybrids constructed as described above can be used to transfer known human chromosomes singly into any mammalian recipient. The complete human chromosome complement can thereby be screened for senescence-inducing activity. The following procedure, which is a modification of that described above, has been employed with newly immortalized Syrian hamster dermal (SHD) cells as recipients. As discussed above for HDFs, conditions for micronucleation must be optimized for each human chromosome "donor" cell line [see also Cuthbert et al. (1995)].

Protocol 5.1

Reagents and Materials

Selection medium: DMEM15 with 600 U/ml hygromycin B

Protocol

(a) Recipient SHD cell lines are replated at a density of 3×10^6 cells/9-cm dish and incubated overnight (the density is critical and dependent on the cell type). Microcell fusions are performed as described above.

(b) Replate cells in selection medium at 1×10^6 cells/9-cm dish 24 h postfusion.
(c) Begin to examine hybrid colonies for growth suppression/senescence after 1–2 weeks in selection (depending on the growth rate of the recipient cell line). In the case of transfers involving human chromosomes with antiproliferative activity, isolate rare actively proliferating colonies (potential segregants) after 2 weeks and expand into mass culture for isolation of DNA in preparation for PCR-STS deletion analysis (see below).

The senescence-inducing effects of the human X chromosome may be studied after microcell transfer directly from human diploid fibroblasts into *hprt*⁻ derivatives of SHD lines; the *hprt* gene is employed as the selectable marker (conferring resistance to HAT). These experiments are performed using the direct microcell transfer protocol described above. The *hprt*-recipient cell lines must be tested for spontaneous reversion to HAT resistance before deciding to use them in experiments. We have isolated two nonrevertible SHD lines for this purpose: 5XH11/6TG4 and 5NL2/6TG1 [Trott et al., 1995a; Cuthbert et al., in press].

Selection protocol

Protocol 5.2

Reagents and Materials

- *Complete HAT medium:* DMEM15 with hypoxanthine, 10^{-4}M, aminopterin, 10^{-6}M, thymidine 5×10^{-5}M.

Protocol

(a) After 24-h recovery from microcell fusion, replate cells at 5×10^5 cells/9-cm dish in complete HAT medium.
(b) Refeed dishes with fresh selection medium every 3–4 days and, beginning at day 5, monitor colonies for senescence over a 6–8-week period.

6. SUB-CHROMOSOMAL LOCALIZATION OF CANDIDATE SENESCENCE-INDUCING GENES BY PCR-STS DELETION MAPPING OF HYBRID SEGREGANTS

A wealth of genetically mapped PCR-detectable microsatellite markers (STSs; see Section 1. Background) covering the entire human genome, is now available [Gyapay et al., 1994]. These markers can be used very successfully to pinpoint the subchromosomal location of antiproliferative/senescence genes identified in the above experiments [N. England et al., submitted]. Even with the most powerful antiproliferative response, rare segregants arise which have lost the growth suppressor function. In a high proportion of cases this correlates with nonrandom deletion of human genetic material.

Protocol 6

Reagents and Materials

- *Reaction mixture (per 10 μl):*

Hybrid DNA	40 ng
Primer	1.0 μM
dNTPs, each	100 μM
$MgCl_2$	0.75mM
Taq polymerase	0.2 U

All in 0.5×Promega Thermo DNA polymerase buffer

- 2.5% agarose
- TBE buffer (see Protocol 4.2)

Protocol

(a) Isolate DNA from putative segregant cell lines using established methods [Sambrook et al., 1989].

(b) Select a set of markers providing good even coverage of the human chromosome being studied (or, in the first instance, a specific region of the chromosome if previous experiments have suggested a candidate location for the senescence gene) and synthesize or purchase PCR primer sets (Research Genetics).

(c) Perform PCR reactions for each STS marker in turn. The following standard set of conditions will usually give good results, but some variations may be required for certain primer sets where amplification of rodent sequences is a problem (uncommon).

(d) Perform reactions in a final volume of 10-μl reaction mixture.

(e) Denature (94°C for 5 min and 80°C for 8 min) prior to performing PCR amplifications. Use the following PCR conditions: 35 cycles of 94°C (40 s), 56°C (30 s), 72°C (30 s), followed by a final extension cycle of 72°C for 5 min.

(f) Resolve PCR amplification products on 2.5% agarose gels in TBE buffer.

7. BRIEF SUMMARY OF RESULTS

7.1. Generation of Newly Immortalized SHD Cell Lines

Using the protocols described here, more than 100 immortal SHD cell lines have been generated and cryopreserved as freshly immortalized stocks. A variety of carcinogenic agents have been found to be effective as immortalizing agents, including polycyclic hydrocarbons and their diol epoxide metabolites, aliphatic alkylating agents, and ionizing radiation (both high-LET and low-LET). This work, which includes a detailed cytogenetic characterization of the newly immortalized lines, has been published elsewhere [Newbold et al., 1982; Trott et al., 1995a]. The most effective immortalizing agent identified thus far has proved to be soluble nickel ($NiCl_2$). The molecular basis of its superior potency is as yet unclear, although there is some evidence that nickel may cause the functional inactivation of a senescence gene (or genes) via an epigenetic mechanism [Trott et al., 1995a].

7.2. Construction of Human Monochromosome Hybrid "Donor" Panels

The direct microcell transfer approach, when combined with the use of retroviral vectors to tag chromosomes efficiently, has proved to be an extremely effective method of generating human : rodent monochromosome "donor" hybrids, in which the single foreign (human) chromosome is highly stable and transferable intact to a wide variety of recipient cell types. Employing the techniques outlined in this chapter, a well-characterized panel of such hybrids has recently been constructed representing the complete human autosomal complement plus the X chromosome [Cuthbert et al., 1995].

7.3. Mapping Human Antiproliferative Genes

The above human monochromosomal hybrid donor panel has been used to screen the complete human chromosome complement for the presence of antiproliferative (including senescence) genes. During the construction of the panel in the mouse A9 cell line, it was found that two human chromosomes, 1 and 9, were not tolerated by these cells presumably because they carry cell growth suppressor genes. Consequently, donors for human chromosomes 1 and 9 were isolated using, as recipients, alternative cell lines that proved refractory to the growth suppressive effects. Transfer of individual human chromosomes to newly immortalized SHD cell lines confirmed the antiproliferative properties of chromosomes 1 and 9 and, in addition, has led to the identification of similar activity associated with human chromosomes 19 and X [N. England et al., submitted]. Furthermore, the cellular effects of introducing the latter human chromosome into SHD lines closely resembled replicative cell senescence. Three of the genes responsible for the induction of growth–arrest have been mapped by STS deletion analysis (of rare hybrid segregants) to chromosome 1q (1q25, and possibly 1q42), 9p (9p21), and Xp (tentative). In the case of 9p21, fine deletion mapping of previously uninformative segregants, employing additional markers between D9S162 and D9S171, has provided strong evidence that the proposed tumour suppressor gene p16^{INK4} (MTS1) is responsible for cell growth-arrest induced by this human chromosome [N. England et al., submitted; Nobori et al., 1994; Kamb, et al., 1994].

8. THE FUTURE: MOLECULAR CLONING STRATEGIES FOR SENESCENCE GENES USING APPROACHES BASED ON FUNCTIONAL ASSAYS

Microcell transfer technology has, when used in conjunction with functional assays, proved to be an extremely powerful technique for identifying growth suppressor genes and tumor suppressor genes. In some cases (see above) the method has permitted the subchromosomal

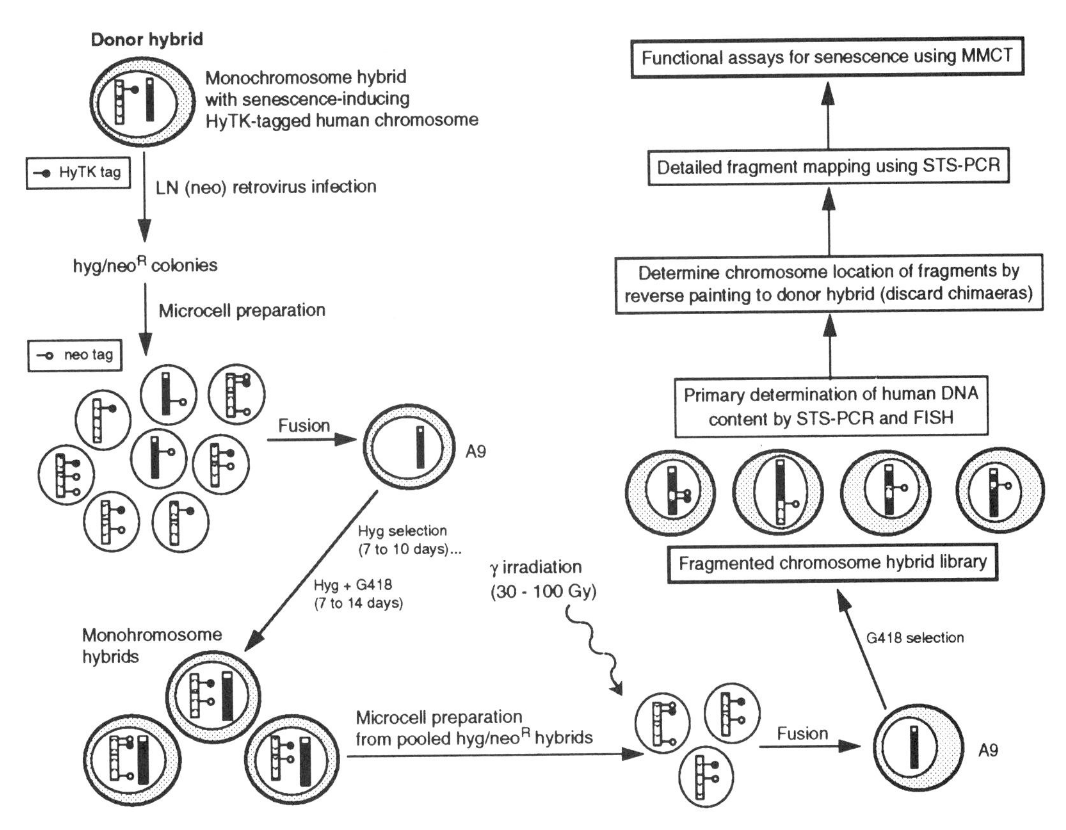

Fig. 3.4. Scheme for the construction of subchromosomal transferrable fragment (STF) hybrids for use in the genetic analysis of cellular senescence. Each STF carries a defined tagged fragment of human genetic material incorporated by recombination into a mouse A9 chromosome.

location of the gene to be established. This work has provided a firm foundation for the development of strategies, again based on function, leading to the molecular cloning of novel tumour suppressor genes and senescence genes. The procedure being used in our laboratory has involved the construction of hybrid subpanels of chromosomes of interest (e.g., human chromosomes 1, 19, and, especially, X) in which a defined tagged subchromosomal transferable fragment (incorporated into a rodent chromosome) is the only human material present in the rodent cell background. A summary of the protocol for generating such hybrids (known as STFs) is shown in Fig.3 4 (see also Koi et al. [1993]).

STFs [3–5 Mbp (megabase pairs) in size] derived from a human monochromosome hybrid with antiproliferative activity, are screened for growth suppressive properties by microcell transfer into the appropriate SHD recipient. STF "donor" hybrids with positive activity can then be used to generate Alu-PCR probes for screening libraries of 650-Mbp-insert yeast artificial chromosomes (YACs). YACs selected in this way are then assayed for growth suppressive properties after transfer to responsive SHD cells by protoplast fusion. This procedure avoids many of the uncertainties and pitfalls associated with positional cloning approaches.

ACKNOWLEDGMENTS

Much of the experimental work leading to the development of the techniques and isolation of hybrids described in this article was supported by grants from the Cancer Research Campaign, the European Commission (Environment Programme), and the Association for International Cancer Research.

REFERENCES

Bai L, Mihara K, Kondo Y, Honma M, Namba M (1993): Immortalization of normal human fibroblasts by treatment with 4-nitroquinoline oxide. Int J Cancer 53: 451–456.

Bobrow M, Cross J (1974): Differential staining of human and mouse chromosomes in interspecific cell hybrids. Nature 251: 77–79.

Carter NP, Ferguson-Smith MA, Perryman MT, Telenius H, Pelmear AH, Leversha MA, Glancy MT, Wood SL, Cook K, Dyson HM, Ferguson-Smith ME, Willatt LR (1992): Reverse chromosome painting: a method for the rapid analysis of aberrant chromosomes in clinical cytogenetics. J Med Genet 29:299–307.

Chen TR. (1977): *In situ* detection of mycoplasma contamination in cell cultures by fluorescent Höechst 33258 stain. Exp Cell Res 104:255–259.

Cole CG, Goodfellow PN, Bobrow M, Bentley DR (1991): Generation of novel sequence tagged sites (STSs) from discrete chromosomal regions using Alu-PCR. Genomics 10: 816–826.

Counter CM, Avilion AA, LeFeuvre CE, Stewart NG, Greider CW, Harley CB,

Bacchetti S (1992): Telomere shortening associated with chromosome instability is arrested in immortal cells which express telomerase activity. EMBO J 11: 1921–1929.
Counter CM, Hirte HW, Bacchetti S, Harley CB (1994): Telomerase activity in human ovarian carcinoma. Proc Natl Acad Sci USA 91: 2900–2904.
Cuthbert AP, Trott DA, Ekong RM, Jezzard S, England NL, Themis M, Todd CM, Newbold RF: Construction and characterization of a highly stable human:rodent monochromosomal hybrid panel for genetic complementation and genome mapping studies. Cytogenet Cell Genet: 71:68–76.
Fan Y-S, Davis L, Shows TB (1990): Mapping small DNA sequences by fluorescence *in situ* hybridization directly on banded metaphase chromosomes. Proc Natl Acad Sci USA 87: 6223–6227.
Freshney, R.I. (1994): "Culture of Animal Cells, a Manual of Basic Technique." New York: Wiley-Liss, 204–206.
Goldstein S (1990): Replicative senescence: the human fibroblast comes of age Science 249: 1129–1133.
Greider CW (1990): Telomeres, telomerase and senescence. Bioessays 12: 363–369.
Gyapay G, Morisette J, Vignal A, Dib C, Fizames C, Millasseau P, Marc S, Bernardi G, Lathrop M, Weissenbach J (1994): The 1993-94 Gènèthon human genetic linkage map. Nature Genet 7: 246–339.
Hensler PJ, Annab LA, Barrett JC, Pereira-Smith OM (1994): A gene involved in control of human senescence on human chromosome 1q. Mol Cell Biol 14: 2291–2297.
Kamb A, Gruis NA, Weaver-Fedhouse J, Liu Q, Harshman K, Tavtigan SV, Stockert E, Day RS, Johnson BE, Skolnik MH (1994): A cell cycle regulator potentially involved in genesis of many tumour types. Science 264: 436–440.
Koi M, Johnson LA, Kalikin LM, Little PFR, Nakamura Y, Feinberg AP (1993): Tumor cell growth arrest caused by subchromosomal transferable DNA fragments from chromosome 11. Science 260: 361–364.
Lui P, Siciliano J, Seong D, Craig J, Zhao Y, de Jong PJ, Siciliano MJ (1993): Dual Alu polymerase chain reaction primers and conditions for isolation of human chromosome painting probes from hybrid cells. Cancer Genet Cytogenet 65: 93–99.
Lupton SD, Brunton LL, Kalberg VA, Overell, RW (1991): Dominant positive and negative selection using a hygromycin phosphotransferase-thymidine kinase fusion gene. Mol Cell Biol 11: 3374–3378.
Mandahl N (1992) Methods in solid tumour cytogenetics. In Rooney DE, Czepulkowski BH (eds): Human Cytogenetics, a Practical Approach, Vol II: Oxford: IRL Press.
Miller AD, Buttimore C (1986): Redesign of retrovirus packaging cell lines to avoid recombination leading to helper virus production. Mol Cell Biol 6: 2895–2902.
Miller AD, Rosman GJ (1989): Improved retroviral vectors for gene transfer and expression. Biotechniques 7: 980–990.
Newbold RF (1985): Multistep malignant transformation of mammalian cells by carcinogens: induction of immortality as a key event. In Barrett JC, Tennant RW (eds): "Carcinogenesis: A Comprehensive Survey," Vol 9. New York: Raven Press, pp 17–28.
Newbold RF, Overell RW (1983): Fibroblast immortality is a prerequisite for transformation by EJ c-Ha-*ras* oncogene. Nature 304: 648–651.
Newbold RF, Overell RW, Connell JR (1982): Induction of immortality is an early event in malignant transformation of mammalian cells by carcinogens. Nature 299: 633–635.
Nobori T, Miura K, Wu DJ, Lois A, Takabayashi K, Carson DA (1994): Deletions of the cyclin-dependent kinase-4 inhibitor gene in multiple human cancers. Nature 368: 753–756.
Sambrook J, Fritsch EF , Maniatis T (1989): "Molecular Cloning, a Laboratory Manual," 2nd ed. Cold Spring Harbor, NY: Cold Spring Harbor Laboratory Press, 3 vols.
Saxon PJ, Stanbridge EJ (1987): Transfer and selective retention of single specific hu-

man chromosomes via microcell-mediated chromosome transfer. Meth Enzymol 151: 313–325.

Trott DA, Cuthbert AP, Overell RW, Russo I, Newbold RF (1995a): Mechanisms involved in the immortalization of mammalian cell by ionizing radiation and chemical carcinogens. Carcinogenesis 16: 193–204.

Trott DA, Cuthbert AP, Todd CM, Newbold RF (1995b): Novel use of a selectable fusion gene as an "in–out" marker for studying genetic loss in mammalian cells. Mol Carcinogenesis: 12:213–224.

APPENDIX: MATERIALS AND SUPPLIERS

Cell Culture and Microcell-Mediated Chromosome Transfer

Materials	Suppliers
Aminopterin	Sigma Chemical Co.
Antibiotics	Life Technologies, Gibco BRL
Centrifugation flasks, 25 cm^2 [Cat. 1-52094A]	Nunclon
Colcemid	Sigma Chemical Co.
Collagenase, grade 1A	Sigma Chemical Co.
Cytochalasin B	Sigma Chemical Co.
Dimethyl sulphoxide (DMSO)	Sigma Chemical Co.
Dulbecco's modified Eagle's medium (DMEM)	Life Technologies, Gibco BRL
Fetal calf serum	Life Technologies, Gibco BRL
Ganciclovir (under agreement)	Syntex Pharmaceutical licenced
Geneticin G418	Gibco BRL
Höescht 33258 (bizbenzimide)	Sigma Chemical Co.
Hygromycin B	Calbiochem
Hypoxanthine	Sigma Chemical Co.
Mycoplasma PCR kit	Stratagene
Newborn calf serum	Life Technologies, Gibco BRL
$NiCl_2$	Sigma Chemical Co.
Phytohaemagglutinin PHA-P	Sigma Chemical Co.
Polybrene (hexadimethrine bromide)	Sigma Chemical Co.
Polycarbonate filters (25-mm 8-, 5-, and 3-μm pore sizes) and holders	Nucleopore
Polyethylene glycol (PEG) Mol. Wt. 1000	Sigma Chemical Co.
Pronase	Sigma Chemical Co.
Thymidine	Sigma Chemical Co.
Tissue culture plastics	Nunclon
Trypsin (bovine pancreas, TRL)	Worthington Enzymes
Tunicamycin	Calbiochem

Characterization of Hybrids

Materials	Suppliers
Antifade-containing mounting solution	Vector Laboratories
Azure B	Sigma Chemical Co.
Biotin conjugated antiavidin antibody	Oncor, Vector
Biotin or digoxigenin-labeled total human DNA probe	Oncor, Vector

APPENDIX (*Continued*)

Characterization of Hybrids (*Continued*)

Materials	Supplies
Cot 1 DNA	Gibco BRL
Eosin Y	Sigma Chemical Co.
Fluorescein isothiocyante–avidin	Oncor, Vector
Formamide	Sigma Chemical Co.
Giemsa	BDH
Propidium iodide	Sigma Chemical Co.
Salmon sperm DNA	Sigma Chemical Co.

Polymerase Chain Reaction

Materials	Suppliers
Agarose	Gibco BRL
PCR primer sets	Research Genetics
Taq polymerse	Promega or Perkin Elmer

4

Development of Immortal Human Fibroblast Cell Lines

L.V. Mayne,* T. N. C. Price, K. Moorwood, and J. F. Burke

Trafford Centre for Medical Research, University of Sussex, Falmer, Brighton BN1 9RY, England (L.V.M., T.N.C.P., K.M.). Biochemistry Group, School of Biological Sciences, University of Sussex, Falmer, Brighton, BN1 9QJ, England (J.F.B.).

Culture of Immortalized Cells, Edited by R. Ian Freshney and Mary G. Freshney.
ISBN 0-471-12134-7

I. INTRODUCTION

Human dermal fibroblasts (HDFs) are robust cells that have been exploited for studies of cell biology including oncogenic transformation, aging and senescence, and as a tool for depicting the cellular basis of genetic disease. Their popularity as an *in vitro* system stems from the ready availability of skin samples and the relative ease with which these cells can be cultured. However, a major limitation on the exploitation of fibroblasts for genetic, physiological, and biological studies is their limited *in vitro* lifespan. Fibroblasts isolated from skin biopsies initially grow very rapidly, forming vigorous cultures that can be subcultured once or twice a week once they reach confluence. These cultures do not grow indefinitely; after about 60 population doublings (PD) the appearance of the cells begins to change and the growth rate of the culture diminishes [Hayflick and Moorhead, 1961]. The culture then degenerates and is characterized by aging or senescent cells that may be multinucleate and large, often with a very spread cytoplasm. Despite the degenerative appearance of these cultures, the cells often remain viable for prolonged periods, although there is no net increase in cell numbers.

Fibroblasts are an important model system for studying *in vitro* aging and for developing the generic technologies for cell immortalization [Huschtcha and Holliday, 1983]. Along with B-lymphocytes, they are probably the most frequently immortalized human cell types. The methods for immortalization are, therefore, well developed but not necessarily well understood. By far the most successful and most frequently used method for deriving immortal human fibroblasts is through the expression of SV40 T-antigen. Unlike rodent cells, human cells rarely, if ever, spontaneously immortalize, as their lifespan is under tight genetic control. The development of the immortalized phenotype requires a number of, as yet unknown, genetic changes. SV40 T-antigen expression does not lead directly to immortalization but initiates a chain of events that results, with a low probability estimated at about 1 in 107 [Shay and Wright, 1989; Huschtcha and Holliday, 1983], in an immortalized derivative appearing.

The immediate effect of SV40 T-antigen expression in fibroblasts is to promote cell division, and changes in gene expression and cell morphology. Ultimately, it cannot prevent cells from senescing, but does lead to an extension of *in vitro* lifespan estimated at 20% [Lomax et al., 1978; Ide et al., 1984]. At the end of their extended lifespan, these cultures enter a degenerative phase known as "crisis"; widespread cell death, aberrant cell division, and gradual deterioration of the culture with loss of viable cells is observed. SV40-transformed cells in crisis differ in many ways from senescent primary cells, although some features appear in common.

Within any SV40-transformed culture, there is a low but finite possibility that secondary genetic changes will occur that will relieve the restrictions imposed on lifespan; if enough cells are cultured, it is likely that an immortalized derivative will appear. The length of time it takes to establish a continuous cell line is dependent on the timing of these genetic events. If they occur early in culture life, then crisis may be short; if they occur later, crisis may be longer. While it is clear that lifespan is genetically controlled, the precise nature and number of genes involved is not yet known. Studies of malignant disease have identified a number of recessive tumor or growth suppressor genes such as *Wt1* and retinoblastoma (*Rb*) [Weinberg, 1990]. Genetic studies on SV40-immortalized cells are consistent with immortalization requiring a loss of function of genes that behave like tumor suppressors [Pereira-Smith and Smith, 1988]. The transformed phenotype in Wilm's tumor and retinoblastoma requires loss or inactivation of both alleles of the tumor suppressor gene and, if immortalization of SV40-transformed cells requires a two hit process to inactivate both alleles of a tumor suppressor gene, then this may account, in part, for the apparent low frequency of immortalization of these cells.

Two critical functions of SV40 are likely to be essential for immortalization to take place. The first is its ability to extend the *in vitro* lifespan. By simply increasing the length of time cultures remain viable and proliferating, SV40 must increase the probability that the requisite genetic alterations for overcoming the restrictions of lifespan will occur. The second important function of SV40 T-antigen is its ability to destabilize the genome. SV40 T-antigen affects chromosome stability and ploidy, and these changes may favor genetic alterations, such as allele loss, which may promote immortalization. The effects of SV40 on prolonging *in vitro* lifespan and destabilizing the genome are likely to be a direct consequence of the interaction of SV40 large-T antigen with the tumor suppressor proteins, p53 and Rb. SV40 large-T antigen has a strong affinity for p53 and can successfully sequester all the available cellular p53, preventing its normal function in regulating the cell cycle. Deregulation of the cell cycle permits DNA replication under unfavorable conditions with subsequent DNA instability [Lane, 1992]. T-antigen also binds the retinoblastoma protein, and this interaction is believed to be critical to the phenotypic changes seen (Bartek et al., 1992).

The method described here for immortalization of human dermal fibroblasts is based on the use of SV40 T-antigen and relies on DNA transfer methods to artificially introduce the T-antigen gene into primary cell cultures. All the vectors described below express both the large-T and small-t antigen as alternative transcripts from the SV40 T-antigen gene. SV40-transfected cells are selected directly under appropriate culture conditions and the surviving cells are subcultured to give rise to a precrisis SV40-transformed cell population. These cells are cultivated continuously until they reach the end of their proliferative lifespan, when they inevitably enter crisis. They must then be nurtured with care and sufficient cells cultured to give a reasonable chance of an immortalized derivative appearing. The general method and points for consideration when setting out to establish immortalized cultures are described below.

2. PRIMARY CELL CULTURE

The success of any gene transfer method is highly dependent on the quality of the starting cells. It is essential that cells be routinely screened for the presence of mycoplasma and maintained in fresh medium with regular subculturing. Vigorously growing healthy cultures are more efficient recipients in gene transfer experiments. We routinely maintain our fibroblast cultures in Eagle's MEM with 15% fetal calf serum (FCS). FCS is prescreened before use for its ability to support fibroblast growth by measuring [^{3}H]-thymidine uptake into the cultured cells. Cells are routinely subcultured at confluence with a split ratio of 1:2 or 1:3 using 0.05% trypsin, 0.02% EDTA in PBSA. Cell stocks are frozen in liquid nitrogen in sterile Nunc ampoules at a concentration of 2×10^6 cells/ml in medium containing 10% sterile DMSO with a minimum of 0.6 ml per ampoule. Ampoules are precooled to −70°C in a Nalgene freezing container with isopropanol before storing in the liquid nitrogen freezer. For further details of basic cell culture techniques, see Freshney [1994].

2.1. Culture Conditions for Human Dermal Fibroblasts

Conditions for growing and handling HDF are well described in Freshney [1994]. We routinely maintain HDF in Eagle's Minimal Essential Medium (ICN/Flow) with 2mM glutamine, and supplemented with 100 U/ml penicillin, 100 μg/ml streptomycin, and 15% fetal calf serum. Transfected cultures are maintained in this medium until postcrisis derivatives appear. Once an immortal culture has been established, the serum concentration can be reduced to 10% and, if desired, the fetal calf serum can be replaced with newborn calf serum.

3. GENE TRANSFER

Transfection experiments can be performed in either flasks or plates, although plates offer easy access if you wish to select individual clones rather than to subculture the culture as a whole. The number of cells for a transfection experiment varies depending on the surface area of the culture vessel used. In our experiments, we use one or two 175-cm^2 flasks with 6×10^5 cells per flask or three 9-cm plates with 2×10^5 cells per plate. Additional plates maybe set up for control experiments. Cells should be trypsinised and seeded 24–48 h before the addition of DNA. For calcium phosphate precipitation, the cells must be subconfluent, and we recommend no more than 70-80% cover of the dish or flask when the cells are transfected.

3.1. Transfection Methods

There are three well-established methods for introducing DNA into human cells: calcium phosphate precipitation, electroporation, and lipofection. Fibroblasts are, in general, relatively easy to transfect successfully and, in our experience, each of these methods gives rise to a similar frequency of transfectants. There is little to choose between these methods, although electroporation requires expensive apparatus and the lipofection solutions are relatively expensive to buy. We routinely use the calcium phosphate precipitation method; the solutions are easy to make up and inexpensive, and the method is simple to perform.

3.2. T-Antigen Expression Vectors

A range of T-antigen expression vectors are available and the essential differences between these vectors are: the choice of dominant selectable marker gene, the source of the promoter that drives T-antigen expression, and alternative forms of the T-antigen itself. The presence of a dominant selectable marker gene allows direct selection for cells that have successfully taken up the vector during transfection. These genes confer resistance to a number of selective toxic agents. The most frequently used dominant selectable marker genes are *gpt, neo,* and *hyg B*. While, in our experience, selection for *gpt* is effective in human fibroblasts, G418 (*neo*) and hygromycin (*hygB*) are much more effective and easier to use.

Efficient constitutive expression can be achieved in human fibroblasts from a range of common promoters, including the Rous sarcoma virus LTR, the immediate early promoter of cytomeglovirus, and the origin region–endogenous SV40 T-antigen promoter. The majority of human SV40-immortalized fibroblast cell lines have been established with either SV40 virus or constructs such as pSV3*neo* that express T antigen from the endogenous promoter. We recommend the use of pSV3*neo* [Southern and Berg, 1982; Mayne et al., 1986] for constitutive expression of T-antigen.

The high level of T-antigen expression seen with most constitutive vectors extends cellular lifespan and promotes genetic events that may give rise to the immortal phenotype but may, at the same time, have deleterious effects, including loss of contact inhibition of growth, reduced serum requirements, changes in gene expression, and altered cell morphology. Many of these deleterious effects can, however, be prevented or minimized through the use of conditional expression vectors [Price et al., 1994]. Previous studies with temperature-sensitive, or inducible, T-antigen expression vectors demonstrated an absolute requirement for the presence of T-antigen to maintain cell viability in SV40-immortalized cultures but did not indicate how much T-antigen was required. Recent studies from our laboratory [Price et al., 1994] have shown that extension of precrisis lifespan and maintenance of the immortal phenotype in postcrisis cells appears to be independent of T-antigen concentration. Very low T-antigen levels are equally effective at maintaining these functions. In contrast, many of the deleterious effects of T-antigen were caused by high levels of T-antigen and could be eliminated by reducing T-antigen levels. Thus, it was possible to restore an apparently normal phenotype to SV40-immortalized human cells by down regulating T-antigen expression. Low-level T-antigen expression was achieved in this study through the use of a novel expression vector (p735.6) employing a modified mouse metallothionein I promoter. This promoter carried a small deletion that reduced basal level gene expression in the absence of any inducing agent. By growing cells without addition of the inducing agent, the low level transcription of the T-antigen gene produced sufficient T-antigen protein to maintain cell viability but not enough to influence the cell phenotype. Induction of high levels of T-antigen produced the typical cell morphology and growth properties associated with the SV40-transformed phenotype.

Different strategies are available for controlling levels of gene expression, and these include the oestrogen-receptor [Kumar et al., 1986], the mouse metallothionein promoter [Price et al., 1994] and temperature-dependent mutations in the T-antigen protein [Radna et al., 1989]. The first two methods control expression at the level of transcription, while the third permits functional inactivation of the protein at different temperatures. A major consideration in the use of controllable expression vectors is the ability to achieve a graded response as T-antigen is essential to maintain cell viability in SV40-immortalized cells, and eliminating T-antigen altogether leads to senescence and death. The trick is to maintain sufficient protein to permit cell survival but not enough to influence other cellular parameters. Few vector systems offer this level of control and, to our knowledge, only p735.6 has been clearly demonstrated to provide adequate control [Price et al., 1994]. Use of temperature-sensitive vector constructs is further complicated by the poor growth of human fibroblasts at the nonpermissive temperature.

3.3. Preparation of Cells for Transfection

Cell cultures should be maintained under optimal conditions with frequent medium changes (1–2 per week). Trypsinise cells 24–48 h before transfection, and seed no more than $2–2.5 \times 10^5$ per 9-cm dish or $5.5-6.8 \times 10^5$ per 175-cm^2 flask. Cells should be subconfluent when transfected with an even 70–80% cover of the dish or flask. The final volume of medium should be 10 ml on a 9-cm dish and 30 ml in a 175-cm^2 flask.

3.4. Transfecting DNA into HDF

It is important to optimise gene transfer conditions in order to maximise the initial number of transfectants. Once you have ensured optimal cell culture conditions, trial transfection experiments to measure transient expression levels are invaluable. When DNA is introduced into human cells, it is not immediately integrated into the genome but can nonetheless be expressed. These transient levels of expression peak between 48 and 72 h after treatment with calcium phosphate-precipitated DNA and then decline as DNA is lost from the cells. Only a proportion of the cells that were competent for DNA uptake and immediate expression of the DNA are able to integrate the sequence into the genome. When integration fails to occur, the sequences are lost either by degradation or simply diluted out as the cells divide. Transient expression levels are not an absolute indicator of subsequent stable transfection frequencies, but they are relatively easy to perform and allow a measure of the success of the gene transfer technique.

Transient expression assays require DNA constructs with reporter genes. Ideally, these vectors should employ the same promoter for the reporter gene as the vector you are using for T-antigen expression in order to act as a suitable control for subsequent experiments. There are a range of reporter genes available to measure the efficiency of transient expression. In particular, genes such as CAT, firefly luciferase, and β-galactosidase produce proteins that can be assayed relatively simply. An alternative method for measuring transient expression, which utilizes the same vector for stable expression, can also be employed. There are a range of antibodies available that permit direct visualisation of T-antigen protein by fluorescent cytochemistry. Cells transfected with SV40-T-antigen can be stained with T-antigen antibody 48-72h after transfection and the number of cells and the intensity of staining gives a direct measure of transient gene expression.

For the calcium phosphate precipitation method, the DNA precipitate is added directly to the culture medium. The volume of DNA precipitate applied to the cultures should never exceed $\frac{1}{10}$th of the total volume, where transfections use small amounts of DNA, application of low volumes of precipitate have not significantly affected the relative transfection frequency, in our hands. Fibroblasts from different individuals vary in their sensitivity to calcium phosphate precipitates, some cells will tolerate only a 6 h exposure, while others can be left overnight. In order to test the success of subsequent selection under the conditions of the experiment, we routinely prepare mock precipitates made without DNA and administer these to a set of control plates.

The DNA precipitate/culture medium is removed by aspiration after 6–18 h exposure at 37°C in a standard CO_2 humidified incubator. The cells are then left undisturbed until 48 h from the start of the experiment (i.e., the point when the calcium phosphate precipitate was added to the cells). At this time, the appropriate selective agent is added (see Appendix for suggested levels) to both the treated and control plates. If the experiment was performed on cell culture plates, handling is avoided from now on to reduce the risk of contamination. Handling is limited to the control plates and one or two of the DNA-treated plates. Examination of sealed flasks does not significantly increase the risk of contamination.

3.4.1. Calcium phosphate–DNA coprecipitation

The calcium phosphate precipitation method relies on the formation of a DNA precipitate in the presence of calcium and phosphate ions. The DNA is first sterilized by precipitating in ethanol and resuspending in sterile buffer. It is then mixed carefully with calcium and the solution is added very slowly, with mixing, to a phosphate solution. When making the precipitate, it is important to note that optimal gene transfer occurs when the final concentration of DNA in the precipitate is 20 μg/ml. For mixing, we recommend the use of two-hand held pipette aids. With two pipettes, one for blowing bubbles of sterile air into the mixture and the other for carefully adding the DNA/calcium in a dropwise fashion to the phosphate solution, good mixing results in an even precipitate. The DNA precipitate sticks very firmly to glass, so we recommend the use of plastic pipettes and containers. It is necessary to leave the mixture to develop for 30 min before it can be added to the cell cultures.

Protocol 3.4.1.

Materials and Reagents, Sterile

• HEPES buffer:	HEPES	12.5mM
	pH 7.12	
• 10× CaHEPES:	$CaCl_2$	1.25M

HEPES 125mM
pH 7.12

- 2× HEPES-buffered phosphate (2×HBP):
  Na_2HPO_4 1.5mM
  NaCl 280mM
  HEPES 25mM
  pH 7.12
- NaOAc: NaOAc 3M
  pH 5.5
- Tris-buffered EDTA (TBE):
  Tris HCl 2mM
  EDTA 0.1 mM
  pH7.12
- Absolute ethanol
- *G418*: 20 mg/ml in Hepes buffer pH 7.5, sterilized by filtration through a 0.2μm membrane and stored in small aliquots at −20°C.
- *Hygromycin:* 2 mg/ml in purified water, filter-sterilized with a 0.2 μm membrane and stored in small aliquots at −20°C.

Protocol

(a) Estimate the amount of DNA required for the transfection
 (i) Use a maximum of 20 μg of your T-antigen vector (without carrier DNA) for each plate and 60 μg vector DNA per flask.
 (ii) Include an additional 20 μg DNA as it is not always possible to recover the full expected volume after preparation of the precipiate.

(b) Prepare a sterile solution of the vector DNA.
 (i) Precipitate DNA with $\frac{1}{10}$th volume 3M NaOAc pH 5.5 and 2.5 volumes of ethanol.
 (ii) Mix well, ensuring that the entire inside of the tube has come into contact with the ethanol solution.
 (iii) Leave on ice briefly (5 min).
 (iv) Centrifuge for 15 min at 15,000 rpm in a microfuge to collect the precipitate.
 (v) Gently remove the tube from the centrifuge and open in a sterile cell culture cabinet.
 (vi) Remove the supernatant by aspiration, taking care not to disturb the pellet-ensure that the ethanol is well drained.
 (vii) Allow the pellet to air-dry in the cell culture hood until any traces of ethanol have evaporated.

(viii) Resuspend your DNA pellet in TBE to give a final concentration of 0.5 mg/ml. It may be necessary to vortex the tube to release the pellet from the side of the tube.

(ix) Incubate the tube at 37°C for 5–10 min with occasional vortexing to ensure that the pellet is well resuspended.

Note: Do not use higher TBE concentrations for resuspending your DNA, as this can interfere with the formation of the DNA precipitate.

(c) Prepare the DNA calcium-phosphate precipitate.

(i) Calculate the total volume of precipitate required. The final concentration of the DNA in the precipitate should be 20 μg/ml, and you will need 1 ml for each 9-cm plate and 3 ml for each 175- cm^2 flask. Remember to make an extra 1 ml of precipitate to ensure recovery of sufficient volume for all your cultures-some loss of volume will occur during preparation of the precipitate.

(ii) Dilute DNA/TBE in 12.5mM HEPES to give 20 μg/ml in final mix.

(iii) Add 10× $CaCl_2$, $\frac{1}{10}$th of final mix volume.

(iv) Add dropwise, with mixing, to an equal volume of 2×HBP.

For instance, for 9, 9cm dishes at 1 ml/dish, with 1 ml spare, (10 ml mix):

(a)	DNA/TBE	0.5ml
(b)	12.5mM HEPES	3.5ml
(c)	10× $CaCl_2$	1.0ml
(d)	Add dropwise to 2×HBP	5.0ml

Final concentrations in mix:		
DNA		20μg/ml
$CaCl_2$	0.125M	
Na_2PO_4	0.75mM	
NaCl	140mM	
HEPES	12.5mM	
pH	7.12	

Note: Remember to use plastic pipettes and tubes when preparing DNA calcium-phosphate precipitates as the precipitate sticks very firmly to glass. Mixing of the DNA-calcium solution into the phosphate solution is best achieved by vigorously blowing sterile air down a pipette into the phosphate solution while gently adding the DNA-calcium solution from a second pipette. Use 1–5-ml pipettes depending on the volume of precipitate being made. As the DNA-calcium solution is being added to the phosphate solution, a light, even, precipitate begins to form. This is quite obvious and gives a milky appearance when complete.

(v) After all the DNA-calcium has been added to the phosphate, replace the lid on the tube, invert gently once or twice, and leave to stand at room temperature for 30 min.

(vi) It is important to make a mock DNA precipitate without DNA. This allows you to assess the effectiveness of your selection conditions. A mock precipitate can be made exactly as described above, but the DNA is replaced with additional HEPES buffer. Mixing the calcium and phosphate together, as for the real precipitate, still produces a visible precipitate despite the absence of DNA.

(d) Add 1 ml of your DNA or mock precipitate to plates or 3 ml to flasks. Make sure that the volume of precipitate is no more than $\frac{1}{10}$th the total volume of the culture medium already on the cells.

(e) Leave the precipitate on the cells for a minimum of 6 h but no more than overnight (16 h). For fibroblasts from some individuals, exposure to calcium and phosphate for more than 6 h may be toxic.

(f) Remove the calcium phosphate precipitate by aspiration.

There is no need to further wash the cells or, in our experience, is there any need for further treatment with either DMSO or glycerol. Under optimal conditions, further treatment with glycerol or DMSO does not improve transfection frequencies.

(g) Leave the flasks and/or plates in the incubator until 48 h from the start of the experiment, and then add selective agents.

The agent that you add will depend on the vector used for transfection. Vectors carrying the *neo* gene confer resistance to G418 (Gibco-geneticin) and vectors carrying the *hyg b* gene confer resistance to hygromycin B (Boehringer-Mannheim). All fibroblasts, in our experience, require 100–200 μg/ml G418 or 10–20 μg/ml hygromycin B to kill the cells gradually over a period of a week.

(h) Change the medium on the transfected plates:
   (i) Dispense the total volume of medium required into a suitable sized sterile flask and add G418 or hygromycin B from the concentrated stocks to give the correct final concentration.
   (ii) Gently swirl to mix.
   (iii) Aspirate the medium from the plates and/or flasks and replace with the selective medium.
   (iv) Return the plates/flasks to the incubator.

(i) Monitor the effects of the selective medium on a daily basis by examining under the microscope. When a significant number of cells have lifted and died, replace the medium with fresh selective medium. Selection should be maintained at all times.

(j) Continue to replace the medium until the background of cells has lifted and died. The mock transfected plates that have not been transfected with DNA should have no viable cells remaining after 7–10 days. If there are cells remaining, then the selection has not worked adequately and it may be necessary to raise the concentration of selective agent.

(k) Once the background of cells has died, it is no longer necessary to routinely change the medium on the cells. They should then be left undisturbed in the incubator for 4–6 weeks to allow the transfected cells to grow and form colonies.

(l) Once colonies arise, pick individually using cloning rings [Freshney, 1994] or bulk together by trypsinizing the whole dish or flask.

(m) Freeze aliquots of cells at the earliest opportunity and regularly thereafter.

Once transfectants have been expanded into cultures and ample stocks frozen in liquid nitrogen, it is necessary to keep the culture going for an extended period of time until it reaches crisis. Such long-term cell culture work is always vulnerable to losses arising from contamination. To minimize the risk from fungal contaminants, we routinely add amphotericin B (Fungizone—Gibco) to our cultures at 2.5 μg/ml.

(n) Subculture routinely until cultures reach the end of their *in vitro* lifespan. As the cells approach crisis, the growth rate often slows. Cultures require careful nurturing and gentle handling. As the cultures begin to degenerate and cell division ceases, it is no longer necessary to subculture the cells. However, if heavy cell debris begins to cling to the remaining viable cells, it may be advisable to trypsinize the cells to remove the debris. Either return all the cells to the same vessel, or use a smaller one to compensate for the cell death that is occurring. In general, the cells grow and survive better if they are not too sparse. With patience, care and the culture of sufficient cells from your freezer stocks, you should, in most cases, obtain a postcrisis line.

(o) When healthy cells begin to emerge, allow the colonies to grow to a healthy size before subculturing, then begin to subculture again. Do not be tempted to put too few cells into a large flask.

(p) Freeze an ampoule of cells at the earliest opportunity and continue to build up a freezer stock before using the culture.

(q) Because of the risk of isolating a contaminating cell line introduced from other cultures in the laboratory, it is essential that you confirm the origin of your postcrisis culture. The best available method is based on DNA fingerprinting.

(r) As a final check that your postcrisis line is truly immortal, we recommend selecting and expanding individual clones from the culture. This has the additional benefit of providing a homogeneous culture derived from a single cell. Details on cloning methods are given in Freshney [1994].

4. POSTTRANSFECTION CARE OF CULTURES

The effectiveness of the selective agent must be monitored. The levels of selective agent are chosen to produce a gentle kill over a period of about a week. Fierce selection leading to immediate kill reduces the number of successful transfectants. Alternatively, levels that are too low may permit growth of clones that are not expressing the transfected vector. The majority of cells on the DNA-treated plates should die within 7 days, those cells remaining should be the successful transfectants. If the level of selection was correct, then no viable cells should remain on the mock-treated control plates at this time. Cell debris is removed by changing the medium and the plates are now left in the incubator for 3–6 weeks to allow colonies to form. Selection is continued and further medium changes are not required unless significant cell debris remains on the plates.

The amount of time for the transfected colonies to appear depends, to some extent, on the success of the experiment, with colonies appearing earlier when the success rate is high. Robust colonies of cells with the characteristic SV40-transformed appearance should be apparent within 6 weeks. Individual clones can now be selected using cloning rings or bulk cultures prepared by trypsinisation of the complete plate or flask. Ampoules of cells should be frozen down at the earliest opportunity and records kept of each subculture.

If a large number of cultures are being maintained, it is not necessary always to count the number of cells obtained after each trypsinisation. A record of the split ratios is sufficient (i.e., the proportion of cells reseeded each time).

It is advisable to freeze ampoules of cells routinely to build up both a stock of transfected cells and to save time if cultures are lost because of contamination.

Medium should be changed 1–2 times per week (depending on cell numbers) and selection continued throughout the experiment. The culture should be trypsinized when it reaches confluence. Cells expressing high levels of T-antigen lose contact inhibition and will continue to grow to high densities, often piling up to the point where cells underneath are no longer in contact with the culture medium. Under these conditions, cells will die and shed into the medium.

Various studies have demonstrated that the expression of T-antigen increases culture lifespan. On average, transfected fibroblasts grow for approximately 60 PD in total, and the posttransfection lifespan of your culture will depend on the age of the starting fibroblast culture.

It is not recommended to use very-late-passage cells for these experiments as older, poor-growing cells are often poor recipients for gene transfer. Young cultures are excellent recipients, but if the culture is very young, it extends the length of time in culture necessary to reach crisis after transfection with T-antigen. We have routinely used cells previously passaged 7–15 times for these experiments.

Figure 4.1 shows a typical growth curve for a skin fibroblast culture expressing T-antigen. As discussed above, T-antigen does not directly immortalize cells, and after a prolonged period of rapid growth, the culture enters crisis. Although the majority of fibroblasts will have an *in vitro* lifespan of around 60 PD, fibroblasts from a few individuals fall outside this range. In a few instances, cells have grown for more than 100 PD posttransfection, and this has led some to assume that their precrisis cultures were in fact immortal. In our experience, all SV40-transfected fibroblasts enter crisis, although there may be considerable variability in the timing of crisis between individual cultures.

There is also considerable variability in the nature and length of crisis between individuals and we have interpreted this as reflecting the timing of the various genetic changes that occur during the immortalization process. We do not yet know whether the cells that ultimately become immortal must themselves undergo crisis, or if the crisis that is observed is simply a phenomenon occurring in the majority of the members of the population that have not accumulated the necessary

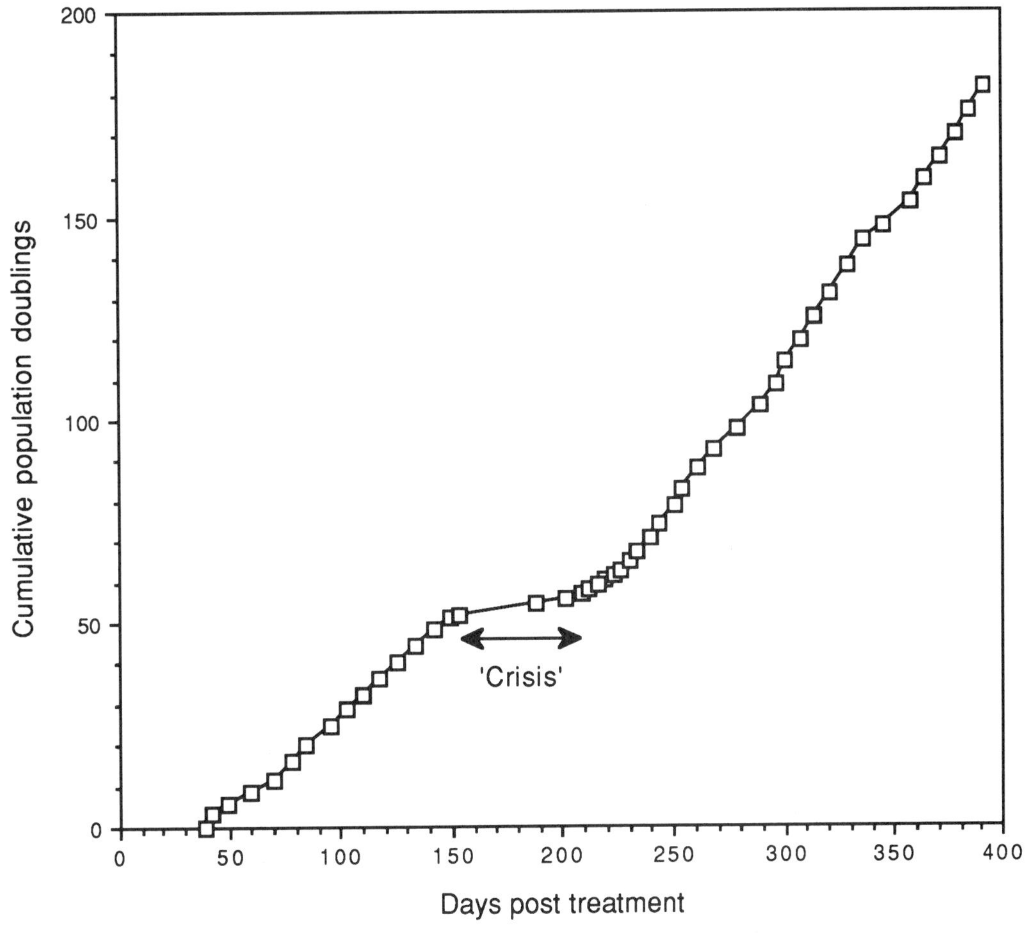

Fig. 4.1. Typical growth curve for SV40-transfected HDF. Cumulative population doublings are plotted against time. Although details will vary between experiments, depending on the culture used, the curve should be very similar to the one shown. In a few instances, crisis may be minimal, and only a brief slowing of the growth rate is observed. HDFs were transfected with a T-antigen expressing vector at day 0 and selected in either G418 or hygromycin B, depending on the vector used. Following selection, individual clones were isolated approximately 4 weeks later. These clones were expanded in culture and grew at an exponential rate for about 150 days, when the culture demonstrated the characteristic signs of crisis. In general, robust and healthy cells can be seen among the remaining non-viable, degenerate cells. Gradually, these cells expand to give a new robust culture that grows exponentially, and, in most cases, indefinitely.

genetic changes to permit immortal growth. In some cases, crisis is very short, with few cells showing overt signs of senescence, and there may only be a brief slowing of the rate of growth of the culture. One interpretation of this observation may be that one or a few cells within the population acquired the necessary genetic changes early in the culture life, such that at the time of crisis, a considerable proportion of the population did not undergo crisis and net population growth only slowed as the immortal cells overgrew the deteriorating mortal cells. In most cases, crisis is a marked event, with the majority of cells showing signs of deterioration and a net loss of viable cells. On average, crisis lasts 3–6 months and the culture may deteriorate to the point where very few, if any, obviously healthy cells are present. In some cases, crisis has lasted 12 months and viable, immortal derivatives have arisen from very degenerate cultures.

As cultures deteriorate and cell debris accumulates in crisis, it becomes difficult to know how to handle the cells. The medium often becomes very acid and the cell debris will stick to both the culture vessel and the remaining viable cells. Often, apparently healthy viable clones begin to emerge, only to subsequently fade and die. Once a culture has entered crisis, you can then reliably predict the timing of crisis for parallel cultures stored in liquid nitrogen. This allows one to retrieve earlier ampoules from the freezer and to build up a number of flasks sitting at the threshold of crisis. As these parallel cultures enter crisis and begin to lose viability, these flasks can be pooled. In some cases, there will be an adequate cover of cells in the flask, but high levels of cellular debris. In these cases, we often recommend replating the cells. The cells are trypsinized, and centrifuged and the pellet returned to the original flask. The flask may be rinsed several times with trypsin to remove any adhering cell debris before returning the cells.

With careful nurturing and ample freezer stocks, the majority of cultures will give rise to immortal derivatives. The first sign of a culture emerging from crisis is usually the appearance of one or more foci of apparently healthy, robust cells with the typical appearance of SV40-transformed cells. On subculturing, these foci expand to give a healthy, regenerating culture. In many cases, the early emerging postcrisis cells grow poorly, but the growth properties improve with further subculturing, and subcloning helps to select individual clones with better growth properties. It is important to freeze an ampoule of your new cell line as early as possible. Our working definition for an immortal line is that the culture has undergone a minimum of 100 PD posttransfection and has survived subsequent subcloning.

As the appearance of an immortal derivative within these cultures is a relatively rare event, it is essential that any postcrisis cell lines that emerge be checked to ensure that they were derived from the original starting material and are not the result of contamination from other immortal cell lines in the lab. DNA fingerprinting presents an excellent method for validating lines. If you do not have the technology inhouse, it can be contracted out.

ACKNOWLEDGMENTS

This work was supported under the LINK Genetic Engineering Programme with funds from the SERC/BBSRC and DTI, Celltech Research Ltd., Glaxo Group Research (Greenford) Ltd. and Unilever

Research, Colworth. L.V.M. was a recipient of a Wellcome Trust Senior Research Fellowship (Basic Biomedical Sciences) and an R.M. Phillips Senior Research Fellowship into Ageing.

REFERENCES

Bartek J, Vortesek B, Grand RJA, Gallimore PH, Lane DP (1992): Cellular localisation and T antigen binding on the retinoblastoma protein. Oncogene 7: 101–108.

Freshney RI (1994): "Culture of Animal Cells, Manual of Basic Technique," 3rd ed. Chichester, UK: Wiley.

Hayflick L, Moorhead PS (1961): The serial cultivation of human diploid cell strains. Exp Cell Res 75: 585–621.

Huschtscha LI, Holliday R (1983): Limited and unlimited growth of SV40-transformed cells from human diploid MRC-5 fibroblasts. J Cell Sci 63: 77–99.

Ide T, Tsuji Y, Nakashima T, Ishibashi S (1984): Progress of aging in human diploid cells transformed with a tsA mutant of simian virus 40. Exp Cell Res 150: 321–328.

Kumar V, Green S, Staub A. Chambon P (1986): Localisation of the oestradiol-binding and putative DNA-binding domains of the oestrogen receptor. EMBO J 5: 2231–2236.

Lane DP (1992): p53, guardian of the genome. Nature 358: 15–16.

Lomax CA, Bradley E, Weber JL, Bourgaux P (1978): Transformation of human cells by temperature-sensitive mutants of simian virus 40. Intervirology 9: 28–38.

Mayne LV, Priestly A, James MR, Burke JF (1986): Efficient immortalization and morphological transformation of human fibroblasts with SV40 DNA linked to a dominant marker. Exp Cell Res 162: 530–538.

Pereira-Smith, OM Smith JR (1988): Genetic analysis of indefinite division in human cells: identification of four complementation groups. Proc. Natl. Acad. Sci. USA 85: 6042–6046.

Price TNC, Moorwood K, James MR, Burke JF, Mayne LV (1994): Cell cycle progression, morphology and contact inhibition are regulated by the amount of SV40 T antigen in immortal human cells. Oncogene 9: 2897–2904.

Radna RI, Caton Y, Jha KK, Kaplan P, Pardenas J, Zainul B, Traganos F, Ozer HL (1989): Growth of immortal simian virus 40 tsA-transformed human fibroblasts is temperature dependant. Mol Cell Biol 9: 3093–3096.

Shay JW, Wright WE (1989): Quantitation of the frequency of immortalization of normal human diploid fibroblasts by SV40 large T antigen. Exp Cell Res 184: 109–118.

Southern PJ, Berg P (1982): Transformation of mammalian cells to antibiotic resistance with a bacterial gene under control of the SV40 early region promoter. J Mol App Genet 1: 327–341.

Weinberg RA (1990): The retinoblastoma gene and cell growth control. Trends Biochem Sci 15: 199–202.

APPENDIX: MATERIALS AND SUPPLIERS

Materials	Suppliers
Eagles MEM	Gibco BRL, Life Technologies
Trypsin	Gibco BRL, Life Technologies
EDTA	Gibco BRL, Life Technologies
Glutamine	Gibco BRL, Life Technologies
Penicillin/streptomycin	Gibco BRL, Life Technologies
G418	Gibco BRL, Life Technologies
HEPES	Gibco BRL, Life Technologies
Amphoteracin B	Gibco BRL, Life Technologies
Hygromycin B	Boehringer Mannheim

5

Immortalization of Human Epidermal Keratinocytes by SV40

Mark L. Steinberg

Biochemistry Division, The City College of New York, Convent Ave. and 138th Street, New York, New York 10031.

Culture of Immortalized Cells, Edited by R. Ian Freshney and Mary G. Freshney.
ISBN 0-471-12134-7

1. INTRODUCTION

Because most human cancers are derived from epithelial cells, there is substantial interest in developing model systems of immortalized cells for studying both normal and abnormal epithelial cell growth and differentiation *in vitro.* However, unlike epithelial cells of rodent origin, spontaneous immortalization of normal human keratinocytes in culture is an extremely rare event [Baden et al., 1987]. In contrast, the early genes of the oncogenic virus, SV40 are highly efficient in inducing immortalization and, unlike other transforming agents such as carcinogenic compounds or other viruses, have no cytotoxic effects. Since the first report of SV40-induced immortalization in human epidermal keratinocytes [Steinberg and Defendi, 1979] SV40 has been widely used as an immortalizing agent for a variety of human epithelial cell types.

The immortalizing activity of SV40 is generally studied from the standpoint of the oncogenic properties of the virus, but the viral, transformants have also been shown to express many of the phenotypic properties of the untransformed host. For this reason SV40-immortalized human epithelial cells have great value as cell lines, providing large quantities of material for studying epithelial function from both normal and pathological tissue *in vitro*. For example, various aspects of normal functioning have been demonstrated in SV40-transformed epithelium from mammary gland [Rudland et al., 1989; Rudland and Barraclough, 1990], bronchial tissue [Ke et al., 1988], eye [Coca-Prados and Wax, 1986], epidermis [Banks-Schlegel and Howley, 1983; Ponec et al., 1985; Vermeer et al., 1986], and thyroid gland [Lemoine et al., 1989]. In our laboratory lines of SV40-transformed cells have been utilized to study synthesis of various antigens associated with nononcogenic epidermal disease states, including epidermolysis acquisita antigen [Okada et al., 1989], lupus-associated SSA/Ro antigen [Miyagawa et al., 1988], and pemphigoid antigen [Okada et al., 1986].

The progressive nature of the process of transformation brought about by SV40 in human epithelial cells has been extensively characterized. Following the introduction of the viral genes, the cells pass through distinct stages defined in terms of a series of changes in properties of growth and differentiation (Fig. 5.1) [Defendi et al., 1982; Steinberg and Defendi, 1983, 1985; Steinberg et al., 1986]. Thus, growth dependence on high serum concentrations and anchorage to solid substrates are markers of transformation which are temporally dissociated from one another. Growth independence from high serum concentrations is manifest quite early—within the first four passages postinfection (about 40 cell generations)—but anchorage independence is never seen until permanent cell lines are established in the period following crisis (>10–15 serial passages; see Steinberg and Defendi [1979], Defendi et al. [1982], and Steinberg and Defendi [1983]. Breakdown of junctional communication, as detected by transfer of microinjected fluorescent dye [Steinberg and Defendi, 1981; Steinberg and Defendi, 1982], clonal growth in the absence of feeder layers [Steinberg and Defendi, 1979; Defendi et al., 1982; Steinberg and Defendi, 1983], and reduction in the production of terminally differentiated squames, as measured by histochemical staining and cornified cell envelope formation [Steinberg and Defendi, 1979; Defendi et al., 1982; Steinberg and Defendi, 1983] all appear in the precrisis stage. Redistribution of fibronectin cables over the cell surface occurs at about the time cells enter crisis [Defendi et al., 1982; Edelman et al., 1985].

In addition to anchorage independent growth, markers of the late phase of the transformation process include altered actin filaments and the ability to form undifferentiated carcinoma-like tumors in nude mice [Steinberg et al., 1986]. A similar time-dependent progression to tumorigenicity has also been demonstrated in bronchial epithelial cells [Reddel et al., 1993] and hepatocytes [Woodworth et al., 1988]. There are also changes in the expression of the keratin genes in that, over time following infection, there is a gradual loss of expression of genes coding for the stratified epithelial keratins while, conversely, expression of genes for the simple epithelial keratins becomes progressively induced. The changes in keratin expression have been shown to be due to altered transcription of the keratin genes [Steinberg and Defendi, 1985; Morris et al., 1985]. Reexpression of the normal epidermal keratins can be seen in cells treated with 5-azacytidine, but this inducibility is eventually lost [Okada et al., 1984]. These findings relate to the clinical observation that ectopic expression of fetal keratins has also been shown to accompany malignant transformation of epithelial cells *in vivo* and is used as a tool in tumor diagnosis [Stieber et al., 1993; Suo et al., 1993; Katabami et al., 1993; Guelstein et al., 1993; Mayall et al., 1993]. Formation of differentiated squames is also inducible at high cell densities [Steinberg and Defendi, 1979, 1981] or after cell fusion [Steinberg and Defendi, 1982] during the early, but not the late stages.

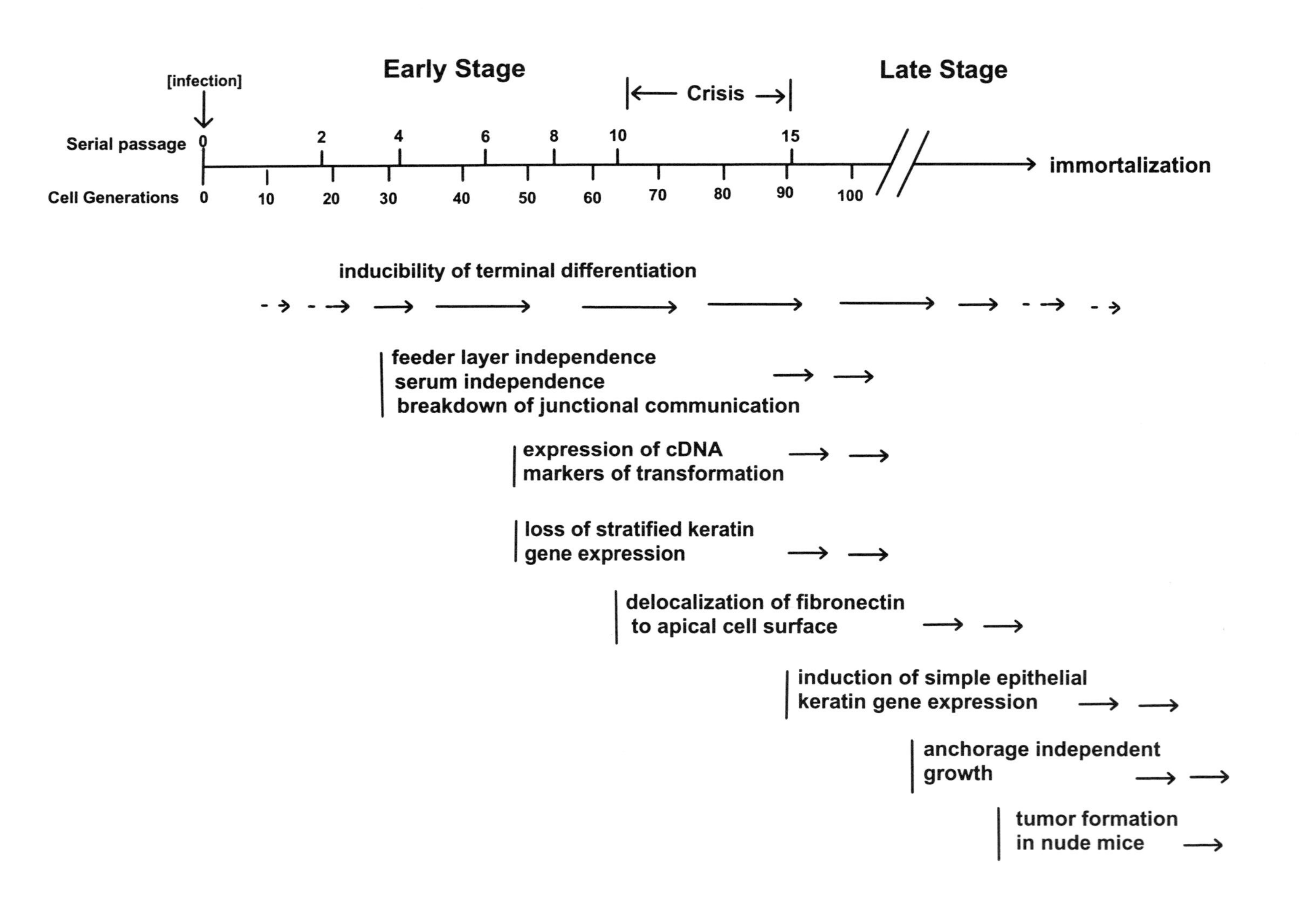
Early Stage
Late Stage
[infection]
Crisis
Serial passage
0
2
4
6
8
10
15
Cell Generations
0
10
20
30
40
50
60
70
80
90
100
immortalization
inducibility of terminal differentiation
feeder layer independence
serum independence
breakdown of junctional communication
expression of cDNA
markers of transformation
loss of stratified keratin
gene expression
delocalization of fibronectin
to apical cell surface
induction of simple epithelial
keratin gene expression
anchorage independent
growth
tumor formation
in nude mice

2. PREPARATION OF MEDIA AND REAGENTS

2.1. Stock Solutions

2.1.1. Hydrocortisone stock solution, 0.7mM

(1) Weigh out sufficient powdered hydrocortisone for a final concentration of 0.25 mg/ml.
(2) Dissolve in ultrapure water (UPW) by stirring overnight.
(3) Sterilize by filtration.

Hydrocortisone stock is stable for at least 6 months at 4°C, although aliquots may be stored frozen at −20°C for longer storage if desired.

2.1.2. 7.5% sodium bicarbonate

(1) Dissolve 7.5 g of sodium bicarbonate in 100 ml of UPW.
(2) Sterilize by vacuum filtration.
(3) Add dropwise with swirling when culture medium becomes acidic; adjust to pH 7.4 by eye using the phenol red in the medium as an indicator.

2.1.3. EGF stock for primary cultures

EGF is available as a sterile, lyophilized powder in a sealed ampule that should stored at 4°C until ready for use.

(1) Reconstitute an ampule containing 100 μg EGF with 1 ml of sterile UPW and dissolve by gentle swirling.
(2) Aliquot and store reconstituted EGF at −20°C, stable for several months.

2.1.4. Cholera toxin

(1) Reconstitute lyophilized cholera toxin with UPW to a final concentration of 1 mg/ml.
(2) Store at 4°C.

Fig. 5.1. Schematic representation of changes in properties of growth and differentiation over time following infection of human epidermal keratinocytes by SV40. The chart represents the composite observations on eight independently derived cell lines. The number of cell generations was calculated on the basis of plating efficiencies and the number of cells harvested at confluence from cultures at each passage (Defendi et al., 1982). The relationship shown between serial passage assumes a constant split ratio of 1:3. Growth independence from feeder layer support (feeder layers were used in the early protocol) and high serum concentrations, altered keratin gene expression, and anchorage-independent growth have all been described. Expression of markers of terminal differentiation begins to diminish after the first few serial passages but remains inducible under various conditions including fusion [Steinberg and Defendi, 1982], treatment with 5-azacytidine [Okada et al., 1984], and TPA [Mufson et al., 1982], and methocel suspension [Steinberg and Defendi, 1983] up to at least the 50th serial passage.

2.2. Media

2.2.1. DMEM from powder

(1) Weigh out powdered DMEM medium for the quantity of medium desired (13.7 g/liter for 1× DMEM; 27.4 g/liter for 2×DMEM).
(2) Dissolve in UPW by stirring for approximately 1 h.
(3) Add 3.7 g of sodium bicarbonate for each liter of medium.
(4) Adjust pH to 7.4 with HCl.
(5) Sterilize the liquid medium by vacuum filtration.
(6) Prepare complete medium by the adding, per liter:
 (a) 2 ml of hydrocortisone stock (see Section 2.1.1) (0.5 μg/ml (1.4μM) final concentration)
 (b) 10 ml of 100× antibiotic concentrate
 (c) 110 ml of fetal calf serum (final concentration 10%)
 Fetal calf serum may be added to a final concentration of up to 20% for cells that may be initially slow growing.

2.2.2. Culture medium for primary cultures

(1) Add 20 μl of cholera toxin stock solution for each liter of medium desired (final concentration 20 ng/ml) to between 5–10 ml of normal culture medium.

(2) Sterilize with a syringe filter and add back to medium.

(3) Remove 0.2 ml of EGF stock using a sterile syringe add directly to one liter of complete DMEM (20 ng/ml).

Because EGF activity in medium diminishes fairly rapidly, fresh medium for primary culture should be made about every 8 weeks or more frequently.

2.3. Trypsin

0.25% 1:250 crude trypsin in PBSA.

2.4. Staining Solutions

2.4.1. Acid fuchsin stock

(1) Dissolve 0.5 g of powdered acid fuchsin in 100 ml of deionized water.

(2) The stock solution is stable for several months at room temperature.

2.4.2. Aniline blue-orange G stock

(1) Dissolve 1.0 g of phosphotungstic acid in 100 ml deionized water.

(2) Add 2.0 g of orange G with stirring until the solution is clear,

(3) Add 0.5 g aniline blue and stir until the blue crystals are dissolved.

The stock solution is stable for several months at room temperature.

3. PRIMARY CULTURES OF EPIDERMAL KERATINOCYTES

Until relatively recently, culture systems for the long-term growth and maintenance of normal human epithelial cells were unavailable. However, in 1975, Rheinwald and Green reported a method of cell culture based on the use of irradiated 3T3 cell feeder layers which permits the growth and subculture of fully differentiated cultures from single cells [Rheinwald and Green, 1975]. While this innovation gave researchers, for the first time, the ability to grow epithelial cells in practical quantities, variability in the lifetime and feeder capabilities of different batches of 3T3 cells are drawbacks inherent in this method. We have found that a modification of the explant technique used by Taichman [L. Taichman, State University of New York, personal communication] provides more reliable results and greater cell yield without the use of feeder layers.

3.1. Source of Tissue

In our experience, foreskin specimens obtained from routine neonatal circumsions provide the best source of tissue for culture, but skin obtained from virtually any anatomical location can be used as long as it is stored in sterile physiological medium (e.g., serum-free culture medium) at 4°C immediately following surgical removal. We have used skin from forearm obtained by punch biopsy, facial epidermis from plastic surgery, and abdominal skin removed from fresh cadavers with good results. Nevertheless, as a general rule, the older the age of the donor, the poorer the growth capacity *in vitro*. Skin specimens derived from the more highly keratinized areas of the body (e.g., the palmar and plantar surfaces) should also be avoided owing to the low yield of proliferative cells.

3.2. Primary Cell Culture

Protocol 3.2

Reagents and Materials (Sterile)

- Serum-free medium: DMEM (Section 2.2.1) without serum
- Primary culture medium (see Section 2.2.2)
- Scissors, flame-sterilized
- Petri dishes, 9 cm
- Pasteur pipettes

Protocol

(a) Trim skin specimens of subcutaneous fat and mince into small pieces (roughly about 2–4 mm^3 per edge) in serum-free medium using flame-sterilized scissors.
(b) Spread the slurry uniformly over the bottom of a culture plate or flask such that the tissue pieces are evenly distributed.
(c) Remove the excess liquid carefully from around each of the tissue pieces, using a flame-drawn Pasteur pipette, so that each piece is damp, but essentially free of surrounding liquid.
(d) Allow the tissue pieces to adhere to the plastic substrate for usually anywhere from 45 min to 1 h.
(e) Flood the culture vessel with fresh primary culture medium.

Longer drying times may be required to ensure that most of the pieces remain adherent after culture medium is added. We have used drying times of up to 3 h with good viability; the basal cells of the epidermis are protected from drying out under these conditions because of the overlying stratum corneum and the underlying layer of dermal tissue. Halos of epidermal cell outgrowths surrounding the attached tissue pieces can be seen in anywhere from several days to 2 weeks, although there is considerable variability in the degree of outgrowth. Microscopic examination of the individual explants can be used to identify fibroblast outgrowths that localize in small colonies around the explant peripheries. Surprisingly, fibroblast outgrowth is generally quite limited in explants. Where fibroblasts are observed, aspiration of the colonies can be used to remove them quite effectively. Cultures are then maintained under standard culture conditions.

4. INTRODUCTION OF IMMORTALIZING VIRAL SEQUENCES

The immortalizing properties of SV40 derive from the early gene region (map positions 31 $\rightarrow$ 2599, counterclockwise; Fig. 5.2) which also accounts for the oncogenic properties of the virus; the early genes

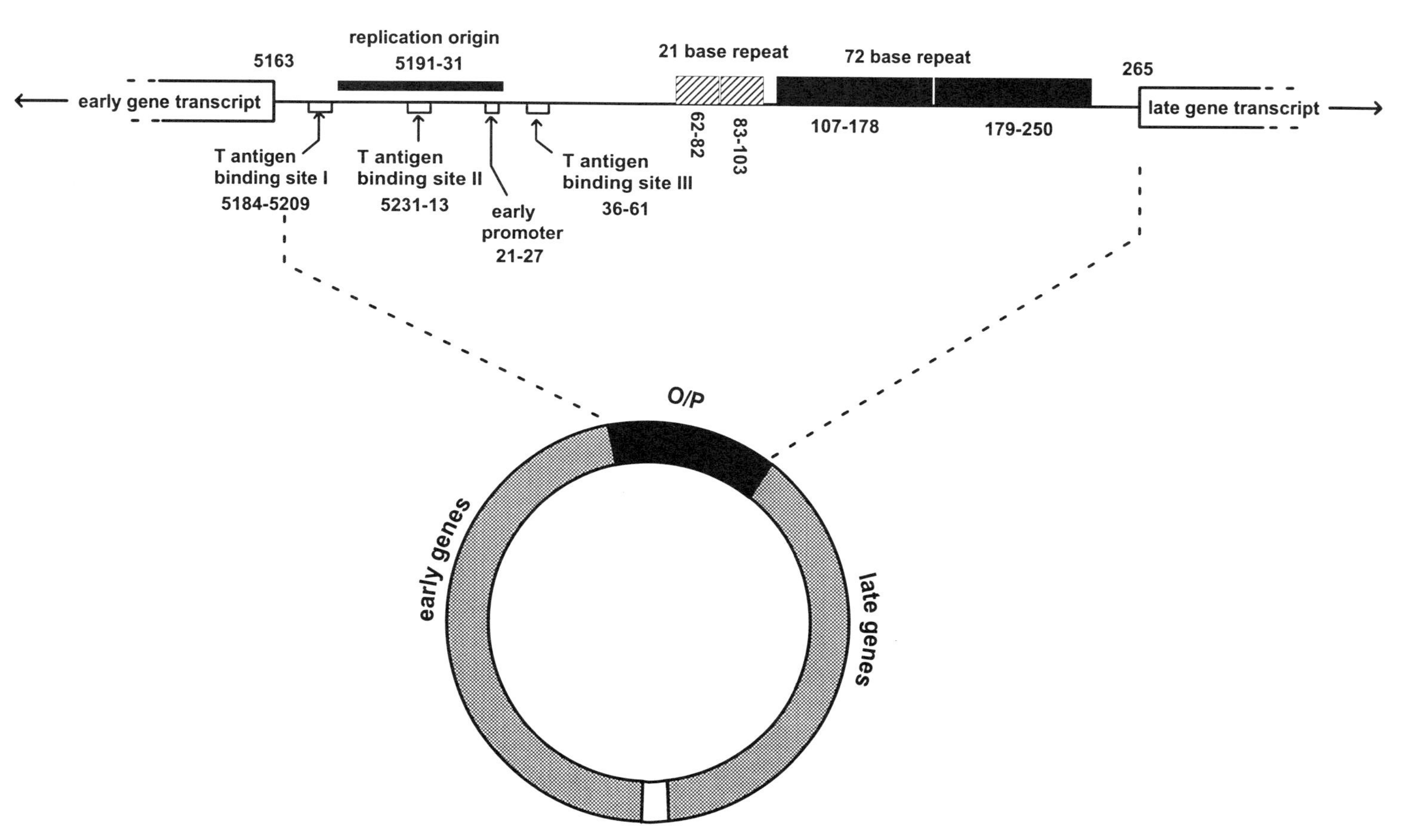

Fig. 5.2. Diagram of the SV40 genome and the main control elements in the origin/promoter (O/P) region. The early gene region codes for two proteins (designated T and t antigens) whose transcripts both arise from the same primary transcript via alternative splicing. The early gene primary transcript is initiated in the counterclockwise direction within the O/P region with coding sequences beginning at nucleotide 5163. Transcription is regulated by T-antigen binding at three distinct sites within O/P located downstream from two enhancer elements comprised of 21 and 72 base repeats, respectively. The oncogenic and immortalizing properties of the virus derive entirely from the early genes. The structural features of the viral genome and the relevant references are described in the SV40 GenBank entry (locus SV4CG; accession V01380).

have been shown to be necessary and sufficient for complete transformation of cells *in vitro* (see Fanning, [1992] and Fanning and Knippers, [1992] for reviews of T-antigen structure and function). Since human keratinocytes do not support viral replication *in vitro,* circumvention of the lytic properties by isolation of the early genes or inactivation of the late gene region is unnecessary. Thus infection of cells using intact SV40 virus can be used to bring about immortalization and is, in fact, the most efficient and reliable means of introducing the immortalizing genes. For this reason infection is much preferable to transfection in cases where immortalization *per se* is the goal. However, where the introduction of cloned viral sequences containing specific alterations is desired, transfection is the only alternative. Transfection of subfragments containing alterations in the SV40 early genes, including mutations conferring temperature-sensitive gene expression [Banks-Schlegel and Howley, 1983, Stamps et al., 1994] or deletions inactivating the viral origin of replication [Berry et al., 1988; Gruenert et al., 1988], has been used with good success to immortalize various types of cultured human epithelial cells *in vitro.* The complete anotated nucleotide sequence of SV40 is available from GenBank (Fig. 5.2). Software for restriction mapping and other types of analyses that can be used for experimental manipulation of the viral genes is available from a number of sources.

4.1. Preparation of SV40 Stocks

Protocol 4.1.

Reagents and Materials

- Confluent monolayers of AGMK or CV-1 monkey cells.
- Sterile stock of SV40 virus at $\sim 10^4$–10^5 plaque-forming units (PFU)/10^7 cells; a low multiplicity of infection (MOI) should always be used in order to prevent the accumulation of virions containing defective genomes.

Protocol

(a) Add the virus to the cells to initiate infection.
(b) At 2–3 days postinfection, check the cells daily for the appearance of cytopathic effects (CPEs) involving vacuolization and detachment of cells from the monolayer.
(c) Pour off the supernatant medium and remove cellular debris by centrifugation 800g when the CPE is extensive (generally 5–7 days postinfection).
(d) Aliquot the clarified medium and used as a viral stock for infection.

A titer of about 10^7 PFU/ml can be used as a working approximation for stocks produced in this way. The viral stocks are stable indefinitely when stored at −80°C.

4.2. Preparation of SV40 DNA for Transfection

Viral infection using a minimum of 10^8 AGMK or CV-1 monkey cells is carried out in the same manner as for preparation of viral stocks described above.

Protocol 4.2

Reagents and Materials

Sterile

- 10^8 AGMK or CV-1 monkey cells
- Phosphate buffered saline (PBSA)
- Sodium deoxycholate
- Nonidet P40 (NP40)
- Rubber policeman
- Ultracentrifuge tubes
- Reagents for CsCl centrifugation [Sambrook et al., 1989]
- Sonicator probe

Nonsterile

- Ultracentrifuge
- SW27 rotor
- Sonicator

Protocol

(a) Collect culture medium when CPE is extensive.

(b) Clarify by centrifugation at 800g.

(c) Ultracentrifuge for 4 h at 27,000 rpm (~100,000g) in an SW27 rotor, or equivalent, in order to pellet the virions.

(d) Scrape the remains of the cell monolayers from the culture vessel in any balanced salt solution or phosphate buffered saline (PBSA) using a rubber policeman.

(e) Pellet the cell slurry by tabletop centrifugation (800g) and combine with the virion pellet.

(f) Suspend the combined pellet at 5 ml per 10^8 cells in PBSA. The volume of salt solution should be kept to a minimum in order that the virion concentration will be sufficiently high to generate a visible band after CsCl density gradient centrifugation.

(g) Add sodium deoxycholate and NP40 to the suspended pellet to give 1% of both.

(h) Sonicate 5–10 times for 5 s each (with intervals to avoid overheating).
(i) Clarify this suspension by centrifugation at 10,000g, or faster.
(j) Band viral particles in CsCl and isolate the viral DNA from the banded virus exactly as described by Sambrook et al. [1989] for bacteriophage DNA.

4.3. Viral Infection

Viral infection is generally carried out in suspension immediately following trypsinization, although infection at low efficiency has been observed even when virus is added directly to cells in monolayer. In either case, it is important that the cells to be infected be subconfluent and preferably during the most rapid phase of growth; postconfluent cultures tend to be heavily keratinized and are less susceptible to viral infection.

Protocol 4.3

Reagents and Materials

- *Complete culture medium:* DMEM with 10% FCS
- Trypsin (see Section 2.3)
- Viral stock (see Section 4.1)

Protocol

(a) Trypsinise keratinocyte culture at middle to late log phase.
(b) Pellet cells by centrifugation at 800g.
(c) Resuspend at a concentration of approximately 5×10^6 cells/ml in complete culture medium.
(d) Mix with viral stock at an MOI of at least 10^8 PFU/10^6 cells in a sealed tube for 2 h at 37°C.
(e) Plate at least 5×10^4 cells/cm^2 under standard culture conditions.

4.4. Transfection of Viral DNA

Precipitate-based methods employing both calcium and strontium phosphate have been used to deliver DNA into various types of human epithelial cells [Banks-Schlegel and Howley, 1983; Brash et al., 1987; Munger et al., 1989; Reddel et al., 1988]. However, the success of these methods is highly variable and is particularly inefficient when dealing with keratinized epithelium such as epidermis. More recently, a liposome-mediated transfection technique using a synthetic cationic lipid (DOTMA; *N*-1-(2,3-dioleyloxy)propyl]-*N*,*N*,*N*-trimethyl ammonium chloride) has been developed by Felgner et al. [1987]. This reagent and a modified version are both available commercially (Lipofectin and LipofectAMINE respectively; Life Technologies, Inc.). Transfection according the manufacturer's instructions is straightforward.

Protocol 4.4

Reagents and Materials (Sterile)

- Keratinocyte monolayers, middle to late log phase
- *Serum-free medium:* DMEM without serum
- *Complete medium:* DMEM with 10% FCS
- Lipofectin or Lipofectamine
- Polystyrene tubes, 15 ml

Protocol

(a) Mix DNA and lipid in a polystyrene tube:
  (i) 1–20 μg DNA in 50 μl sterile UPW
  (ii) 30–50 μg lipofectamine reagent in 50 μl sterile UPW
  for each 60-mm plate.
(b) Allow liposome complexes to form over a period of 15 min at room temperature.
(c) Add the liposome suspension to cell monolayers in serum-free medium.
(d) Incubate cultures at 37°C for 10–12 h.
(e) Replace the serum-free medium with complete medium.

5. SELECTION AND MAINTENANCE OF CELL SUBLINES

Regardless of which technique is used to introduce the transforming viral sequences into cells, the percentage of cells initially expressing the viral sequences is quite small. Even for infection, which is the most efficient technique, we have found that the percentage of cells expressing SV40 T antigen within the first week as measured by immunofluorescence is generally only about 2% (Fig. 5.3) [Steinberg and Defendi, 1979, 1983]. However, the viral-infected cells are rapidly self-selecting and soon predominate over their normal counterparts in culture. This is due to at least three factors: (1) enhanced growth rate, (2) increases in plating efficiency, and (3) the viral-induced loss of senescence (see also Fig. 5.3); in uninfected cells, senescence-related loss of growth potential is generally manifest within a period of 5–10 serial passages. Thus, the number of T-antigen expressing cells is found to increase rapidly over time, typically constituting a majority

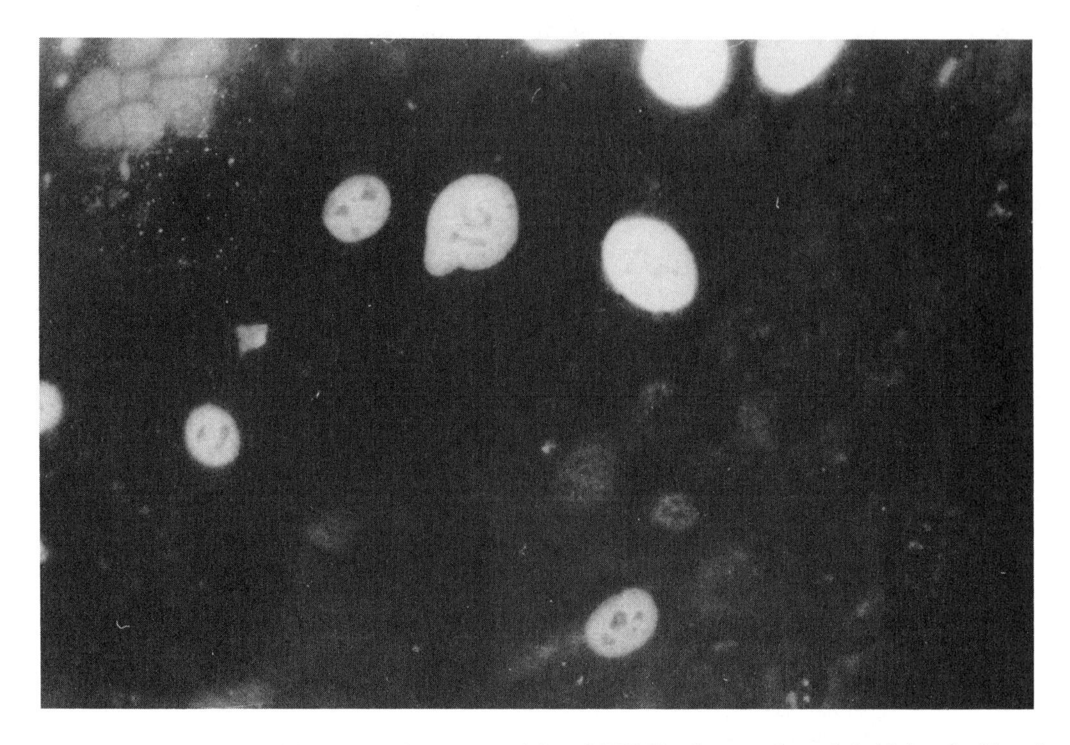

Fig. 5.3. Immunofluorescent staining of SV40 T antigen; a culture of viral-infected epidermal keratinocytes (line 130) at the 2nd serial passage. Note that, at this stage, most cells do not stain for T-antigen and that there is substantial heterogeneity in the level of nuclear staining. The percentage of T-antigen-positive cells increases only slightly between platings but undergoes marked increases when the cultures are split. This is attributed to the higher plating efficiency of the viral- infected subpopulation and the loss of uninfected cells by senescence. Generally, T antigen negative cells are completely absent from the cell population after about the 5th serial passage. (×700.)

of the cell population within 2–3 serial passages, a period representing approximately 30 cell generations under normal culture conditions. For this reason, there is usually no reason at this stage to apply selection to the cell population. Surprisingly, the infected cells also possess the ability to undermine, and thus effectively remove, contaminating colonies of fibroblasts that may have escaped removal from the primary cultures, even those that have been inadvertently infected or transfected and express the SV40 T antigen.

5.1. Crisis

The initial period of viral-induced growth stimulation is almost invariably followed by a period of growth crisis in which growth slows and cells exhibit a characteristic type of cytopathic effect involving gross enlargement, vacuolization and detachment of the affected cells from the culture substrate (Fig. 5.4). Human fibroblasts infected by SV40 have also been shown to go through a period of crisis prior to the establishment of cell lines [Girardi et al., 1965]. The severity of the crisis has been found to vary widely from one cell line to the next. In its most severe form, the cells are progressively, and completely, destroyed within a period of about 10 days following the first signs of CPE. Interestingly, when the number of cell divisions is carefully counted, the onset of the crisis period appears to occur at approximately the same number of cell generations (Fig. 5.1) as would precede senescence in uninfected cultures, suggesting that crisis may be the manifestation of normal senescence in an abortively transformed subpopulation of cells.

Crisis not only poses the major obstacle to obtaining immortalized cell lines but, because the onset of crisis occurs at least several months following the introduction of virus, it can also lead to a substantial loss of time and material. While there appears to be no completely effec-

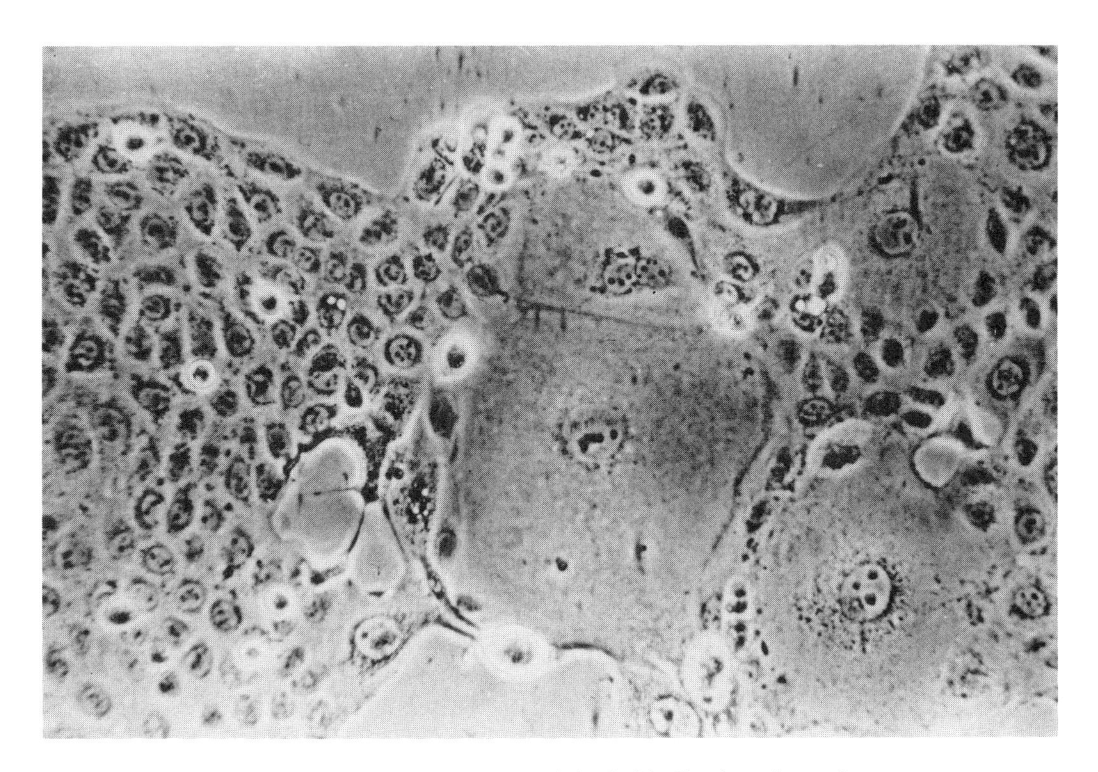

Fig. 5.4. Cells showing cytopathic effects (CPEs) characteristic of crisis. The photomicrograph is of line 130 SV40-infected keratinocytes at the 10th serial passage. Cells showing CPE are highly enlarged with linear dimensions greater than 10 times normal. Perinuclear vacuolization and multinucleation develop as the cells enlarge. The affected cells rapidly detach from the plastic substrate. (×280.)

tive means of preventing crisis, we have found that by maintaining cultures at high cell densities during the crisis period, the loss of many cell lines can be prevented. In some cases where CPE has been severe at the outset, we have been able to rescue cultures by immediately transferring the cells into smaller culture vessles as a means of increasing the cell density. Cultures may attain a steady state lasting for up to 2 weeks where there is no net growth as cells lost to CPE are replaced in equal number. For cell lines that ultimately survive, the crisis period usually lasts for a period of several weeks. The end of crisis is marked by the appearance of colonies representing clonal populations containing the immortalized cells. The crisis period represents an important demarcation between the early and late phases of the transformation process; a number of properties associated with the transformed phenotype including clonal growth (see below) appear only after crisis (Fig. 5.1).

5.2. Clonal Growth

SV40 infected keratinocytes show substantial phenotypic heterogeneity, which is evident even at an early stage in the immortalization process. There is, for example, a striking heterogeneity in the level of expression of the keratin proteins [Steinberg and Defendi, 1985; Okada et al., 1984] and anchorage independent growth [Defendi et al., 1982; Naimski and Steinberg, 1985]. For this reason it may be desirable to derive clonal sublines in which expression of certain properties is maximized for a specific purpose(s). During the precrisis period cell growth is dependent on relatively high population densities and cultures split at less than about 10^3 plated cells/cm^2 may fail to grow even in the presence of conditioned medium or high concentrations of fetal calf serum. For this reason dilution techniques involving low plating densities, which are normally used for the isolation of colonies arising from single cells, cannot be applied to the cell populations in the precrisis phase. The ability to grow at clonal densities (~1–10 cells/cm^2) is a property which appears abruptly after crisis [Steinberg and Defendi, 1983]. At this time clonal populations of cells can be easily derived by standard techniques where cells are grown either on plastic substrates or under anchorage independent conditions involving growth in suspension in semisolid media.

5.2.1. Cloning on plastic substrates

Isolation of clonal subpopulations on standard plastic cell culture substrates can be accomplished in a straightforward manner by endpoint dilution.

Protocol 5.2.1

Reagents and Materials

- Trypsin
- Complete medium
- Microwell plates: 96-well microtitration plates

Protocol

(a) Resuspend trypsinized cells in fresh medium at a final concentration of approximately 100 cells/ml.
(b) Plate into tissue culture microwell plates at 0.1ml of cell suspension per microwell.
(c) Examine the wells microscopically on the following day and record the location of wells containing 1–3 plated cells.

A small number of clonal populations derived from single cell wells can usually be obtained in one step by this method, but two rounds of selection of populations derived from microwells with more than one cell will yield a much larger sample of clones if a large sample is desired. We have utilized this technique to derive clonal subpopulations of viral-infected keratinocytes that either do, or do not, stably express a vimentin cytoskeleton.

5.2.2. Protocol for anchorage-independent cloning

We use a simplified version of the agar suspension technique originally described by MacPherson [1973] to derive anchorage independent clones of virally transformed cells.

Protocol 5.2.2

Reagents and Materials

- *Agar,* 1%: Prepare stocks of 1% agar by autoclaving Difco Bacto agar in UPW. Aliquot and store for future use.
- 2× culture medium (see Section 2.2.1): 2×DMEM with 20% FCS
- *Agar medium:* Prepare on the day of use by combining an aliquot of melted agar at 44°C with an equal volume of 2× culture medium, and keep at 44°C.
- *Agar underlays:* Dispense 4 ml of the agar medium into each 60-mm plate to create an underlay.

Protocol

(a) Trypsinise cells and resuspend in fresh culture medium.
(b) Dilute in complete medium to 5× the final cloning concentration.

(c) Combine complete medium (at 20°C), with an equal volume of the agar medium (at 44°C), and then 0.2-ml cell suspension, to give a final volume of 1 ml, and then, immediately, layer over the agar underlay.
(d) Incubate the plates in a CO_2 incubator for periods ranging up to about 2 weeks.
(e) Feed cells by adding 0.5 ml of normal culture medium to the cultures every 3–4 days; old medium can be gently removed from the top agar layer with a Pasteur pipette prior to feeding.

Colonies visible to the naked eye (~0.5 mm diameter) can usually be observed by the second week. Colony forming efficiency (CFE) varies considerably from one cell line to the next and exhibits a striking density dependence; at plating densities of less than 10^4 cells/60 mm plate we observe CFEs of less than 2% and often less than 0.5% [Steinberg and Defendi, 1983]. Individual colonies can be picked from the agar layer under a dissecting microscope with a pasteur pipette and then transferred to wells in a multiwell culture tray for growth in monolayer. Since the colonies picked from agar plates are generally surrounded by a layer of agar that interferes with plating of the cells onto plastic substrates, care should be taken to break up each transferred colony by vigorous pipetting. Clones derived by agar selection tend to be less differentiated than clones derived by anchorage-dependent means, and this is a consideration that should be born in mind if maintaining the expression of differentiated properties is important.

6. TERMINAL DIFFERENTIATION IN SV40-INFECTED EPIDERMAL KERATINOCYTES

The expression of various aspects of normal keratinocyte differentiation in SV40-transformed keratinocytes is well documented [Steinberg and Defendi, 1979, 1981, 1982, 1983]. Nevertheless, one must always be aware that expression of the features of terminal differentiation are manifest prior to crisis but become greatly reduced (and eventually undetectable) after long-term culture. Therefore, it is highly recommended that stocks of early passaged cells be frozen as soon as possible after introduction of the virus.

6.1. Markers of Terminal Differentiation

In epidermal keratinocytes infected by SV40, markers of terminal differentiation, including desmosomal junctions, squame production, and the formation of cells bearing cornified cell envelopes, are seen mainly during the precrisis period [Steinberg and Defendi, 1981; Steinberg et al., 1983]. We have employed an orange G-acid fuchsin histochemical stain [Ayoub and Shklar, 1963] as a fast and convenient means of monitoring the level of differentiation at various time points following viral infection. The extent of staining has been used as a quantitative measure of differentiation that correlates with the production of cornified envelopes in which differentiation has been induced at high cell densities [Steinberg and Defendi, 1979, 1981, 1983] or by cell fusion [Steinberg and Defendi, 1982].

6.1.1. Histochemical staining

Protocol 6.1.1

Reagents and Materials

- Coverslip cultures of keratinocytes
- PBSA
- 10% formalin in PBSA
- Deionized water
- Acid fuchsin solution (see Section 2.4)
- Aniline blue-orange G solution (see Section 2.4)
- 95% ethanol
- Absolute ethanol
- Xylene
- Permount

Protocol

(a) Rinse coverslip cultures of keratinocytes in phosphate-buffered saline (PBSA).

(b) Fix in 10% formalin in PBSA for 10 min.

(c) Stain

  (i) Rinse the coverslips several times in deionized water

  (ii) Transfer to acid fuchsin solution for 3 min

  (iii) Transfer to aniline blue-orange G solution for 45 min.

  (iv) Rinse the coverslips 3× in 95% ethanol to remove residual stain.

  (v) Dehydrate in several changes of absolute ethanol.

  (vi) Clear in xylene.

  (vii) Mount on slides with Permount.

6.2. Keratin Gene Expression

SV40 immortalized keratinocytes have great utility as model systems for studies of epithelial differentiation and transformation. Expression of the keratin intermediate filament genes can be used as a molecular marker of epithelial differentiation, which also becomes altered according to a predictable pattern after infection by SV40.

Following crisis, cultures of virally infected epidermal keratinocytes pass through a transient period of phenotypic instability in which there is a marked variability in the level of keratin expression [Steinberg and Defendi, 1985]. During this time there is a striking heterogeneity in the level of expression of the keratin proteins, as is clearly seen when cultures are stained with polyclonal keratin antisera, by immunofluorescence, which reveals the presence of many nonstaining variant cells (Fig. 5.5). Clonal analyses of unselected cell populations on palladium island grids has shown that the variant cells spontaneously arise and disappear at random, precluding the possibility that stable variant clones can be derived [Steinberg and Defendi, 1985].

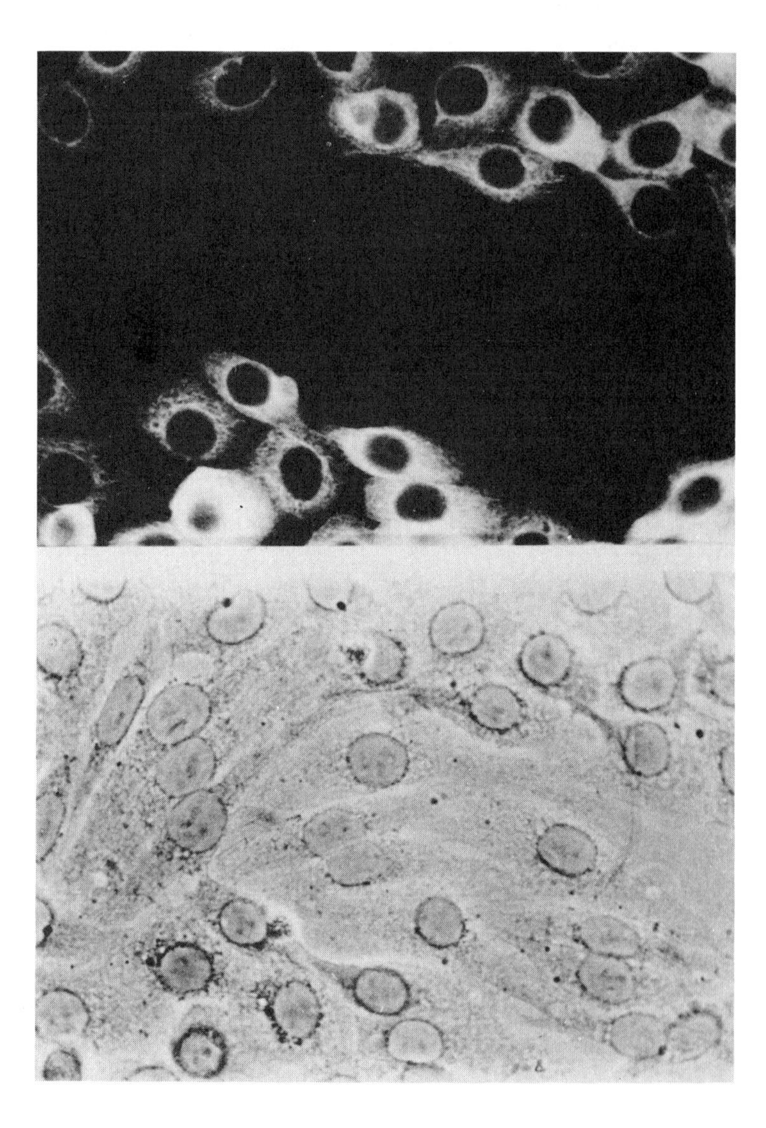

Fig. 5.5. Phenotypic instability manifest in levels of keratin expression in a postcrisis culture of SV40-infected keratinocytes. The photomicrograph shows a coverslip culture of line 130 SV40-infected keratinocytes at the 14th serial passage stained by immunofluorescence using a rabbit polyclonal antiserum [Steinberg and Defendi, 1985]. A patch of variant cells in the center of the field are virtually unstained giving the appearance of a clonal population. The same field in phase contrast is also shown. (×450.)

Detailed analyses of individual keratins have shown that there is a loss of the stratified epithelial keratins K5 and K6, which can be observed as early as about the 8th serial passage, although transcripts for these genes remain detectable for far longer; at least up to the 98th serial passage [Morris et al., 1985]. Expression of keratins K14, K16, and K17 also persists into the postcrisis period but is eventually lost [Hronis et al., 1984; Okada et al., 1984; Morris et al. 1985; Steinberg and Defendi, 1985; Steinberg et al., 1986]. In contrast, ectopic expression of keratins, K8, K13, and K18 characteristic of simple epithelium is permanently induced just after crisis [Hronis et al., 1984; Okada et al., 1984; Morris et al. 1985; Steinberg and Defendi, 1985; Steinberg et al., 1986].

7. MAINTENANCE OF THE IMMORTALIZED STATE: DEPENDENCE ON CONTINUED EXPRESSION OF THE VIRAL EARLY GENES

In SV40-infected cells the viral sequences are initially present in the form of full-length unintegrated viral genomes at high copy number. However, over time there is an accumulation of aberrant subgenomic fragments with an accompanying loss of the unintegrated sequences such that, ultimately, the viral sequences are retained only as integrated subfragments in low copy number [Defendi et al., 1982; Naimski and Steinberg, 1985]. In one subline in which the viral sequences were found to be completely contained on two BamHI fragments, sequence analysis revealed the presence of a full-length integrated SV40 genome and two head-to-tail integrants containing the viral early and late promoter elements (Fig. 5.6); the viral sequences were all found to contain numerous acquired mutations. On the other hand, the cells never become cured of fragments containing the viral early gene sequences and so it appears that continued growth in vitro remains dependent on the presence of the viral early genes. Because the cell population is constantly undergoing growth selection in culture, we also view the mutations in the early genes as possibly representing more oncogenic forms that may confer a growth advantage(s) on the cells that contain them.

ACKNOWLEDGMENTS

This work was supported by grants RR03060 and RR08168 from the National Institutes of Health and a PSC-CUNY award from the State of New York.

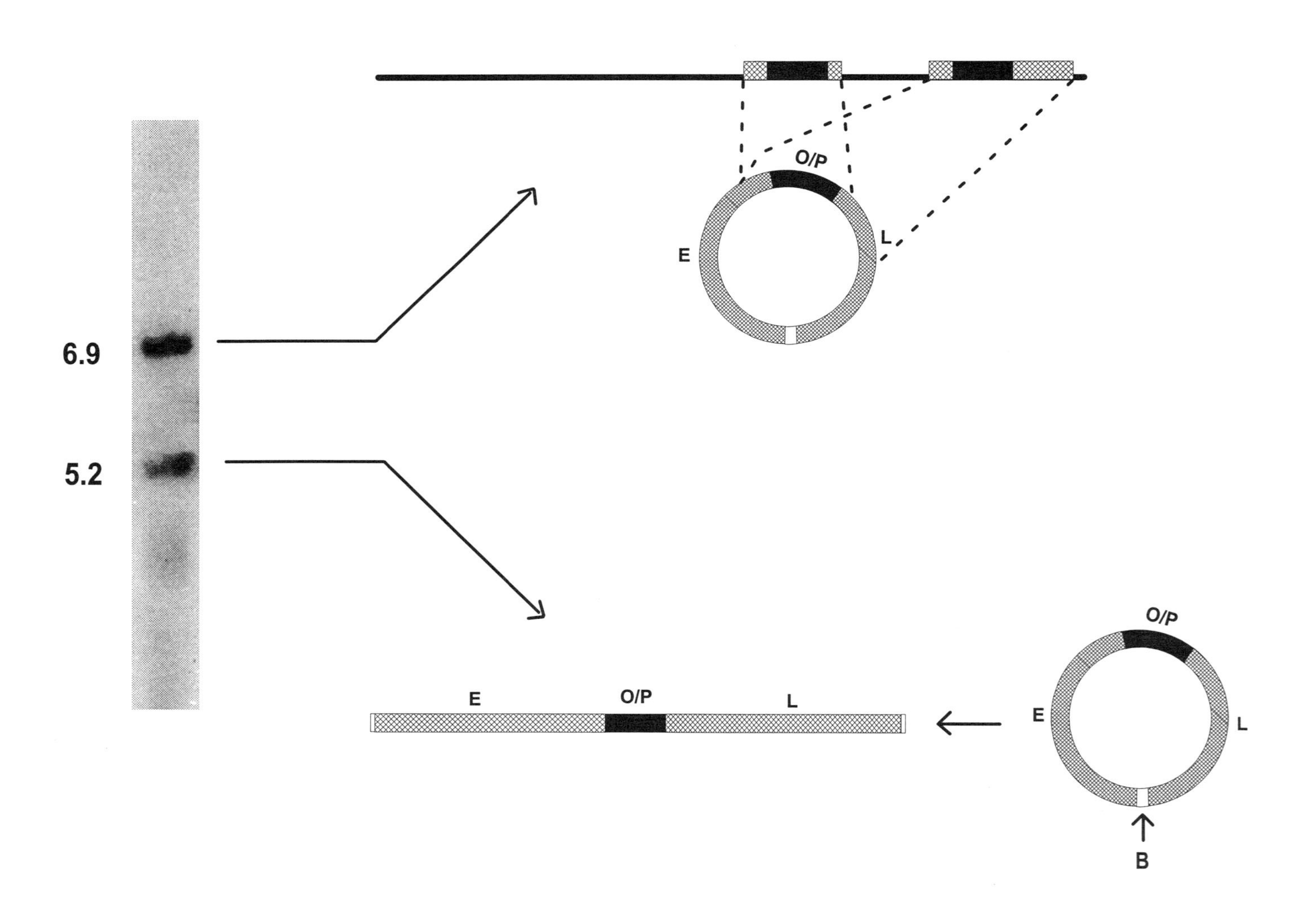
6.9
5.2
O/P
E
L
E
O/P
L
O/P
E
L
B

APPENDIX: MATERIALS AND SUPPLIERS

Materials	Suppliers
Acid fuchsin	Fisher Scientific
Agar, Bacto Agar	Difco
Aniline blue	Fisher Scientific
Cholera toxin	Sigma Chemical Co.
DOTMA	Boehringer Mannheim
Dulbecco's modified minimal essential medium (DMEM)	Fisher Scientific
Epidermal growth factor (EGF)	Sigma Chemical Co.
Fetal calf serum	Gemini Bioproducts
Hydrocortisone, H0888	Sigma Chemical Co.
Lipofectamine	Gibco BRL Life Technologies
Nonidet P40 (NP40)	Sigma Chemical Co.
Orange G	Fisher Scientific
Permount	Fisher Scientific
Phosphotungstic acid	Sigma Chemical Co.
Sodium deoxycholate	Sigma Chemical Co.
Sonicator	Heat Systems–Ultrasonics, Inc
SV40 virus, AGMK and CV-1 cells	ATCC
Trypsin	Gibco BRL Life Technologies

REFERENCES

Ayoub P, Shklar G (1963) A modification of the mallory connective tissue stain as a stain for keratin. Oral Surg Oral Med Pathol., 16: 580–581.

Baden HP, Kubilus J, Kvedar JC, Steinberg ML, Wolman SR (1987): Isolation and characterization of a spontaneously arising long-lived line of human keratinocytes (NM 1). In Vitro 23: 205–213.

Banks-Schlegel SP, Howley PM (1983): Differentiation of human epidermal cells transformed by SV40. J Cell Biol 96: 330–337.

Berry RD, Powell SC, Paraskeva C (1988): In vitro culture of human foetal colonic epithelial cells and their transformation with origin minus SV40 DNA. Br J Cancer, 57: 287–289.

Brash DE, Reddel RR, Quanrud M, Yang K, Farrell MP, Harris CC (1987): Strontium phosphate transfection of human cells in primary culture: stable expression of the

Fig. 5.6. Integration of viral fragments in a subline of SV40-transformed epidermal keratinocytes. A Southern blot of BamHI digested genomic DNA hybridized to an SV40 probe is shown at the left. The viral sequences in this subline are almost entirely contained on the two restriction fragments [Steinberg et al., 1989]. Sequence analysis of the fragments shows that the smaller, 5.2-kbp fragment, consists of full-length viral DNA containing a number of point mutations. The larger, 6.9-kbp fragment (GenBank accession U08313, locus HSU08313) contains a segment of the viral promoter/origin region including small portions of the early and late transcripts; there is a direct repeat of this segment with an additional portion of the late gene coding region about 1.6 kbp downstream. Human sequences flanking the integrated viral segments are represented by the solid black line.

simian virus 40 large-T-antigen gene in primary human bronchial epithelial cells. Mol Cell Biol 7: 2031–2034.
Coca-Prados M, Wax MB (1986): Transformation of human ciliary epithelial cells by simian virus 40: induction of cell proliferation and retention of beta 2-adrenergic receptors. Proc Natl Acad Sci, 83: 8754–8758.
Defendi V, Naimski P, Steinberg ML (1982): Human cells transformed by SV40 revisited: the epithelial cells. J Cell Physiol Suppl 2: 131–140.
Edelman B, Steinberg ML, Defendi V (1985): Changes in fibronectin synthesis and binding distribution in SV40-transformed human keratinocytes. Int J Cancer 35: 219–225.
Fanning E, Knippers R (1992): Structure and function of simian virus 40 large tumor antigen. Annu Rev Biochem 61: 55–85.
Fanning E (1992): Simian virus 40 large T antigen: the puzzle, the pieces, and the emerging picture. J Virol 66: 1289–1293.
Felgner PL, Gadek TR, Holm M, Roman R, Chan HW, Wenz M, Northrop JP, Ringold GM, Danielsen M (1987): Lipofection: a highly efficient, lipid mediated DNA-transfection procedure. Proc Natl Acad Sci USA, 84:7413–7417.
Girardi AJ, Jensen FC, Kiprowski H (1965): SV40-induced transformation of human diploid cells: crisis and recovery. J Cell Comp Physiol 65: 69–84.
Gruenert DC, Basbaum CB, Welsh MJ, Li M, Finkbeiner WE, Nadel JA (1988): Characterization of human tracheal epithelial cells transformed by an origin-defective simian virus 40. Proc Natl Acad Sci 85: 5951–5955.
Guelstein VI, Tchipysheva TA, Ermilova VD, Troyanovsky SM (1993): Immunohistochemical localization of cytokeratin 17 in transitional cell carcinomas of the human urinary tract. Virchows Arch B Cell Pathol 64: 1–5.
Hronis TS, Steinberg ML, Defendi V, Sun T-T (1984): The simple epithelial nature of some simian virus-40 transformed human epidermal keratinocytes. Cancer Res 44: 5797–5804.
Katabami M, Fujita H, Honke K. Makita A, Akita H, Miyamoto H, Kawakami Y, Kuzumaki N (1993): Marked reduction of type I keratin (K14) in cisplatin-resistant human lung squamous-carcinoma cell lines. Biochem Pharmacol 45: 1703–1710.
Ke Y, Reddel RR, Gerwin BI, Miyashita M, McMenamin M, Lechner JF, Harris CC (1988): Human bronchial epithelial cells with integrated SV40 virus T antigen genes retain the ability to undergo squamous differentiation. Differentiation, 38: 60–66.
Lemoine NR, Mayall ES, Jones T, Sheer D, McDermid S, Kendall-Taylor P, Wynford-Thomas D (1989): Characterisation of human thyroid epithelial cells immortalised in vitro by simian virus 40 DNA transfection. Br. J. Cancer 60: 897–903.
MacPherson J (1973): Soft agar techniques. In Kruse PF, Patterson MK (eds): "Tissue Culture, Methods and Applications." Academic Press, pp 276–280.
Mayall FG, Goddard H, Gibbs AR (1993): The diagnostic implications of variable cytokeratin expression in mesotheliomas. J Pathol 170: 165–168.
Miyagawa S, Okada N, Kitano Y, Sakamoto K, Steinberg ML (1988): SSA/Ro antigen expression in SV40-transformed human keratinocytes. J Invest Dermatol 90: 342–345.
Morris A, Steinberg ML, Defendi V (1985): Keratin gene expression in SV40 transformed human keratinocytes. Proc Natl Acad Sci 82: 8498–8502.
Mufson RA, Steinberg ML, Defendi V (1982): The effects of 12-O-tetradecanoyl phorbol-13-acetate on the differentiation of SV40-infected human keratinocytes. Cancer Res 42: 4600–4605.
Munger K, Phelps WC, Bubb V, Howley PM, Schlegel R (1989): The E6 and E7 genes of the human papillomavirus type 16 together are necessary and sufficient for transformation of primary human keratinocytes. J Virol 63: 4417–4421.
Naimski P, Steinberg ML (1985): Analysis of stable and unstable viral forms in SV40 infected human keratinocytes. J Cell Bioch 29: 95–103.
Okada N, Kitano Y, Miyagawa S, Sakimoto K, Steinberg ML (1986): Expression of

pemphigoid antigen by SV40 transformed human keratinocytes. J Invest Dermatol 86: 399–401.
Okada N, Miyagawa S, Hiriguchi Y, Kitano Y, Yoshikawa K, Sakamoto K, Steinberg ML (1989): Synthesis of epidermolysis bullosa acquisita antigen by simian virus 40-transformed human keratinocytes. Arch Dermatol Res 281: 1–4.
Okada N, Steinberg ML, Defendi V (1984): Re-expression of differentiated properties in SV40-infected human epidermal keratinocytes induced by 5-azacytidine. Exp Cell Res 153: 198–207.
Ponec M, Lavrijsen S, Kempenaar J, Havekes L, Boonstra J (1985): SV40-transformed (SVK14) and normal keratinocytes: similarity in the expression of low-density lipoprotein, epidermal growth factor, glucocorticoid receptors, and the regulation of lipid metabolism. J Invest Dermatol 85: 476–482.
Reddel RR, Ke Y, Gerwin BI, McMenamin MG, Lechner JF, Su RT, Brash DE, Park JB, Rhim JS, Harris CC (1988): Transformation of human bronchial epithelial cells by infection with SV40 or adenovirus-12 SV40 hybrid virus, or transfection via strontium phosphate coprecipitation with a plasmid containing SV40 early region genes. Cancer Res 48: 1904–1909.
Reddel RR, Salghetti SE, Willey JC, Ohnuki Y, Ke Y, Gerwin BI, Lechner JF, Harris CC (1993): Development of tumorigenicity in simian virus 40-immortalized human bronchial epithelial cell lines. Cancer Res 53: 985–991.
Rheinwald JG, Green H (1975): Serial cultivation of strains of human epidermal keratinocytes: the formation of keratinizing colonies from single cells. Cell 6: 331–343.
Rudland PS, Barraclough R (1990): Differentiation of simian virus 40 transformed human mammary epithelial stem cell lines to myoepithelial-like cells is associated with increased expression of viral large T antigen. J Cell Physiol 142: 657–665.
Rudland PS, Ollerhead G, Barraclough R (1989): Isolation of simian virus 40-transformed human mammary epithelial stem cell lines that can differentiate to myoepithelial-like cells in culture and in vivo. Devel Biol 136: 167–180.
Sambrook J, Fritsch EF, Maniatis T (1989): "Molecular Cloning: A Laboratory Manual," 2nd ed. New York: Cold Spring Harbor Press
Stamps AC, Davies SC, Burman J, O'Hare MJ (1994): Analysis of proviral integration in human mammary epithelial cell lines immortalized by retroviral infection with a temperature-sensitive SV40 T-antigen construct. Int J Cancer 57: 865–874.
Steinberg ML, Defendi V (1979): Altered pattern of growth and differentiation in human keratinocytes infected by SV40. Proc Natl Acad Sci 76: 801-805.
Steinberg ML, Cassai N, Defendi V (1983): A scanning electron microscope study of normal and SV40-infected human keratinocytes. Scanning Electron Microsc I, part 1: 343–349.
Steinberg ML, Defendi V (1981): Patterns of cell communication and differentiation in SV40 transformed human keratinocytes. J Cell Physiol 109: 153–159.
Steinberg ML, Defendi V (1982): Fusion induced differentiation of SV40 transformed human keratinocytes. Exp Cell Res 139: 369–375.
Steinberg ML, Defendi V (1983): Transformation and immortalization of human keratinocytes by SV40. J Invest Dermatol 81: 131s–136s.
Steinberg ML, Defendi V (1985): Altered patterns of keratin synthesis in human epidermal keratinocytes transformed by SV40. J Cell Physiol 123: 117–125.
Steinberg ML, Morris A, Goodman S (1986): Oncogenic properties of human epidermal keratinocytes transformed by SV40. In Bernstein IA, Hirone T, (eds): "Processes in Cutaneous Epidermal Differentiation," New York: Praeger Scientific, pp 287–299.
Steinberg ML, Rossman TG, Morris A, Chen G (1989): Specific high frequency rearrangements induced by MNNG in SV40-infected human keratinocytes. Carcinogenesis 10: 1801–1807.
Stieber P, Bodenmueller H, Banauch D, Hasholzner U, Dessauer A, Ofenloch-Haehnle B, Jaworek D, Fateh-Moghadam A (1993): Cytokeratin 19 fragments: a new marker for non-small-cell lung cancer. Clin Biochem 26: 301–304.
Suo Z, Holm R, Nesland JM (1993): Squamous cell carcinomas. An immunohisto-

chemical study of cytokeratins and involucrin in primary and metastatic tumours. Histopathology 23: 45–54.
Vermeer BJ, Wijsman MC, Mommaas-Kienhuis AM, Ponec M (1986): Binding and internalization of low-density lipoproteins in SCC25 cells and SV40 transformed keratinocytes. A morphologic study. J Invest Dermatol 86: 195–200.
Woodworth CD, Kreider JW, Mengel L, Miller T, Meng YL, Isom HC (1988): Tumorigenicity of simian virus 40-hepatocyte cell lines: effect of in vitro and in vivo passage on expression of liver-specific genes and oncogenes. Mol. Cell Biol 8: 4492–501.

6

Immortalization of Human Bronchial Epithelial Cells

Ruwani De Silva, Elsa L. Moy, John F. Lechner[1] and Roger R. Reddel

Children's Medical Research Institute, 214 Hawkesbury Road, Westmead, Sydney 2145, Australia.
[1]Inhalation Toxicology Research Institute, P.O. Box 5890, Albuquerque NM 87185

Culture of Immortalized Cells, Edited by R. Ian Freshney and Mary G. Freshney.
ISBN 0-471-12134-7

1. INTRODUCTION

For many types of human cells immortalization may be induced *in vitro* by genes of DNA tumor viruses such as simian virus 40 (SV40) or human papillomavirus (HPV) via a two-phase process (reviewed in Bryan and Reddel [1994]). This is illustrated by normal human bronchial epithelial (NHBE) cells transfected with SV40 early region genes (Fig. 6.1). The first phase of "lifespan extension" consists of morphologic transformation accompanied by karyotypic changes and an increased proliferative potential, ranging from 20 to 60 population doublings (PD) beyond the point at which untreated NHBE cells undergo senescence [Reddel et al., 1988; De Silva and Reddel, 1993; De Silva et al., 1994]. The phase of lifespan extension usually ends in culture crisis, during which the number of cells no longer increases and changes in cellular morphology including flattening occur.

In some, but not all, cultures a subpopulation of cells may escape from crisis and form a permanent cell line. This second phase of immortalization occurs at a frequency of 1 in 10^5 [Shay et al., 1993] to 1

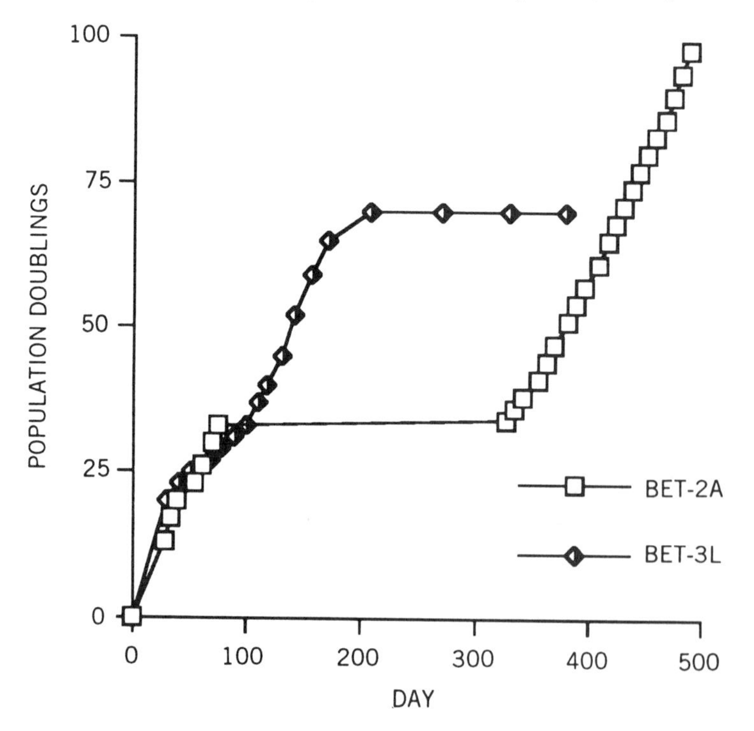

Fig. 6.1. Growth curves of two independent clones of normal human bronchial epithelial (NHBE) cells transfected with pRSV-T, which contains the SV40 early region genes. Cells were transfected at day 0, and individual clones were isolated and serially passaged. Cumulative population doublings were calculated at each passage. Control NHBE cells senesced at approximately 3-5 PD. Clone BET-3L entered crisis and failed to become immortalized [De Silva and Reddel, 1993]. Clone BET-2A escaped crisis to become an immortal cell line [Reddel et al., 1995].

in 10^9 [E. Duncan et al., unpublished data] human epithelial cells containing the viral oncogene(s). Such cultures are often regarded as immortalized when they have reached 100 PD. The SV40 transforming proteins are required in both the first and second phases of immortalization. In the first phase they are both necessary and sufficient, and they are necessary but not sufficient for the second phase, which presumably requires additional genetic change(s). The probability of the additional changes occurring may be increased by both the karyotypic instability induced by the viral proteins and the expanded cell population resulting from lifespan extension. Those cultures where immortalization occurs are referred to as "immortalization-competent," whereas those where immortalization fails to occur are referred to as "immortalization-incompetent." Approximately 1 in 3 bronchial epithelial cell cultures expressing the viral transforming genes are immortalization-competent [De Silva and Reddel, 1993; Reddel et al., [1995].

We describe here a procedure for producing SV40-immortalized human bronchial epithelial (HBE) cell lines. The technique for obtaining monolayer cultures of NHBE cells from explants of large airways tissue is based on the method of Lechner and LaVeck [1985]. Although HBE cell lines may be established following infection of NHBE cells with SV40 virus [Reddel et al., 1988, 1995], this may result in a high SV40 copy number and additional genetic instability due to the presence of the SV40 origin of replication (reviewed in Bryan and Reddel [1994]). We therefore describe the transfection of NHBE cells with an SV40 origin-minus plasmid, via strontium phosphate–DNA coprecipitation [Brash et al., 1987]. Foci of SV40-transformed cells become visible within 3–5 weeks, and these may be individually trypsinized and passaged as separate cultures (Fig. 6.2).

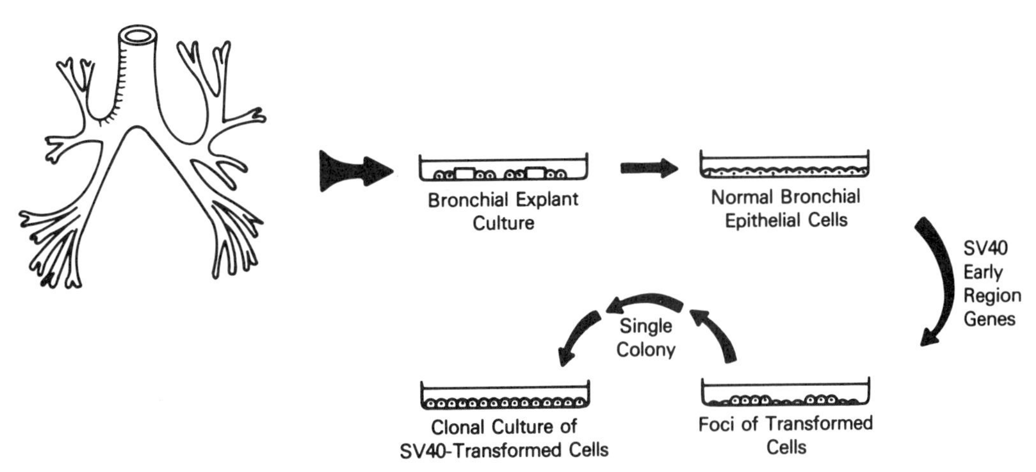

Fig. 6.2. Diagrammatic representation of bronchial explant culture and generation of SV40-transformed HBE cell clones.

Those cultures in which a subpopulation escapes from crisis will usually become established as an immortalized cell line.

2. PREPARATION OF REAGENTS AND MEDIA

2.1. Solutions

Phenol Red

Phenol red	1.0 g
Water (ultrapure water, e.g., distilled–deionized, carbon-filtered, UPW; see Preface) to 200 ml	

HEPES buffered saline (HBS)

HEPES	4.76 g
NaCl	7.07 g
KCl	0.20 g
Glucose	1.70 g
Na_2HPO_4	1.022 g
UPW to	900 ml
0.5% w/v phenol red solution	0.25 ml
Adjust the pH to 7.5 with 1.0M NaOH, and the final volume to 1 liter with UPW. Filter-sterilize the solution (0.2 μm) and store at room temperature.	

Bovine serum albumin (BSA) stock

BSA	100 mg
UPW to	100 ml
Filter-sterilize (0.2 μm) and store at 4°C	

Collagen/fibronectin-coating solution Coating solution is prepared essentially as described in Lechner and LaVeck [1985].

Bovine fibronectin	2 mg
LHC-basal medium (see Section 2.3)	2 ml
Dissolve by heating to 37°C for 1 h.	
To this add BSA stock	20 ml
Vitrogen 100	2 ml
LHC-Basal medium	198 ml
Filter-sterilize the solution (0.2 μm) and store at 4°C.	

Trypsin solution (PET)

PVP stock	
Polyvinylpyvrolidone (PVP)	10 g
HBS to a final volume of	100 ml
Store at −20°C.	
EGTA stock	
EGTA	200 mg
HBS to a final volume of	100 ml
Filter-sterilize (0.2 μm) and store at room temperature.	
Trypsin	
Trypsin XII-S	100 mg
HBS	10 ml
To prepare PET:	
PVP stock	50 ml

EGTA stock	50 ml
Trypsin	10 ml
HBS	390 ml
Filter-sterilize (0.2 μm) and store at −20°C.	

Soybean trypsin inhibitor (SBTI)	
SBTI	1.0 g
HBS	33 ml
Filter-sterilize (0.2 μm) and store at −20°C.	

2× antibiotic freezing medium	
FBS	40 ml
Gentamicin sulfate (50 mg/ml)	0.4 ml
1M HEPES, pH 7.6	4 ml
L15 medium (see Section 2.3),	156 ml
Filter-sterilize (0.2 μm) and store at −20°C.	

2× DMSO freezing medium	
PVP stock (see trypsin solution above)	40 ml
Dimethylsulfoxide (DMSO)	30 ml
1M HEPES, pH 7.6	4 ml
L15 (see Section 2.3)	126 ml
Filter-sterilize (0.2 μm) and store at −20°C.	

Insulin (2 mg/ml)	
Insulin	0.12 g
4mM HCl	60 ml
Store at 4°C, unfiltered.	

Epidermal growth factor (EGF; 5 μg/ml)	
EGF	100 μg
BSA stock (see above)	2.0 ml
HBS	18 ml
Prepare aseptically and store at −20°C.	

Phosphoethanolamine (P) and ethanolamine (E) stock	
Phosphoethanolamine stock, 10^{-2}M	
Phosphoethanolamine	114 mg
BSA, 1 mg/ml	10 ml
HBS	90 ml
Filter-sterilize and store at −20°C	
Ethanolamine stock, 10^{-2}M	
Ethanolamine	60 μl
BSA, 1mg/ml	10 ml
HBS	90 ml
Filter sterilize (0.2 μm) and store at −20°C.	
To make 50 ml PE stock:	
Phosphoethanolamine, 10^{-2}M	0.5 ml
Ethanolamine, 10^{-2}M	0.5 ml
HBS	49 ml
Filter-sterilize (0.2 μm) and store at 4°C.	

Hydrocortisone (3.6 mg/ml)	
Hydrocortisone	72 mg
95% ethanol	20 ml
Store at 4°C.	

Transferrin (5 mg/ml)

Bovine transferrin	200 mg
BSA, 1 mg/ml	4 ml
HBS	36 ml
Filter-sterilize (0.2 μm) and store at −20°C.	

Epinephrine (1 mg/ml)

Epinephrine	100 mg
HCl, 10mM	100 ml
Filter-sterilize (0.2 μm) and store at −80°C.	

Retinoic acid (1 μg/ml)

Retinoic acid	20 mg
DMSO	20 ml

Dilute 1: 1000 in DMSO to form a stock solution of 3.3 μM. Filter-sterilize (0.2 μm) and store at −80°C.

3,3',5-Triiodothyronine (T_3) (1mM)

Triiodothyronine (T_3)	13 mg
50% *n*-propanol	20 ml
Store at −80°C.	

Dilute 1: 10 in 50% *n*-propanol for use as a working stock (0.1mM).

Trace elements Dilute 1 ml of each trace-element stock solution (Table 6.1) together with concentrated HCl (1 ml) in distilled water to a final volume of 1000 ml to make a working stock. Filter-sterilize (0.2 μm) and store at room temperature.

2× transfection-HBS Prepare 2× transfection-HBS from 20× stock solutions (Table 6.2):

Combine 10 ml of each of the 20× stock solutions.
Adjust the pH to the optimal level (Section 3.4.3).
Make total volume up to 100 ml with UPW.
Double-filter (see note below) and store at 4°C.

Table 6.1. Components of the Trace-Element Stock Solution[a]

Element	Formula	Amount (mg) Added to 100 ml Water	Concentration in Stock Solution	Concentration in Medium
Manganese	$MnSO_4$	1.5	100μM	1nM
Molybdenum	$(NH_4)_6Mo_7O_{24}{\cdot}4H_2O$	12.4	100μM	1nM
Nickel	$NiCl_2{\cdot}6H_2O$	1.2	50μM	0·5nM
Selenium	$NaSeO_3$	52	3.0mM	30nM
Silicone	$Na_2SiO_3{\cdot}9H_2O$	1,420	50mM	0·5μM
Tin	$SnCl_2{\cdot}2H_2O$	1.1	50μM	0·5nM
Vanadium	NH_4VO_3	5.9	0.5mM	5nM

[a]Based on Lechner and La Veck [1985].

Table 6.2. 20× Stock Solutions for Transfection-HBS[a]

Additive	g/500 ml distilled water
$Na_2HPO_4 \cdot 7H_2O$	1.88
Dextrose	10.8
NaCl	80.0
KCl	3.72
HEPES	47.6

[a]Based on the addendum to Brash et al., [1987]. Store each of these five stock solutions at −20°C.

Glycerol/1× transfection-HBS

Glycerol	15 g
2× transfection-HBS	50 ml
Make up total volume with UPW to	100 ml
Double-filter and store at 4°C.	

$SrCl_2$ solution

Make 100 ml of a 2M solution of $SrCl_2$ in UPW.
Double filter and store at 4°C.

Note: The 2× transfection-HBS, glycerol/1× transfection-HBS, and $SrCl_2$ solutions used in strontium phosphate transfections are each filtered through two 0.2-μm Nalgene filtration units (i.e., double-filtered). The brand of filtration unit used has a substantial effect on transfection efficiency. Rinse each filter by discarding the initial 4–5 ml of filtrate.

2.2. Plasmid DNA

pRSV-T [Fig. 6.3; (referred to as pRSV-TAg in Sakamoto et al., [1993]). pRSV-T expresses the SV40 early region genes, large-T antigen, small-t antigen, and 17kT, driven by the Rous sarcoma virus 3'-long terminal repeat. Other plasmids containing SV40 early region DNA, preferably without a functional SV40 origin of replication, may also be used. DNA may be prepared for transfection by two cesium chloride ultracentrifugation steps [Sambrook et al., 1989] or by Qiagen column chromatography.

2.3. Cell Culture Media

LHC-Basal Medium (LHC-BM). Dissolve 1 bottle of LHC-BM powder (154.3 g per bottle, which includes stocks 4, and 11 and calcium [Lechner and LaVeck, 1985]), in 8 liters of distilled water. Add $NaHCO_3$ (10 g) and adjust the pH to 7.4 ± 0.05 with 1M HCl or 1M NaOH, and adjust the final volume to 10 liters. Filter-sterilize the medium in a 0.2-μm 10 liter filter unit (e.g., Mediakap™) and store at

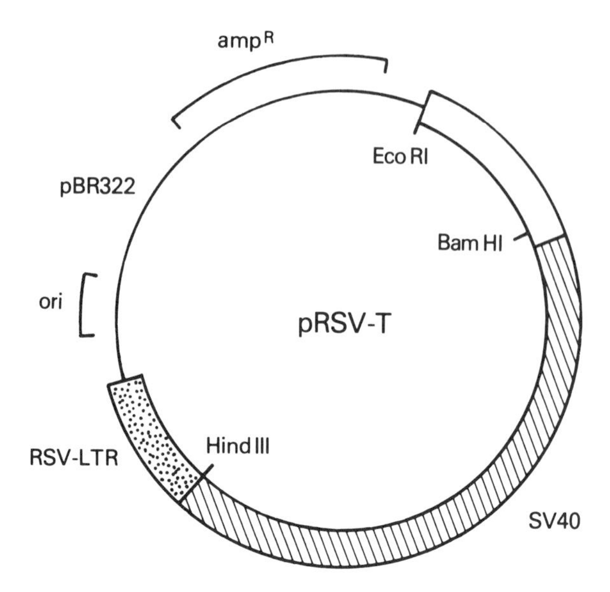

Fig. 6.3. Diagrammatic representation of plasmid pRSV-T [Sakamoto et al., 1993], which expresses the SV40 large and small T antigen genes from the SV40 Rous sarcoma virus 3'-long terminal repeat (RSV-LTR) promoter/enhancer element. The bacterial origin of replication (ori) and ampicillin resistance gene (amp^R) are from the pBR322 plasmid.

4°C in the dark. LHC-basal medium, LHC-8, and LHC-9 (see below) are also available in a liquid form from Biofluids Inc.

LHC-8 and LHC-9 media. LHC-8 and -9 are prepared from LHC-BM essentially as described in Lechner and LaVeck [1985]. The composition of these growth media is described in Table 6.3. Add the appropriate ingredients to LHC-BM (500 ml), 0.2 μm filter-sterilize the medium, and store at 4°C in the dark.

Table 6.3. Preparation of LHC-9 Medium[a]

Ingredient	Volume	Concentration of Stock Solution	Concentration in Medium
LHC-BM	500 ml		
EGF	0.5 ml	0.83μM (5 μg/ml)	0.83nM
Gentamicin	0.5 ml	50 mg/ml	50 μg/ml
Hydrocortisone	0.1 ml	10mM (3.6 mg/ml)	2.0μM
Insulin	1.25 ml	0.35mM (2 mg/ml)	0.87μM
P&E stock[b]	2.5 ml	1mM P and E	0.5μM P and E
Trace elements[c]	5 ml	See Table 6.1	
Transferrin	1.0 ml	62.5μM (5 mg/ml)	0.13μM
Triiodothyronine[c]	50 μl	0.1mM (0.065mg/ml)	10nM
Epinephrine[c]	250 μl	5.4mM (1 mg/ml)	2.7μM
Retinoic acid[c]	50 μl	3.3μM (1 μg/ml)	0.33nM
Bovine pituitary extract[d]	2.5 ml	70 mg/ml	0.35 mg/ml

[a] Based on the method of Lechner and La Veck [1985]. Note that stocks 4 and 11 and calcium do not need to be added to LHC-basal reconstituted from powder.
[b] Phosphoethanolamine and ethanolamine.
[c] Working stock (Section 2).
[d] LHC-8 medium is LHC-9 medium without epinephrine and retinoic acid. LHC-7.5 medium is LHC-8 without bovine pituitary extract and is used for transfections.

Liebowitz 15 (L15) medium. Dissolve L15 powder (14.8 g) in 950 ml distilled water, adjust the pH to 7.6 with 1M NaOH or 1M HCl, and adjust the volume to 1 liter with distilled water, 0.2 μm, filter-sterilize, and store at 4°C.

3. PROTOCOLS

3.1. Culture Conditions

Appropriate biological containment procedures, including the use of a biological containment cabinet, should be observed at all times to avoid operator contact with human tissue samples, and plasmids or cell lines containing tumor virus DNA (see Chapter 2, although local and national rules will apply).

Explant cultures, monolayer cultures of NHBE cells and immortal HBE cell lines are grown in disposable plastic cell culture vessels in a humidified atmosphere of 3.5% CO_2 in air at 36.5°C.

The surface of the culture dish or flask should be coated with a collagen/fibronectin solution (Section 2) as follows.

Protocol 3.1

Reagents and Materials

- Collagen/fibronectin solution (see Section 2)
- LHC-basal medium
- Scalpel

Protocol

(a) Place 2 ml collagen/fibronectin solution in each 100-mm petri dish or 75-cm^2 flask (scale appropriately for culture vessels of other sizes), spread evenly over growth surface of culture vessel, then leave for a minimum of 2 h (maximum 10 days) in the incubator at (36.5–37°C). Flasks may be sealed to prevent drying of the solution.

(b) Remove excess collagen/fibronectin solution by aspiration just before seeding cells into the flask or dish. If the solution has dried, redissolve in LHC-basal medium, then remove by aspiration.

(c) To facilitate adherence of explanted bronchial tissue to culture dishes, use a sterile scalpel to score the culture surface at 8–10 evenly spaced sites (where the tissue blocks will be placed) before coating with collagen/fibronectin solution.

Unlike fibroblasts whose growth is enhanced in the presence of serum, NHBE cell growth is inhibited by serum, which induces cross-linked envelope formation and terminal squamous differentiation [Lechner et al., 1984]. The primary differentiation-inducing serum factor for NHBE cells is contained in the platelet fraction of serum and appears to be transforming growth factor (TGF)-β [Masui et al., 1986]. The clonal growth of SV40-immortalized HBE cell lines may also be inhibited by serum [Ke et al., 1988]. Therefore explant cultures, NHBE cell monolayer cultures, and SV40-transfected HBE cells are cultured in serum-free media, i.e. LHC-8 or -9 (Table 6.3).

3.2. Explant Culture of Bronchial Tissue

Immediate autopsy specimens of normal human bronchus, collected in cold L15 medium (Section 2.3) containing 0.1 mg/ml gentamicin, may be used to derive cultures of NHBE cells.

Protocol 3.2

Reagents and Materials

Sterile

- Scissors
- Forceps
- Scalpels
- 100-mm dishes that have previously been scored and coated with collagen/fibronectin solution (Section 3.1)
- L15 medium
- LHC-9 medium

Protocol

(a) Dissect out the large airways from the rest of the bronchial tree.
(b) Dissect away muscle, fat, and connective tissue from the large airways.
(c) Wash the bronchial tissue 3 times with L15 medium.
(d) Cut the bronchial tissue into approximately 0.3-cm^2 tissue blocks.
(e) Place 8–10 tissue blocks with the epithelial side facing upward into 100-mm scored and coated dishes
(f) Add sufficient LHC-9 medium (Section 2.2) to cover the epithelial surface of the tissue pieces and incubate the dishes in a humidified 3.5% CO_2 incubator at 36.5°C.
(g) Over the next 7–10 days observe for epithelial outgrowths radi-

ating from the tissue blocks. Change the medium approximately twice per week.

(h) Transfer the tissue explants to new scratched and collagen/fibronectin-coated plastic culture dishes for further outgrowths of epithelial cells. Tissue blocks can be so transferred about 3 times. Fibroblast contamination is usually not a problem during the first 2–3 transfers.

(i) Culture the epithelial cells that grow out from the explants for a further 3–5 days, then harvest the cells for passaging or cryopreservation (Section 3.3.2).

3.3. Primary Culture of NHBE Cells

NHBE cells are grown as monolayer cultures in collagen/fibronectin-coated culture vessels in LHC-8 or -9 medium [Lechner and LaVeck, 1985]. Growth medium is changed 2–3 times per week.

3.3.1. Trypsinization of NHBE cells

When the NHBE cells become confluent, they are trypsinized and seeded at a lower density into a new vessel and/or cryopreserved. NHBE cells are particularly susceptible to damage by trypsin, and this may be minimized by the following procedure.

Protocol 3.3.1

Reagents and Materials

- HBS
- PET
- SBTI

Protocol

(a) Remove growth medium, and wash the cells 3 times in HBS (Section 2.1).

(b) Incubate the cells in 2 ml PET (Section 2) per 100-mm dish or 75-cm^2 flask, for the minimum time required for the cells to detach from the culture vessel surface (usually 5–10 min).

(c) Harvest the cells in PBSA containing 10% FBS and centrifuge for 5 min at 200g.

(d) Discard the supernatant and resuspend the cells in 0.15 ml SBTI (Section 2.1) for 3 min prior to cryopreservation or passaging.

3.3.2. Cryopreservation and thawing of NHBE cells. Standard cryopreservation procedures may be used.

Protocol 3.3.2

Reagents and Materials

- 2× Antibiotic freezing medium (see Section 2.1)
- 2× DMSO freezing mixture (see Section 2.1)
- 1-ml cryogenic storage vials
- Insulated freezing box or controlled rate cooler
- Waterbath or bucket at 37°C, with lid
- Growth medium: LHC-9
- Collagen/fibronectin-coated flask or dish

Protocol

(a) Resuspend 2×10^6 cells in 0.5 ml cold 2× antibiotic freezing medium.
(b) Add 0.5 ml cold 2× DMSO freezing medium.
(c) Place cells in a 1-ml cryogenic vial.
(d) Gradually (~1°C/min) decrease temperature to −80°C overnight.
(e) Transfer vials to a liquid nitrogen cryostorage unit for long-term storage.
(f) To thaw cells, remove a vial of cells from the liquid nitrogen cryostorage unit and thaw rapidly in a 37°C waterbath (3–5 min). *Note:* if the vials have been submerged in liquid nitrogen, the water bath must be covered, as vials can inspire nitrogen and explode violently on thawing.
(g) Resuspend the cells slowly, with constant mixing, in 9 ml of growth medium, and centrifuge at 100*g* for 3 min.
(h) Discard the supernatant and resuspend the cells in fresh growth medium before seeding the cells into a collagen/fibronectin-coated tissue culture vessel.

3.4. Strontium Phosphate Transfection of NHBE Cells

Strontium phosphate transfection is a modification of the well-known calcium phosphate transfection procedure, and was introduced to circumvent problems arising from calcium-induced toxicity in NHBE cells [Brash et al., 1987]. It may also be used to transfect other cell types, including immortalized HBE lines. The method detailed here is based on the addendum to the original description by Brash et al. [1987]. Although LHC-8 is acceptable for transfections, for reasons of cost only we omit bovine pituitary extract from LHC-8 and refer to this medium as LHC-7.5.

3.4.1. Preparation of NHBE cells

Protocol 3.4.1

Reagents and Materials

- Hemocytometer
- LHC-7.5 or LHC-8 medium
- LHC-9 growth medium

Protocol

(a) On the day before transfection, harvest and count cells using a hemocytometer.
(b) Seed cells at a concentration of $0.5–1 \times 10^6$ cells per 100-mm dish in 10 ml LHC-9 growth medium.
(c) Return the cells to a 3.5% CO_2, 36.5°C incubator overnight.
(d) Place LHC-7.5 (or LHC-8) medium previously optimized for transfections (Section 4.5.3) into the 3.5% CO_2, 36.5°C incubator in a container with a loosened cap for equilibration overnight.

3.4.2. Strontium phosphate transfection

All volumes are given for a 100-mm dish, and may be scaled appropriately for other culture vessels.

Protocol 3.4.2

Reagents and Materials

Sterile

- HBS
- LHC-basal (LHC-BM)
- LHC-7.5
- LHC-9
- 2× transfection-HBS
- Ultrapure water
- Plasmid DNA
- 15% (w/v) glycerol/1× transfection-HBS solution (Section 2.1)
- Nalgene filters, 0.2 μm
- Polypropylene tubes, 17×100 mm

Nonsterile equipment

- Heating block

Protocol

(a) On the day of transfection, remove the growth medium and wash the cells 3 times with HBS.

(b) Add 10 ml of the preequilibrated LHC-7.5 medium to each dish.
(c) Return the dishes to the incubator and allow to equilibrate for a further 1 h (maximum 3 h).
(d) Prewarm polypropylene tubes to 37°C in a heating block during the 1-h equilibration.
(e) Add the following (in order) to each polypropylene tube (1 tube per dish):
    (i) 2× transfection-HBS, 500 μl (Section 2)
    (ii) Double-filtered water, 440-*x*[see (iii) below] μl (use Nalgene filters, prerinsed with distilled water)
    (iii) Plasmid DNA, 10–20μg per dish: *x* μl
    (iv) 2M $SrCl_2$, 62 μl (Section 2.1)
(f) Briefly agitate each polypropylene tube and pipette up and down once with a sterile 1-ml pipette to mix thoroughly.
(g) Return each tube to the heating block and incubate for precisely 30 s at 37°C
(h) Pipette DNA-PO_4 mixture up and down once more with a 1-ml pipette and add to each dish of cells dropwise.
(i) Rock each dish gently to mix, and return to the incubator for 2 h.
(j) After 2 h observe for a fine dust-like DNA-PO_4 precipitate carpeting the cells.
(k) Discard the medium and glycerol-shock the cells by incubating them for 30 s at room temperature in 3 ml 15% (w/v) glycerol/1× transfection-HBS solution.
(l) Remove and discard the glycerol solution and wash the cells 3 times with LHC-BM at room temperature.
(m) Feed the cells with 10 ml LHC-9 medium and return to the incubator.
(n) Change the growth medium 2–3 times per week and observe over the ensuing 2–6 weeks for colony formation.

3.4.3. Optimization of strontium phosphate transfections

The efficiency of strontium phosphate transfections is critically dependent on the quality of the DNA-PO_4 precipitate, and this is in turn dependent on the pH of the LHC-7.5 (or LHC-8) growth medium and the 2× transfection-HBS. A precipitate that is too fine results in a greatly reduced efficiency, whereas one that is too coarse results in severe cell toxicity. An optimum precipitate is shown in Fig. 6.4. The pH of the transfection-HBS should be optimized for each batch of LHC-7.5 medium. Since variations in CO_2 concentration occur among different incubators, it is preferable to use the same incubator throughout.

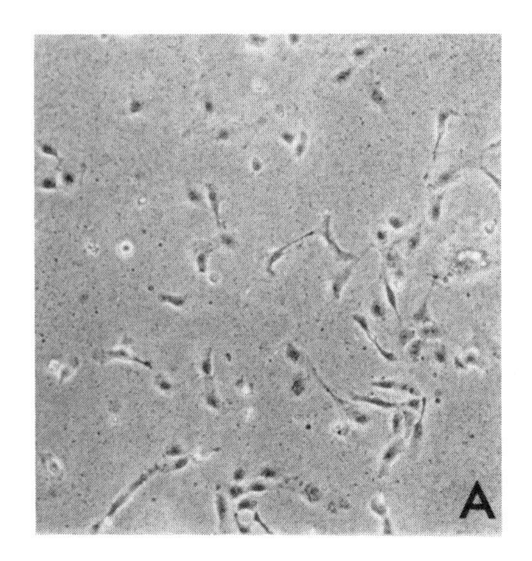

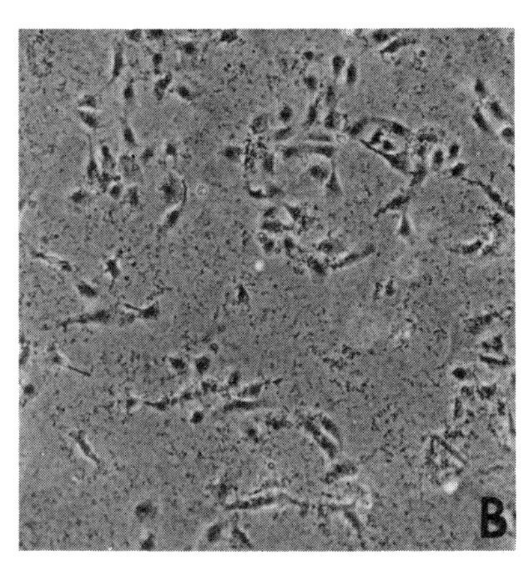

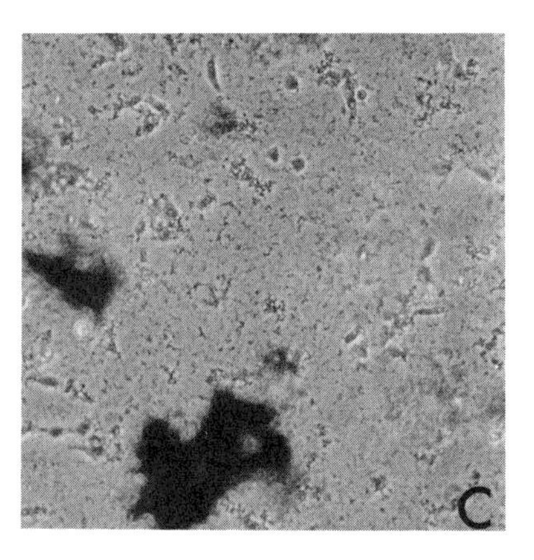

Fig. 6.4. Phase-contrast photomicrographs of NHBE cells during strontium phosphate transfection [Brash et al., 1987]. Optimal precipitate formation at pH 7.3 (A); suboptimal precipitate formation at pH 7.4 (B); excessive toxic coarse precipitate formation at pH 7.5 (C).

Protocol 3.4.3

Reagents and Materials

- 2× transfection-HBS
- Plasmid DNA

Protocol

(a) Make up 2× transfection-HBS at a range of pH values (e.g., at 0.05 intervals between 7.20 and 7.80).
(b) Use these to transfect plasmid DNA into test cells.

The optimum transfection-HBS pH may be assessed on the basis of quality of precipitate (Fig. 6.4) or expression of a reporter gene such as chloramphenicol acetyltransferase.

3.5. Isolation of Transformed Colonies

NHBE cells expressing the SV40 early region genes become visible as foci of rapidly proliferating morphologically altered cells on a background of normal cells. Individual foci, each having arisen from a single cell, may be isolated with cloning cylinders, and serially passaged as clonal SV40-transformed cell strains. Since approximately 1 in 3 SV40-transformed human epithelial cell cultures are immortalization-competent, to obtain at least one immortalized HBE line it is advisable to isolate a minimum of 3, and preferably 5-10 foci.

Protocol 3.5

Reagents and Materials

Sterile

- HBS
- PET
- PBSA with 10% FBS (PBS10FB)
- SBTI
- Growth medium: LHC-9
- Cloning cylinders: coat the floor of a glass petri dish with a layer of vacuum grease, embed the base of the cloning cylinders into the vacuum grease, place the lid on the petri dish, and autoclave.
- Forceps
- Six-well collagen/fibronectin-coated plates
- 25- and 75-cm^2 collagen/fibronection-coated culture flasks

Protocol

(a) Isolate individual foci as follows. Remove growth medium from the cells and wash the dish 3 times with 10 ml HBS (Section 2.1).

After removing the HBS, use sterile forceps to place a sterile glass cloning cylinder around each focus using the autoclaved vacuum grease to attach the cloning cylinder to the dish.

(b) Harvest the cells within the cloning cylinder by trypsinization (using approximately 0.2 ml PET solution), and collect the detached cells in 0.8 ml PBS10FB.

(c) Centrifuge for 5 min at 200g and discard the supernatant.

(d) Resuspend the cells in SBTI as described in Section 3.3.1.

(e) Transfer the cells from each focus to one well of a six-well collagen/fibronectin-coated cell culture plate.

(f) Add 2 ml growth medium to give a total volume of 3 ml per well.

(g) When confluent, harvest the cells by trypsinization with PET and transfer to a collagen/fibronectin-coated 25-cm^2 tissue culture flask and, when confluent again, from there to a collagen/fibronectin-coated 75-cm^2 flask for subsequent passaging.

3.6. Generation of Immortalized Cell Lines

Each clonal cell population is serially subcultured in 75-cm^2 flasks.

Protocol 3.6.1

(See previous sections, as indicated, for reagents and materials.)

(a) Prior to the onset of crisis, passage the cells (see Section 3.3.1) such that they become confluent again after 7 days; this usually means a split ratio of 1:8 or 1:16.

(b) Cryopreserve the cells (see Section 3.3.2) at every other passage until crisis is reached. Once a population of cells reaches crisis, no net proliferation occurs, although the population can remain viable for an extended period of time (up to 9 months).

(c) During culture crisis, less frequent medium changes may be required (e.g., once per week).

Immortalization of SV40-transfected epithelial cells occurs at a frequency of approximately 1 in 10^5 to 1 in 10^9 transformed cells. Therefore, in order to obtain an immortalized cell line it is advisable to have a minimum of 5×10^7 cells (equivalent to 5–10 confluent 150-cm^2 flasks) in culture by the onset of crisis.

Protocol 3.6.2

(See previous sections, as indicated, for reagents and materials.)

(a) Thaw cryopreserved cells approximately 4 passages prior to crisis.

(b) Expand cell numbers so that there are 5–10 150-cm^2 flasks by the time crisis is reached.
(c) Maintain the crisis cultures such that each flask is at >50% confluence. This improves the efficiency of immortalization. As the cell population decreases, passage the cells once or twice into progressively fewer and/or smaller culture vessels such that the cells remain at approximately 70% confluence

In some cultures an immortal cell clone will appear, recognizable as a small colony of dividing cells on a background of cells in crisis. This can be serially passaged to give rise to an immortalized cell line. A population of human cells is often regarded as being immortalized when it has reached 100 PD. The morphology of normal, senescing, and SV40-transformed cells is illustrated in Fig. 6.5.

4. VARIATIONS ON THE METHOD

NHBE cells may also be cultured from bronchial biopsies [De Jong et al., 1993] or bronchial brushings [Kelsen et al., 1992].

Other transfection methods, such as liposome-mediated gene transfer, may be used for NHBE cells [Willey et al., 1991; Viallet et al., 1994].

Human papillomavirus (HPV) E6 and E7 genes may be used as an alternative to those of the SV40 early region. The HPV-16 genes also immortalize human epithelial cells via a two-step process and at approximately the same frequency as SV40 [De Silva et al., 1994]. Transfection of NHBE cells with a plasmid containing the HPV-16 E6 and E7 genes linked to the human β-actin promoter did not result in focus formation [De Silva et al., 1994]. Isolation of clonal populations of HPV-16 transformed cell strains therefore requires either cotransfection with a selectable marker such as the neomycin resistance gene and subsequent selection with G418 or mass culture of a population of HPV-16 transfected cells followed by limiting dilution.

When subculturing SV40-transfected HBE cell lines, FBS may be replaced by BSA in the solution used to harvest cells after trypsinization to avoid selection of subpopulations that are resistant to the inhibitory effects of serum.

5. CHARACTERIZATION OF THE CELLS

5.1. Immunostaining for SV40 T Antigens

The expression of the SV40-transforming proteins is confirmed on methanol fixed cells using the PAb419 antibody at a dilution of 1:50 from the stock solution, followed by a labeled secondary antibody,

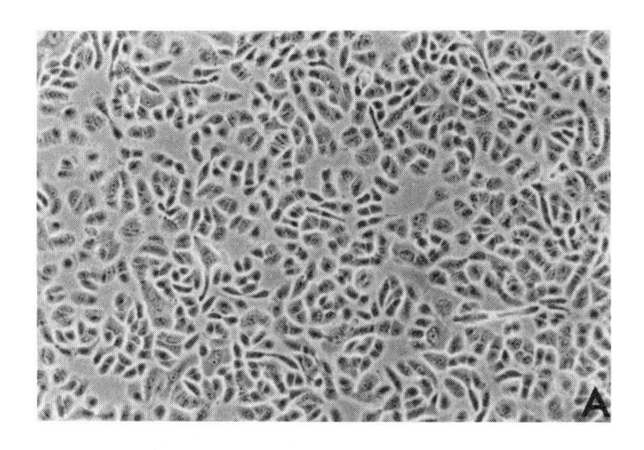

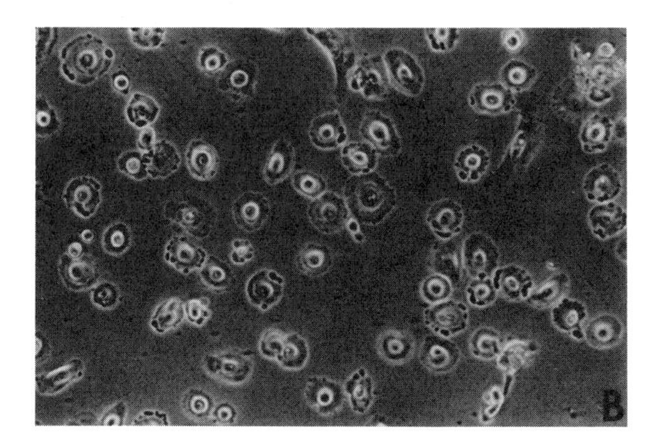

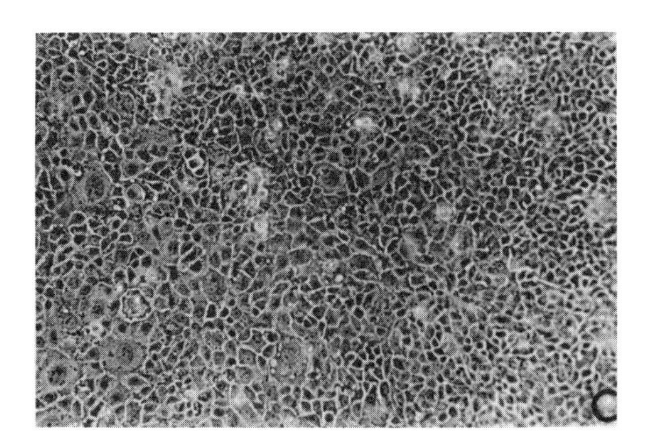

Fig. 6.5. Phase-contrast photomicrographs of (A) NHBE cells (B) senescing NHBE cells (C) a focus of SV40 T-antigen transformed NHBE cells. (×120.). Reproduced from Reddel et al. [1988} with permission of the publisher.

such as FITC-conjugated goat antimouse IgG. SV40 T-antigen immunostaining is characteristically nuclear with nucleolar sparing (Fig. 6.6A).

5.2. Confirmation of Epithelial Origin of Cells

The epithelial origin of NHBE cells and derived HBE cell lines may be confirmed by immunofluorescent staining with an antibody to an

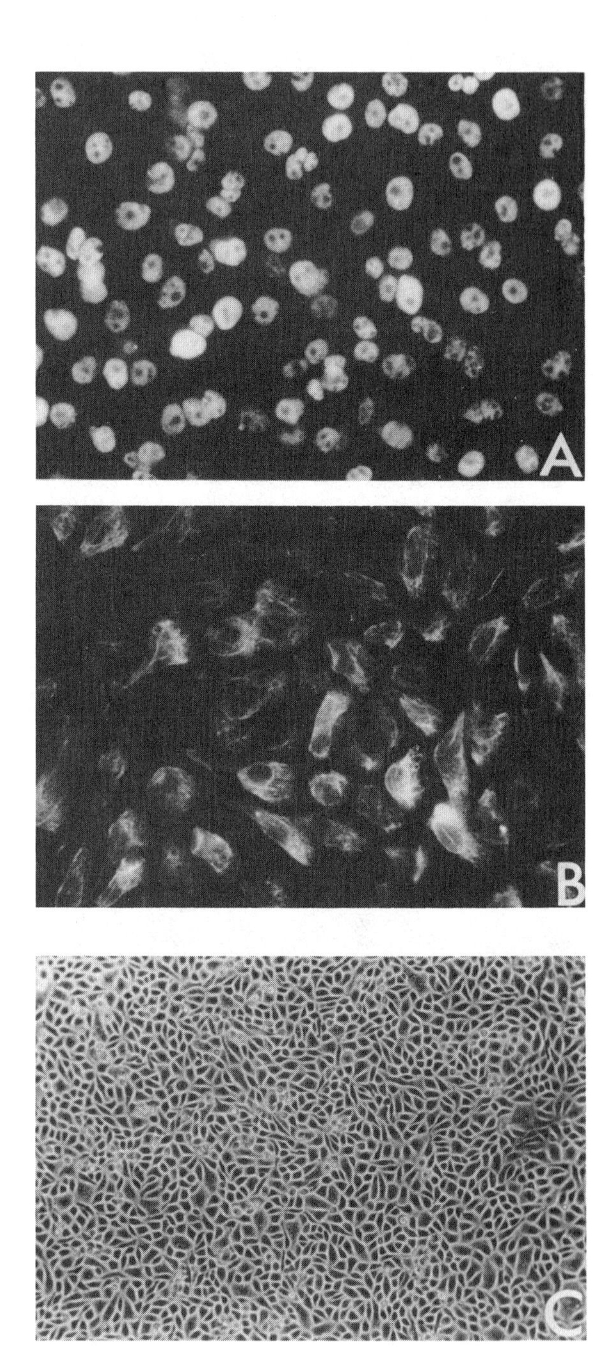

Fig. 6.6. Indirect immunofluorescent staining of immortalized HBE cells for (A) SV40 large and small T antigens; (B) keratin 18. Phase-contrast photomicrography of immortalized HBE cells (C). (×120.)

appropriate keratin, such as Ab-1, which recognizes keratin 18 (Fig. 6.6B).

5.3. DNA Fingerprinting

Confirmation that immortalized cell lines are derived from the parental cell strain is essential because of the particular vulnerability of cells in culture crisis to cross-contamination with immortalized cells, and may be achieved through DNA fingerprinting. We have mostly used single-locus DNA fingerprinting with probes recognizing highly polymorphic loci (e.g. Duncan et al. [1993]), although any DNA fingerprinting method would be acceptable.

ACKNOWLEDGMENTS

The procedures described have arisen out of work commenced in the laboratory of Dr. Curtis Harris, at the U.S. National Cancer Institute. Research in the authors' laboratory at the Children's Medical Reseach Institute is supported by the New South Wales Cancer Council.

REFERENCES

Brash DE, Reddel RR, Quanrud M, Yang K, Farrell MP, Harris CC (1987): Strontium phosphate transfection of human cells in primary culture: stable expression of the simian virus 40 large-T-antigen gene in primary human bronchial epithelial cells. Mol Cell Biol 7: 2031–2034.

Bryan TM, Reddel RR (1994): SV40-induced immortalization of human cells. Crit Rev Oncogenesis 5: 331–357.

De Jong PM, Van Sterkenburg MAJA, Kempenaar JA, Dijkman JH, Ponec M (1993): Serial culturing of human bronchial epithelial cells derived from biopsies. In Vitro Cell Dev Biol 29A: 379–387.

De Silva, R, and Reddel R.R. (1993): Similar simian virus 40-induced immortalization frequency of fibroblasts and epithelial cells from human large airways. Cell Mol Biol Res 39: 101–110.

De Silva R, Whitaker NJ, Rogan EM, Reddel RR (1994): HPV-16 E6 and E7 genes, like SV40 early region genes, are insufficient for immortalization of human mesothelial and bronchial epithelial cells. Exp Cell Res 213: 418–427.

Duncan EL, Whitaker NJ, Moy EL, Reddel RR (1993): Assignment of SV40-immortalized cells to more than one complementation group for immortalization. Exp Cell Res 205: 337–344.

Ke Y, Reddel RR, Gerwin BI, Miyashita M, McMenamin M, Lechner JF, Harris CC (1988): Human bronchial epithelial cells with integrated SV40 virus T antigen genes retain the ability to undergo squamous differentiation. Differentiation 38: 60–66.

Kelsen SG, Mardini IA, Zhou S, Benovic JL, Higgins NC (1992): A technique to harvest viable tracheobronchial epithelial cells from living human donors. Am J Respir Cell Mol Biol 7: 66–72.

Lechner JF, Haugen A, McClendon IA, Shamsuddin AM (1984): Induction of squamous differentiation of normal human bronchial epithelial cells by small amounts of serum. Differentiation 25: 229–237.

Lechner JF, LaVeck MA (1985): A serum-free method for culturing NHBE cells at clonal density. J Tiss Cult Meth 9: 43–48.

Masui T, Wakefield LM, Lechner JF, LaVeck MA, Sporn MB, Harris CC (1986): Type β transforming growth factor is the primary differentiation-inducing serum factor for normal human bronchial epithelial cells. Proc Natl Acad Sci USA, 83: 2438–2442.
Reddel RR, Ke Y, Gerwin BI, McMenamin MG, Lechner JF, Su RT, Brash DE, Park JB, Rhim JS, Harris CC (1988): Transformation of human bronchial epithelial cells by infection with SV40 or adenovirus-12 SV40 hybrid virus, or transfection via strontium phosphate coprecipitation with a plasmid containing the SV40 early region genes. Cancer Res, 48: 1904–1909.
Reddel RR, De Silva R, Duncan EL, Rogan EM, Whitaker NJ, Zahra DG, Ke Y, McMenamin MG, Gerwin BI, Harris CC (1995): SV40-induced immortalization and *ras*-transformation of human bronchial epithelial cells. Int J Cancer, 61: 199–205.
Sakamoto K, Howard T, Ogryzko V, Xu N-Z, Corsico CC, Jones DH, Howard B (1993): Relative mitogenic activities of wild-type and retinoblastoma binding-defective SV40 T antigens in serum-deprived and senescent human diploid fibroblasts. Oncogene 8: 1887–1893.
Sambrook J, Fritsch EF, Maniati, T (eds.) (1989): "Molecular Cloning: A Laboratory Manual," 2nd ed. New York: Cold Spring Harbor Laboratory Press.
Shay JW, Van Der Haegen BA, Ying Y, Wright WE (1993): The frequency of immortalization of human fibroblasts and mammary epithelial cells transfected with SV40 large T-antigen. Exp Cell Res. 209: 45–52.
Viallet J, Liu C, Emond J, Tsao M-S (1994): Characterization of human bronchial epithelial cells immortalized by the E6 and E7 genes of human papillomavirus type 16. Exp Cell Res 212: 36–41.
Willey JC, Broussoud A, Sleemi A, Bennett WP, Cerutti P, Harris CC (1991): Immortalization of normal human bronchial cells by human papillomavirus 16 or 18. Cancer Res 51: 5370–5377.

APPENDIX: MATERIALS AND SUPPLIERS

Materials	Suppliers
Ammonium metavanadate, A2286	Sigma Chemical Co
Bovine pituitary extract, 680-30281E	Gibco-BRL, Life Technologies
Bovine serum albumin (BSA), 735086	Boehringer Mannheim
Cloning cylinders, 10-mm external diameter, 2090-01010	Bellco Biotechnology
Collagen (Vitrogen 100), PC0701	Celtrix Laboratories
Dextrose, 17-0340-01	Pharmacia LKB
Dimethylsulfoxide (DMSO), D2650	Sigma Chemical Co
Disodium hydrogenphosphate, 10249	BDH/Merck
EDTA, 10093	BDH/Merck
EGTA, E3889	Sigma Chemical Co.
Epidermal growth factor (EGF), E4127	Sigma Chemical Co.
Epinephrine, 151065	ICN Biomedicals
Ethanol (Spectrasol grade), 1170	Ajax Chemicals
Ethanolamine, E0135	Sigma Chemical Co.
Fetal bovine serum (FBS), 15-010-0500V	Trace Biosciences
Fibronectin, F4759	Sigma Chemical Co.
FITC-conjugated goat antimouse IgG (H+L), stock concentration: 1.0 mg/ml, 115-015-062	Jackson Immunoresearch Labs
G418 (Geneticin™), 11811	Gibco-BRL, Life Technologies
Gentamicin, G1397	Sigma Chemical Co.
Glucose, G7520	Sigma Chemical Co.
Glycerol (Analytical grade), 242	Ajax Chemicals

Appendix *(Continued)*

Materials	Suppliers
HEPES, H9136	Sigma Chemical Co.
Hydrochloric acid, 10307	BDH/Merck
Hydrocortisone, H0888	Sigma Chemical Co.
Insulin, I6634	Sigma Chemical Co.
LHC-basal medium with stocks 4, 11, and calcium, 441	Biofluids
Liebowitz 15 (L15) medium, L4368	Sigma Chemical Co.
Manganese sulfate, M1144	Sigma Chemical Co.
Mediakap™ filtration units, 101 0.2 μm, ME2M-10B-12S	Microgon Inc.
Molybdic acid, M1019	Sigma Chemical Co.
Mouse monoclonal anticytokeratin antibody Ab-1 to keratin 18, IF04, stock concentration: 100 mg/ml	Oncogene Science Inc.
Mouse monoclonal antibody PAb419 to SV40 large- and small T antigens, DP01, stock concentration: 100 mg/ml	Oncogene Science Inc.
N-Propanol, 10345	BDH/Merck
Nalgene filtration units, 150 ml 0.2 μm, 125-0020	Nalge Company
Nickel chloride, N6136	Sigma Chemical Co.
Phenol red, 20090	BDH/Merck
Phosphate buffered saline (PBS), Ca^{2+}/Mg^{2+}-free (= PBSA), 11-075-0500V	Trace Biosciences
Phosphoethanolamine, P0503	Sigma Chemical Co.
Plasmid DNA, *pRSV-T,* was obtained from Dr. Bruce Howard, Laboratory of Molecular Growth Regulation, Building 6, Room 416, NICHHD, National Institutes of Health, Bethesda, MD 20892.	
Polypropylene tubes, 2059	Becton Dickinson
Polyvinylpyrrolidone (PVP), 20611	U.S. Biochemical Corp.
Potassium chloride, 383	Ajax Chemicals
Qiagen Plasmid Mega kit, 12181	Qiagen Inc.
Retinoic acid, R2625	Sigma Chemical Co.
Sodium bicarbonate, S5761	Sigma Chemical Co.
Sodium chloride, 10241	BDH/Merck
Sodium hydroxide, 10252	BDH/Merck
Sodium metasilicate, S4392	Sigma Chemical Co.
Sodium selenate, S5261	Sigma Chemical Co.
Soybean trypsin inhibitor (SBTI)	Sigma Chemical Co.
Stannous chloride, S9262	Sigma Chemical Co.
Strontium chloride, 14029	BDH/Merck
Transferrin (bovine), T8027	Sigma Chemical Co.
Triiodothyronine, T6397	Sigma Chemical Co.
Trypsin (bovine), T2271	Sigma Chemical Co.
Vacuum grease, D1400	Dow Corning

7

Monkey Kidney Epithelium

J. B. Clarke

Cell Resources Department, CAMR, Porton Down, Salisbury, Wiltshire, SP4 0JG, England

Culture of Immortalized Cells, Edited by R. Ian Freshney and Mary G. Freshney.
ISBN 0-471-12134-7

1. INTRODUCTION

Immortalized kidney cells are of considerable interest for studies in renal physiology and biochemistry, and also in pharmacology, toxicology, and pathology. They may also be useful for detection and production of a number of viruses. In particular, immortalized monkey kidney cells are now proving to have similar sensitivity to human viruses to the primary cells that are currently used extensively for human virus detection and isolation. Kidneys are relatively large, even in small laboratory animals, and are readily dissected out for processing. In larger species such as monkeys, they may be perfused *in situ* with enzyme solution prior to removal, which gives higher efficiencies of cell recovery and cell suspensions of higher viability than can be obtained by simple removal and disruption.

Kidneys exhibit a complex microanatomy of nephrons and tubules, organized into distinct segments and consisting of cell types that are morphologically and functionally distinct. If the intended studies require a specific type of cell, it is desirable to enrich the initial cultures for this, prior to immortalization. Specific segments of the nephrons and collecting ducts can be isolated by microdissection. A less technically demanding procedure is to culture cells from different regions of the kidney. For instance, dissociation of the outer cortex can generate cultures enriched with cells expressing functional characteristics of the proximal tubule. Density gradient methods may also be used to isolate similarly enriched cultures.

Primary kidney cells in general grow well for three or more subcultures, and many of the cells retain differentiated characteristics over this period. There are therefore usually no special difficulties in bringing primary kidney cell cultures to a suitable condition for successful transfection. Primate kidney cells can be successfully transformed by plasmids containing the SV40 large T-antigen. As monkeys are a natural host of SV40, the plasmids should be origin-deleted, to ensure the absence of any extrachromosomal replication.

2. MATERIALS AND EQUIPMENT

2.1. Solutions and Reagents

2.1.1. Citrate/trypsin solution

This is prepared in two stages.

Base solution

Sodium chloride	8.0 g
Potassium chloride	0.2 g
Disodium hydrogen phosphate	1.0 g
Potassium dihydrogen phosphate	0.2 g
Trisodium citrate	6.5 g

Make up to 1 liter with double-distilled deionised water at pH 7.3.

Final ingredients

Glucose	20 g
Lactalbumin hydrolysate	20 g
Trypsin	25 g

Add these three ingredients to 1 liter of citrate buffer, final pH 6.6 (± 0.1). Sterilize by filtration through a 0.22-m filter.

2.1.2. Hanks/trypsin solution

This is prepared in two stages

Base solution

Sodium chloride	8.0 g
Potassium chloride	0.4 g
Potassium dihydrogen phosphate	0.06 g
Disodium hydrogen phosphate (anhydrous)	0.096 g
Magnesium chloride· $6H_2O$	0.1 g
Magnesium sulfhate· $7H_2O$	0.1 g
Calcium chloride· $2H_2O$	0.186 g
Sodium hydrogen carbonate	0.35 g
Glucose	1 .0 g

Make up to 1 liter with double-distilled deionized water, final pH 7.2. Sterilize by filtration through an 0.22-μm filter and bottle in 375-ml volumes.

Final ingredients

Trypsin	5% (w/v), pH 7.2
DNAse	0.2% (w/v)
Disodium carbonate	5% (w/v), pH 8.2 ±0.1
Penicillin/streptomycin, 1000 IU/mg	5 mg/ml

Dissolve individually in ultrapure water (UPW) and sterilize by filtration through 0.22 μm filter.

Add the final ingredients to 375-ml bottles of Hanks' buffer on the day of use:

Trypsin	25 ml
DNase	8.5 ml
Disodium carbonate	6.0 ml
Penicillin/streptomycin	4.0 ml

2.1.3. Buffers

Prepare the following stock solutions. Dissolve the weighed ingredients and make up to the indicated final volume with deionised double-distilled water.

Stock Solution	*Concentration*	*Weight (g)*	*Final Volume (ml)*
Tris pH 7.5	2.0M	4.84	20
EDTA	0.2M	0.744	10
NaCl	5.0M	29.2	100
KCl	1.0M	0.75	10
$CaCl_2$.2H2O	2.0M	2.94	10
$MgCl_2$	0.1M	0.203	10
$NaHPO_4$	1.0M	3.58	10
$CaCl_2$(anhydrous)	2.0M	2.22	20

Prepare the following buffer solutions, using the prepared stock solutions.(Add HEPES as a weighed solid)

TE(P) Buffer

Constituent	*Stock Concentration*	*Volume of Stock (ml)*	*Final Concentration*
Tris	2.0M	50	1mM
EDTA	0.2M	25	0.05mM
Distilled water to		100ml	

HBSS(2X) Buffer (pH 7.12)

Constituent	*Stock*	*Volume/wt. of stock*	*Final Concentration*
NaCl	5M	5.6 ml	280mM
Na_2HPO_4	1M	0.15 ml	1.5mM
HEPES	(solid)	1.192 g	50mM
Distilled water to		100 ml	

TBS Buffer (pH 7.4)

Constituent	*Stock Concentration*	*Volume of Stock (ml)*	*Final Concentration*
Tris (pH 7.5)	2M	1.25	25mM
NaCl	5M	2.74	137mM
KCl	1M	0.5	5mM
$MgCl_2$	0.1M	0.5	0.7mM
Na_2HPO_4	1.0M	60	0.5mM
$CaCl_2 \cdot 2H_2O$	2M	35	0.6mM
Distilled water to	100 ml		

Sterilize by filtration through 0.22-μm filters.The buffers and the calcium chloride solutions should be stored at −20°C.

2.2 Cell Culture Medium and Serum

M199 medium
Newborn bovine serum (NBS)
Lactalbumin hydrolysate
Penicillin/streptomycin
Trypsin

2.3 Additional Materials

Anaesthetic (Vetelar)
DNAse
DOTAP, DOTMA

2.4 Sterile Equipment

Glass filter funnel containing a sheet of muslin gauze within a conical steel gauze.
Fine string, cannula (50mm long, 2mm diameter, stainless steel), tubing, and dispenser top for 1 liter bottle (including inlet air filter) (Fig. 7.1)
Trypsinisation column plus tubing, clamps and collecting bells (see Fig. 7.2)
Centrifuge tubes or bottles (200-ml)
Artery clamps, bone cutters, forceps, scalpel blades (disposable), scissors.
Trypsinisation flask
Cloning rings (glass, PTFE, or stainless steel)
Silicone grease.

2.5 Nonsterile Equipment

Animal electric hair shaver
Variable-speed peristaltic pump

3. PROTOCOLS

The simple method of preparing the cells is to remove the kidneys from the animal and trypsinize *in vitro*. The cells are then seeded in growth medium. Different parts of the kidney may be selected for trypsinization to provide cultures enriched for different cell types. A more efficient method of obtaining cells by trypsinizing *in situ* is also described. This procedure gives a 4-fold increase in viable cells com-

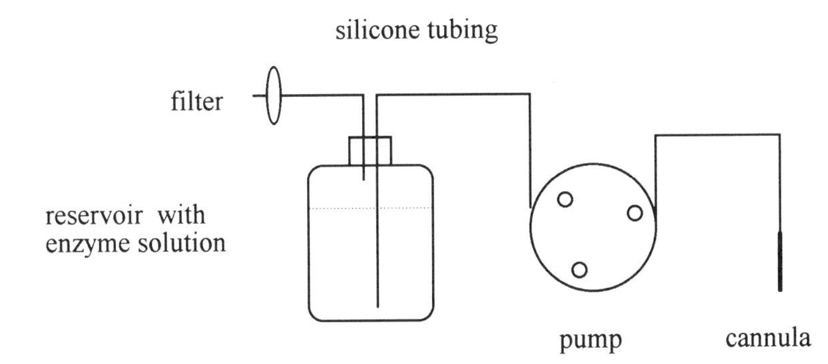

Fig. 7.1. Equipment for *in vitro* kidney perfusion.

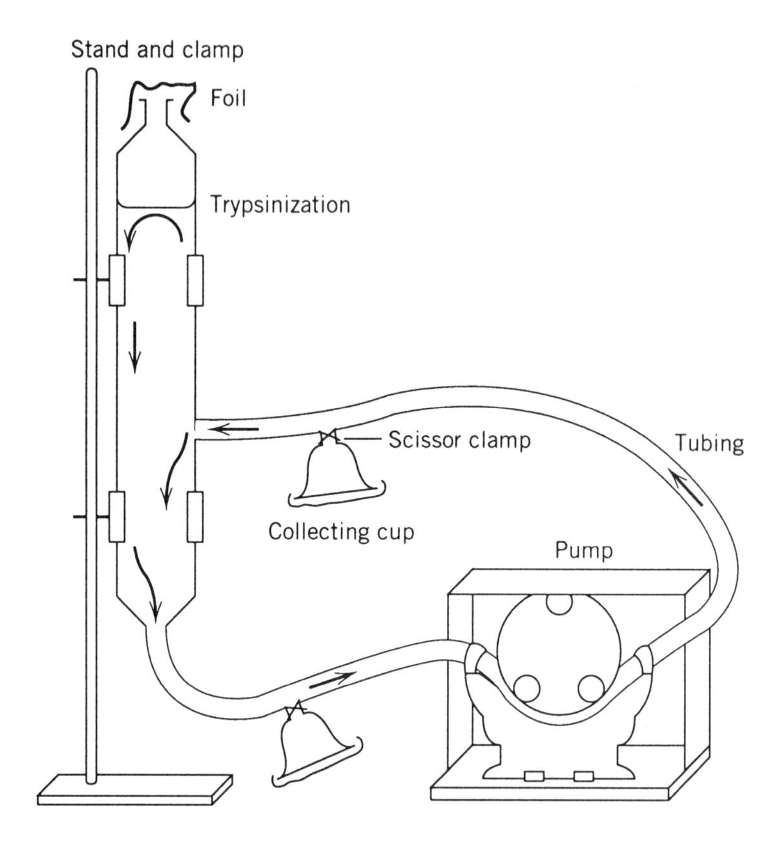

Fig. 7.2. Trypsinization apparatus.

pared to the simple procedure. The technique was developed for use with monkeys. It is adaptable for other species, but is not easy to perform on animals much smaller than rabbits.

3.1. Preparation of Primary Cell Cultures

3.1.1. Simple dissection procedure

Protocol 3.1.1

Reagents and Materials

Sterile

- Scalpels, large-bladed, ~40 mm
- Forceps, rat-toothed
- Scissors
- Nylon bag, autoclavable, 150×300 mm, or dish, 150 mm diameter × 50 mm deep
- Vetelar anesthetic
- Disinfectant, e.g., hypochlorite
- Shaver
- Horizontal laminar flow cabinet

Protocol

(a) Terminally anesthetize the animal by injecting Vetelar, 0.15 ml/g intramuscularly into the thigh. Shave, dip in dilute disinfectant, and place in a dissection cabinet with horizontal sterile airflow.
(b) Using large-bladed scalpel and rat-toothed forceps, cut away the skin over the central area of the torso.
(c) Open the body wall midventrally using sterile scissors. Avoid cutting into the internal organs.
(d) Using fresh sterile scissors, dissect connective and fatty tissue away from around the kidneys, avoiding blood vessels and adrenal glands.
(e) Remove the kidneys and place in a sterile bag or petri dish.

3.1.2. Preparation of single cell suspension after simple dissection

Protocol 3.1.2

Reagents and Materials, Sterile

- Scissors
- Hanks/trypsin solution
- Indented trypsinization flask (Bellco) with bar magnet [Freshney, 1994]
- Magnetic stirrer
- Culture flask, Erlenmeyer flask, or bottle, 250 ml
- Newborn bovine serum, (NBS)
- Medium M199 with 10% NBS and antibiotics (M199/10NB)

Protocol

(a) Cut the kidneys into small pieces (about 1-mm cubes) in a Class III cabinet. Class II may be used for nonprimate tissues.
(b) Suspend the fragmented kidneys in 100 ml Hanks/trypsin solution and pour into the trypsinization flask.
(c) Trypsinize the kidneys by stirring the cell suspension at 37°C for 30 min.
(d) Allow the remaining pieces of kidney to sediment.
(e) Decant the kidney cell suspension into a sterile flask containing 15 ml NBS to inactivate the trypsin.
(f) Resuspend the remaining kidney tissue in 100 ml Hanks/trypsin.
(g) Proceed as in (c) above.
(h) Repeat (d) above. At the end of this stage, a single cell suspension should have been achieved with only fatty tissue remaining in the flask.
(i) Resuspend in prepared and prewarmed M199/10NB with $\geq 10^5$ cells/ml.
(j) Perform flask culture as in Section 3.2.

3.1.3. Perfusion procedure

Protocol 3.1.3

Reagents and Materials

Non-sterile equipment

- Shaver
- Peristaltic pump, 20–100 ml/min
- Dissection cabinet with horizontal sterile airflow

Sterile instruments and materials

- Scalpel, large-bladed, ~40 mm
- Rat-toothed forceps
- Scissors, 2 pairs
- Bone cutters or heavy scissors
- Spencer–Wells forceps
- String
- Cannula, or stainless steel tube, ~50 mm long, 2 mm diameter
- Bag, nylon, autoclavable, 150×300 mm

Reagents

- Citrate/trypsin solution, sterile
- Vetelar anesthetic
- Disinfectant

Protocol

Dissection

(a) Terminally anesthetize the animal by injecting 0.15 ml Vetelar per kilogram intramuscularly into the thigh.
(b) Shave, dip in dilute disinfectant, and place in a dissection cabinet with horizontal sterile airflow.
(c) Using large-bladed scalpel and rat-toothed forceps, cut away the skin over the central area of the torso.
(d) Open the body wall midventrally using sterile scissors. Avoid cutting into the internal organs.
(e) Using fresh sterile scissors, dissect connective and fatty tissue away from around the kidneys, avoiding blood vessels and adrenal glands.
(f) Cut the diaphragm away from the rib cage.
(g) Cut out and remove the lower $\frac{2}{3}$ of the rib cage and sternum, using bone cutters or heavy scissors.
(h) Dissect connective and fatty tissue away from the aorta and vena cava below the kidneys, avoiding the mesenteric artery and vein.
(i) Clamp the exposed aorta and vena cava, below the kidneys and preferably above the mesenteric vessels, using Spencer-Wells forceps.
(j) Clamp the hepatic artery and other arteries branching from the aorta anterior to the kidneys.

Perfusion

(a) Pass a sterile string under the aorta anterior to the kidneys.
(b) Nick the vena cava, posterior to the kidneys but anterior to the clamp.
(c) Pump citrate/trypsin solution through the perfusion apparatus to the tip of the cannula, to prevent air bubbles from entering the kidneys.
(d) Make a cut into the aorta anterior to the kidneys, about $\frac{1}{3}-\frac{1}{2}$ across.
(e) Insert the cannula into the cut in the aorta.
(f) Tie the string tightly around the aorta where it covers the cannula, to seal the cannula in the aorta and prevent back-leakage of the citrate/trypsin solution.
(g) Start perfusion at a relatively slow rate, approximately 30 ml/min, then increase gradually to 60–80 ml/min. Perfuse as much of the citrate/trypsin as possible through the kidneys without allowing air bubbles to enter. Note that the right (lower) kidney often perfuses faster than the left; completion is shown by the organ turning pale gray-beige in color and swelling to 1.5–2.0 times its original size. If the right kidney has perfused significantly more rapidly than the left, clamp the renal artery and vein to the right. This will cause the remaining citrate/trypsin solution pass exclusively through the left kidney and complete the digestion.
(h) Remove the kidneys into a sterile bag. Care is required as the kidneys will be soft and slippery as a result of the enzyme action.
(i) Inspect the lungs, spleen, and liver for overt signs of disease and parasites. Reject the kidneys if such signs are found.

3.1.4. Preparation of single cell suspension after perfusion

Protocol 3.1.4

Reagents and Materials

Sterile reagents

- Hanks' BSS (HBSS)
- Hanks/trypsin
- M199 medium + 10% NBS and antibiotics at 37°C (M199/10NB)

Sterile Instruments and materials

- Large-bladed (~40-mm) and small-bladed (~20-mm, pointed) scalpels
- Sterile muslin gauze
- Flask, 250–500ml containing 15 ml newborn calf serum (NBS)

Sterile equipment

- Trypsinization column (see Fig. 7.2).
- Centrifuge tubes or bottles, 200 ml

Nonsterile equipment

- Peristaltic pump
- Centrifuge

Protocol

(a) Cut the perfused kidneys into small pieces (about 1 mm) in a Class II cabinet (Class III for primate kidneys).
(b) Dilute the fragmented kidneys in Hanks/trypsin solution then pour into the trypsinization column.
(c) Trypsinize the kidneys by pumping the cell suspension through the tubing loop for 20 min.
(d) Allow the remaining pieces of kidney to sediment below the level of the side arm to the collecting cup.
(e) Siphon the kidney cell suspension into a large flask containing 15 ml NBS to inactivate the trypsin.
(f) Further dilute the kidney cell suspension remaining within the column with HBSS.
(g) Repeat the trypsinization process until a single cell suspension has been achieved and all that remains in the column is fatty tissue.
(h) Filter the kidney cell suspension through a sterile muslin gauze into centrifuge pots and centrifuge at 200*g* for 10 min.
(i) Resuspend in prepared and prewarmed M199/10NB.

3.2. Flask Culture

Protocol 3.2

Reagents and Materials

Sterile reagents

- Dulbecco's phosphate-buffered saline without Ca^{2+} and Mg^{2+} (PBSA)
- EDTA/trypsin (5 ml/25-cm^2 flask; 10 ml/75-cm^2)
- Medium: M199/10NB

Sterile flasks

- Culture flasks, 25 or 75-cm^2

Nonsterile equipment

- Cell counter or hemocytometer

Protocol

(a) Resuspend cells in prewarmed M199/10NB, and adjust concentration to $\geq 10^5$ cells/ml.

(b) Dispense 10 ml diluted cells into 25-cm^2 flasks, or 25 ml into 75-cm^2 flasks.
(c) Incubate the flasks in stationary mode at 37°C, in 5% CO_2.
(d) Inspect microscopically at 48-h intervals. The cells grow initially in discrete patches, but these usually coalesce to form a monolayer, which becomes virtually confluent by 7–10 days.
(e) The medium may be changed after 4–6 days, to remove debris, and promote cell growth.
(f) To subculture the cells, decant the spent medium and wash the monolayers once with PBSA.
(g) Add EDTA/trypsin solution to each flask (5 ml/25-cm^2 flask; 10 ml/75-cm^2).
(h) Leave this on the cells for 20 s and then decant, leaving a residual film.
(i) Incubate the flasks at 37°C, and inspect microscopically every 5 min until cell detachment is observed.
(j) Add M199/10NB medium to each flask (10 ml/25-cm^2 flask; 25 ml/75-cm^2 flask), and agitate to detach the cells.
(k) Pool the resulting cell suspensions, dilute 1:2 or 1:3 with complete M199 medium, and dispense into fresh flasks (10 ml/25-cm^2 flask; 25-ml/75 cm^2).
(l) Incubate the flasks at 37°C in stationary mode, in 5% CO_2.

3.3. Transfection of Primary Monkey Kidney Cells: Calcium Phosphate Precipitation Procedure.

3.3.1. Preparation of cells

Protocol 3.3.1

Sterile Reagents and Materials

- PBSA.
- EDTA/trypsin (5 ml/25-cm^2 flask; 10 ml/75-cm^2)
- Medium: M199/10NB
- Flasks, 25- or 75-cm^2

Protocol

(a) Select rapidly dividing cultures.
(b) Decant the spent medium and wash the monolayers once with PBSA.
(c) Add EDTA/trypsin solution to each flask (5 ml/25-cm^2 flask; 10 ml/75-cm^2).
(d) Leave this on the cells for 20 s and then decant, leaving a residual film.
(e) Incubate the flasks at 37°C, and inspect microscopically every 5 min until cell detachment is observed.

(f) Add complete M199 medium to each flask (10 ml/25-cm^2 flask; 25 ml/75-cm^2 flask), and agitate to detach the cells.
(g) Pool the resulting cell suspensions, dilute 1:3 with M199/10NB medium (final concentration 3-5$\times 10^4$- cells/ml), and dispense 10 ml of this diluted suspension into fresh 25-cm^2 flasks.
(h) Incubate the flasks at 37°C in stationary mode.
(i) Inspect daily using a microscope. The cells are ready for use when 50–70% confluent.

3.3.2. *Transfection*

Protocol 3.3.2

Sterile Reagents and Materials

- Medium: M199/10NB
- Plasmid preparation (pUK42; pUKEori-; pUKδt [Kreuzberg-Duffy and MacDonald, 1994]
- TBS
- TE(P) buffer, 500 ml
- 2×HBSS
- 2M $CaCl_2$
- Falcon polypropylene tubes, 15 ml
- 25-cm^2 flasks containing growing cell cultures from Protocol 3.3.1, (i)

Non-sterile equipment

- Vortex mixer

Protocol

(a) Determine concentration of DNA in purified plasmid preparation and dispense volumes containing 10 μg of plasmid DNA into 15-ml Falcon polypropylene tubes, one tube for each 25-cm^2 flask containing cells.
(b) Add TE(P) buffer to the plasmid-containing tubes to give a final volume of approximately 420 μl.
(c) Dispense aliquots of 480 μl 2×HBSS into another set of polypropylene tubes.
(d) Add dropwise 60 μl 2M $CaCl_2$ to DNA-containing tubes; vortex at same time and ensure that each drop is thoroughly mixed before the next is added.
(e) Add dropwise the DNA-$CaCl_2$ solution to the tubes containing 2×HBSS; vortex at same time and ensure that each drop is thoroughly mixed before the next is added.
(f) Allow preciptate to form for 30 min at room temperature.
(g) Add preciptate suspension dropwise to the medium in the 25-

cm^2 flask cultures, with the flask horizontal (distribute evenly). Use one tube for each 25-cm^2 flask.

(h) Expose cells to DNA-$CaPO_4$ preciptate for 16–24 h. (The precipitate should be observable with microscope after 1 h).

(i) Decant medium and wash with 5 ml prewarmed TBS, then wash with 5 ml prewarmed M199 complete medium.

(j) Add 5 ml M199/10NB and incubate at 37°C, 5% CO_2.

3.2.3. Isolation of colonies

Protocol 3.3.3

Reagents and Materials

Sterile reagents

- Selective agent: this will depend on the resistance gene(s) present on the plasmid in use. A common one is neo, which confers resistance to the cytotoxic antibiotic geneticin (G418). A preliminary titration should be performed to determine the minimum lethal dose of the agent for the cells in use, and a concentration selected slightly above this. Excessive concentrations will kill even transfected cells. G418 is often used at 500 μg/ml.
- Trypsin, 0.25% in PBS
- Medium: M199/10NB

Sterile materials

- 4- or 6- well tissue culture plates
- 24-well plates
- 25- and 75-cm^2 flasks
- Plastic, 1-ml pipettes
- Forceps
- Cloning rings
- Silicone grease

Nonsterile

- Marker pen, waterproof

Protocol

(a) Inspect cells daily using a microscope.

(b) Subculture into 4- or 6-well tissue culture plates when growth appears well established or cells become confluent.

(c) Inspect as in (a), and when cell growth is well established, add the selective agent.

(d) Inspect the cells regularly, using a microscope. The cells may remain viable for some time, but will eventually cease growing and start to die. If the transfection has been successful, colonies of immortalized cells will start to appear.

(e) The cells should be fed at intervals with fresh medium, replacing part of the old medium only, until colonial growth is well established. After this, fresh medium alone may be used.

(f) If possible, it is best to allow cell colonies to grow until 1–5 mm in diameter before harvesting.

(g) Mark the positions of colonies that appear to have arisen from one growth centre and are well separated from other cells, using a waterproof marker on the base of the tray.

(h) Remove the medium from the cultures.

(i) Using sterile forceps, pick up a cloning ring and dip its base in silicone grease. Ensure that the grease is evenly distributed.

(j) Place the ring around a marked colony and press down, moving slightly to obtain a good seal.

(k) Repeat steps (i) and (j) for the other marked colonies.

(l) Fill the rings with trypsin solution, leave for 20 s, then remove most of the solution, leaving a thin film. Use a new pipette for each ring.

(m) Close the plate and incubate at 37°C. Inspect periodically to monitor the detachment process.

(n) When the cells have detached, fill each ring with complete medium.

(o) Carefully pump the medium up and down to suspend the cells, and transfer the suspension to a well of a 24-well plate. The total medium volume per well should not exceed 1 ml.

(p) Repeat the process with the contents of each ring, using a new pipette each time.

(q) Incubate at 37°C in a humidified atmosphere of 5% CO_2/95% air.

(r) When sufficient cells are present (~50% of the area of the well is covered), subculture into 25-cm^2 flasks and subsequently into 75-cm^2 flasks.

(s) Stocks should be frozen in liquid nitrogen as soon as a sufficient number of cells are available.

4. ALTERNATIVE PROCEDURES

In addition to the standard calcium phosphate precipitation procedure described, transfection of primate kidney cells may also be achieved with DEAE-Dextran or by *lipofection,* using polycationic detergents such as DOTMA or DOTAP. Experience with monkey kidney cells suggest that these reagents give high levels of transient plasmid expression, but low levels of stable transfection, when compared with calcium phosphate. While lipofection introduces DNA into larger numbers of cells than the calcium phosphate procedure, it introduces smaller quantities, and the amount of DNA transfected may be particularly important for a significant chance of incorporation into the genome in monkey kidney cells. Electroporation is a potentially useful procedure, but the present author has not used it with these cells.

5. SAFETY CONSIDERATIONS

Primates may be infected with viruses and other agents that can infect humans. Animal house procedures must be designed to screen for such agents. Staff involved in the handling of primates and kidney collection must wear protective clothing and be vaccinated against rabies. Initial disaggregation of primate kidneys should be carried out under ACDP Class III conditions. When the cells are suspended in growth medium, containment may be reduced to Class II. For other species, Class II is sufficient unless there are particular indications of agents that may affect humans (see also Chapter 2).

6. CRITERIA FOR CHARACTERIZATION AND VALIDATION OF CELLS

Primate kidney cells may be validated for species by the procedures of multilocus DNA fingerprinting and isoenzyme analysis. Function of different kidney cell types may be validated by determining the presence of a variety of biochemical and physiological activities and cell surface proteins. For instance, cells from the renal proximal tubule exhibit a specific pattern of hormonal stimulation of cAMP, and express specific brush border enzymes [Kempson et al., 1986]. Cells from the thick ascending loop of Henle exhibit the $Na+$, $K+$, Cl^- cotransport system, an Na, K-ATPase [Eveloff et al., 1980]. They also secrete the Tamm–Horsfall mucoprotein [Sikri et al., 1979].

REFERENCES

Eveloff J, Haase W, Kinne R (1980): Separation of renal medullary cells from the thick ascending limb on Henle's loop. J Cell Biol, 87: 672–681.

Freshney RI (1994): " Culture of Animal Cells, a Manual of Basic Technique." New York: Wiley-Liss.

Kempson SA, McAteer JA, Al-Mahrouq HA, Dousa TP, Dougherty GS, Evan AP (1989): Proximal tubule characteristics of cultured human renal cortex epithelium. J Lab Clin Med, 113: 285–2965.

Kreuzberg-Duffy U, MacDonald C (1994): Analysis of SV40 early region expression in immortalised mouse macrophage lines. In: (Spier RE, Griffiths JB, Berthold W eds): "Animal Cell Technology—Products for Today, Prospects for Tomorrow Oxford: Butterworth-Heinemann, pp 80–82.

Sikri KL, Fosater CL, Bloomfield FL, Marshall RD (1979): Localisation by immunofluorescence and by light and electron-microscopic immunoperoxidase techniques of Tamm-Horsfall glycoprotein in adult hamster kidney. Biochem J 181: 525–532.

APPENDIX: MATERIALS AND SUPPLIERS

Materials	Suppliers
General chemicals	
(Analar, except where stated)	BDH
Cell culture medium and serum.	
Lactalbumin hydrolysate	Gibco BRL, Life Technologies, Ltd.
M199 medium	Gibco BRL, Life Technologies, Ltd.
Newborn bovine serum (NBS)	Gibco BRL, Life Technologies, Ltd.
Penicillin/ streptomycin	Gibco BRL, Life Technologies, Ltd.
Trypsin	Gibco BRL, Life Technologies, Ltd.
Additional materials	
Anaesthetic (Vetelar)	Centaur Services Ltd.
DNAse	Gibco BRL, Life Technologies, Ltd.
DOTAP	Boehringer Mannheim
Sterile equipment	
Cannula, 50 mm long, 2 mm diameter, stainless steel	Propper, Rocket of London
Centrifuge tubes or bottles (200-ml)	Fisons Ltd.
Cloning rings	Bellco
(Glass, PTFE, or stainless steel)	
Instruments	
Artery clamps	Rocket of London
Bone cutters	Rocket of London
Forceps	Rocket of London
Scalpel blades (disposable)	Paragon Razor Co.
Scissors	Rocket of London
Trypsinisation flask	Bellco
Silicone grease	Dow Corning, Hopkin & Williams
Nonsterile equipment	
Animal electric hair shaver	Arnolds Veterinary Products Ltd.,
(Oster Corporation, USA)	
Variable-speed peristaltic pump (types HR100 and 603 S/R)	Watson Marlow

8

Human Hepatocytes

K. Macé, C. C. Harris, M. M. Lipsky[2] and A. M. A. Pfeifer

Nestlé Research Centre, P.O. Box 44, Lausanne, CH-1000, Switzerland (K. M., A. M. A. P.) and NCI, NIH, Bethesda, Maryland (C. C. H.)
[2]University of Maryland, Maryland.

1. BACKGROUND

Cultures of primary human hepatocytes are now frequently used for *in vitro* biochemical and pharmacotoxicological studies and for investigation of pathogenesis of liver disease [Hsu et al., 1985; Butterworth

Culture of Immortalized Cells, Edited by R. Ian Freshney and Mary G. Freshney.
ISBN 0-471-12134-7

et al., 1989; Lopez et al., 1991]. Generally, these experiments are performed on either freshly isolated cells or short-term primary cultures. Although methods have been developed for replicative cultures of many types of adult normal human epithelial cells [Harris, 1987], establishing conditions to support long-term cell cultures of human liver epithelial cells has proven difficult. The difficulties encountered with human hepatocytes have been linked to the differentiated phenotype of these cells and the thereby limited capacity to undergo cell division. In addition, standard culture techniques cannot preserve differentiated functions over long periods. This point is particularly well-illustrated by the loss of cytochrome P450-dependent activities during the first few days of primary culture of liver cells [Guillouzo et al., 1985; Donato et al., 1992a]. The use of extracellular matrix and basal medium supplemented with defined components (e.g., insulin, transferrin, hydrocortisone and EGF) has significantly extended the lifetime of cultured human hepatocytes and improved the maintenance of differentiated functions essential for metabolic studies [Lechner et al., 1989; Gibson-D'Ambrosio et al., 1993; Koebe et al., 1994]. However, the limited access to liver tissue from healthy donors and the functional variability of different tissue isolates underline the significance of having standardized and defined human liver cell models.

Increased or indefinite life-span mediated by chemical, physical, or microbiological agents, has been widely reported with cultured human epithelial cells from different tissues, such as skin, lung, kidney, and bladder [Harris, 1987]. *In vitro* transformation of normal human cells has proven to be more difficult than transformation of rodent cells [DiPaolo, 1983], due, probably, to a relatively greater genomic stability. In addition, the limited capacity of hepatocytes to undergo cell division complicates the immortalization process of this cell type and reduces its efficiency. Several immortalized rat hepatocytes have been established [Isom et al., 1980; Lafarge-Frayssinet et al., 1984; Woodworth et al., 1986; Woodworth and Isom, 1987; Bayad et al., 1992] with some of them retaining differentiated features of normal hepatocytes, including the expression of albumin, transferrin, hemopexin, and glucose phosphatase [Woodworth et al., 1986; Woodworth and Isom, 1987; Bayad et al., 1992]. In contrast, only two studies have reported immortalization of human liver epithelial cells with expression of hepatocyte-specific markers [Pfeifer et al., 1993; Ueno et al., 1993]. Successful immortalization required a limited potential of cells for cell division. Populations of liver cells that can still undergo cell divisions either due to a reduced state of differentiation or via induction of dedifferentiation have been described [Kaighn and Prince, 1971; Lechner et al., 1989; Sell, 1990]. If this observation is correct the establishment of replicative cultures of adult human hepatocytes would involve methods to induce the proliferation of less differentiated or "dedifferentiated" cells. In a second phase, immortalization of this cellular phenotype would be initiated, followed by the application of culture conditions that promote their redifferentiation into hepatocytes.

Therefore, our first goal for achieving immortalization of liver epithelial cells was the development of a liver cell growth medium. This involves the growth of primary human hepatocytes on collagen I/fibronectin-coated flasks in a proliferation-inducing serum-free liver cell medium (LCM) in the presence of medium conditioned with hepatocellular carcinoma cells [Pfeifer et al., 1993]. Proliferating cultures are then immortalized with a high-titer recombinant SV40 large-T-antigen virus, which introduces the immortalizing gene with high efficiency and minimal cell toxicity. This technique allows the reproducible establishment of functional liver epithelial cell lines, designated THLE cells, from nondiseased adult donors and overcomes many of the limitations reported for liver models: extended life-span in contrast to immortalization, tumorigenicity, loss of metabolism, and differentiation functions.

2. PREPARATION OF MEDIA AND REAGENTS

2.1. Perfusion Solutions

2.1.1. Hanks' balanced salt solution (HBSS)

Prepare Ca^{2+}- and Mg^{2+}-free HBSS containing 25mM tricene buffer and 0.5mM EGTA, pH 7.2.

2.1.2. Collagenase–dispase solution

Dissolve 140 mg of collagenase type P, 1 g of dispase, and 10 mg of soybean trypsin inhibitor in 1 liter of phosphate buffered saline (PBSA) containing 50mM HEPES and 0.6mM $CaCl_2$.

2.1.3. L15 complete medium

Supplement L 15 medium with 10% fetal bovine serum (FBS, lot number 30101107), 10 μg/ml insulin, 0.2μM hydrocortisone, and 50 μg/ml gentamicin.

2.2. Cell Culture Reagents

2.2.1. Hepes-buffered saline (HBS)

Dissolve 50mM HEPES in PBS. The prepared solution is commercially available from Biofluids.

2.2.2. Liver cell medium (LCM)

The medium is based on a modification of Ham's F12 medium referred to as PMFR-4 [Lechner et al., 1980], except that arginine is replaced by 0.3mM ornithine and the calcium concentration is reduced to 0.4mM. The medium is supplemented with hormones, growth factors and additives as described in Table 8.1.

2.2.3. DMEM

DMEM basal medium is supplemented with 10% fetal calf serum.

2.2.4. PC-1 medium

PC-1 medium is a serum-free medium used for virus collection.

2.2.5. Hep G2-conditioned medium

Incubate 80% confluent Hep G2 cells (HB 8065, ATCC) in LCM for 3 days. The conditioned medium is filter sterilized before use.

2.2.6. Coating solution

Fibronectin	5 mg
Collagen type 1 (Vitrogen 100), 3 mg/ml	5 ml
Bovine serum albumin, 0.1%	50 ml
PMFR-4 basal medium	500 ml

Filter, 0.22 μm, and store at 4°C.

2.2.7. Geneticin®

Dissolve 200 mg of Geneticin® in 1 ml H_2O. Filter on 0.22 μm and store at -80°C.

Table 8.1. Liver Cell Medium (LCM)

Nutrients	
PMFR-4 modified to contain	0.3mM ornithine
	0.0mM arginine
Hormones and Factors	
Insulin	1.75μM
Transferrin	10.0 μg/ml
Hydrocortisone	0.2μM
Triiodothyronine	50.0nM
Retinoic acid	0.33nM
Epidermal growth factor	5.0 ng/ml
Other Additives	
P/E[a]	0.5μM
Bovine pituitary extract	7.5 μg/ml
Chemically denatured serum	3.0%
Hep G2 conditioned medium[b]	30%
L-Glutamine	2mM
Gentamycin	50 μg/ml

[a]Phosphoethanolamine/Ethanolamine stock
[b]After several cell passages the conditioned medium can be removed

2.2.8. Polybrene solution

Dissolve 1 mg of polybrene (hexadimethrine bromide) in 1 ml of water. Filter on 0.22 μm and store at 4°C.

2.2.9. 2×BBS buffer

N,N-bis-(2-hydroxyethyl)-2-aminoethanesulfonic acid (BES)	50 mM
NaCl	280 mM
Na_2HPO_4	1.5 mM

2.2.10. E-PET

Trypsin	0.02%
Polyvinylpyrrolidine	1%
Ethylene glycol bis	0.02%

In HEPES buffered saline.

3. PROCEDURES

3.1. Isolation and Culture of Hepatocytes

Human liver cells are isolated by a two-step perfusion technique [Hsu et al., 1985; Pfeifer et al., 1993] modified from the initial procedure established by Berry and Friend [Berry and Friend, 1969]. Large pieces of liver or an entire lobe obtained from immediate autopsy of adult donors with no clinical evidence of cancer are used.

3.1.1. Perfusion

Protocol 3.1.1

Reagents and Materials (Sterile)

- Ca^{2+}/Mg^{2+}-free HBSS
- Collagenase-dispase solution (see Section 2.1.2)
- L15 medium with supplements (see Section 2.1.3).
- Pasteur pipettes
- Artery clamps
- Scissors and scalpels
- Sterile gauze mesh size 1, increasing to 6–8
- Tygon tubing, $\frac{1}{4}$ in. (6 mm) long, $\frac{1}{8}$ in. (3 mm) diameter

Protocol

(a) Cannulate the piece of liver via the largest available blood vessel, by using an inverted 5.75in. (150 mm) Pasteur pipette with Tygon tubing attached.

(b) Clamp all other visible large vessels to avoid the flow of solutions.
(c) Place the liver in a sterile container in a waterbath maintained at 37°C.
(d) Start the perfusion by using Ca^{2+}/Mg^{2+}-free HBSS at 37°C for 20–30 min or until the liver perfusate appears clear. "Massage" to aid in disloging blood clots.
(e) Continue the perfusion by using the collagenase-dispase solution for an additional 20–30 min or until there is a noticeable blanching and softening of most of the specimen.
(f) Mince the perfused tissue into small pieces, place it in fresh perfusion solution, and gently agitate in a large spinner flask for 10–20 min.
(g) Filter the resulting solution through successive layers of sterile gauze of decreasing porosity (mesh size 1, increasing to 6–8).
(h) Dilute the filtered cell solution with an equal volume of L15 medium.
(i) Centrifuge at 50g for 10 min.
(j) Remove the supernatant and resuspend the cell pellet in fresh L15 medium.
(k) Repeat the centrifugation two or more times.
(l) Resuspend the final cell pellet in fresh L15 medium.
(m) Incubate an aliquot with trypan blue to determine cell number and viability.

3.1.2. Primary culture

Protocol 3.1.2

Reagents and Materials

- Collagen I-fibronectin (see Section 2.2.5)
- *HBS:* HEPES-buffered saline (see Section 2.2.1)
- LCM (see Section 2.2.2)
- *LCM33CM:* LCM medium containing 33% of conditioned medium (see Section 2.2.5) isolated from human hepatoblastoma Hep G2 cultures (Table 8.1)
- L15 medium (see Section 2.1.3)

Protocol

(a) Seed isolated cells at a moderate cell density of 6000 cells/cm^2 into T-75 culture flasks previously coated for at least 1 h with a mixture of collagen I-fibronectin.
(b) Incubate in a humidified incubator with 3.5% CO_2 at 37°C.
(c) After 4 h, aspirate the medium and unattached cells.

(d) Add 13 ml fresh L15 medium.
(e) After incubation overnight remove the medium and rinse the cells 2–3 times with HBS.
(f) Maintain the primary cultures at 37°C with 3.5% CO_2 in LCM33CM for 1–2 weeks before infection.
(g) Change the medium every 2–3 days.

3.2. Immortalization

3.2.1. Retroviral vector

Use a retroviral vector containing the 2.5-kb BglI/HpaI fragment of the SV40 genome (GenBank accession J02400). This fragment was cloned via BamH1-linkers into the BamH1 site of the pZip Neo SV(X)1 vector [Cepko et al., 1984]. The plasmid has a truncated origin of replication and lacks both the early promoter and the polyadenylation site.

3.2.2. Virus preparation

The high-titer SV40 virus is generated by a modification of the original "ping-pong" infection protocol described by Lynch and Miller involving cocultivation of packaging cell lines with different host ranges [Lynch and Miller, 1991].

Transfection of packaging cell lines and virus collection. Separate $CaCl_2$ transfection [Chen and Okayama, 1987] is performed on $0.5–1 \times 10^6$ ecotropic Psi 2 (gift from the laboratory of R. C. Mulligan) [Mann et al., 1983] and amphotropic PA317 (CRL 9078, ATCC) packaging cell lines grown in DMEM supplemented with 10% FCS, at 37°C, 5% CO_2.

Protocol 3.2.2.1

Reagents and Materials

- DNA/$CaCl_2$ solution: 10 μg of SV40 T-antigen plasmid in 250mM $CaCl_2$
- Sterile 15-ml conical tube
- 2×BBS buffer (see Section 2.2.9)
- 1.5mM Na_2HPO_4
- 10-cm petri dish cultures of Psi-2 and PA317 cells, 70% confluent
- Serum-free PC-1 medium (see Section 2.2.4)
- PBS
- DMEM with 10% FCS

Protocol

(a) Slowly add 500 μl of a DNA/$CaCl_2$ solution to a sterile 15-ml conical tube containing 500 μl of 2×BBS buffer.
(b) Gently mix the solution and incubate for 10–20 min at room temperature until a precipitate appears.
(c) Distribute the precipitate dropwise onto the medium-containing 10-cm plates of Psi 2 and PA317 cells.
(d) Incubate the Psi 2 and PA317 cells for 18 h at 37°C in 3.5% CO_2.
(e) Wash the cells twice with PBS and add fresh medium (DMEM + 10% FCS).
(f) 48 h after transfection, trypsinise and mix the two cell populations together in a 1:1 ratio, 0.5×10^6 cells of each in DMEM with 10% FCS, and incubate for reciprocal ping-pong infection at 37°C, 5% CO_2.
(g) Grow to confluence and split 1:5.
(h) Collect the virus over 18 h from a 70% confluent culture growing in serum-free PC-1 medium.
(i) Collect a second harvest of virus at 80–90% confluence.
(j) Filter through 0.45-μm micropore filter (see Section 3.2.3).
(k) Directly apply the collected virus for virus titration and hepatocyte infection or store frozen at −80°C.

Virus titration

Protocol 3.2.2.2

Reagents and Materials

- NIH 3T3 cells, 8×10^4/60-mm dish
- Viral supernatant from Protocol 3.2.2.1 above
- Polybrene, 1 mg/ml (see Section 2.2.8)
- DMEM10FC: DMEM with 10% FCS

Protocol

(a) Titrate the virus by infecting NIH 3T3 cells with 0.5 ml of 10^{-2}, 10^{-3}, and 10^{-4} dilutions of viral supernatant in the presence of 8 μg/ml polybrene for 2 h at 37°C, 5% CO_2.
(b) Wash the cells with PBS and add fresh DMEM10FC.
(c) 48h later treat the cells with 850 μg/ml Geneticin®.
(d) Calculate the titer (cfu/ml) by multiplying the colonies developing after 10d of Geneticin® selection by the diluting factor.
Expected titer should be in the range 10^5–10^6 cfu (colony-forming units)/ml.

3.2.3. Infection of human hepatocytes

Protocol 3.2.3

Reagents and Materials

- 75-cm^2 flask of replicating liver epithelial cells (corresponding to approximately 1–5$\times 10^6$ cells) (see Section 3.1.2)
- 0.45-μm-filtered high-titer virus (10^6 cfu/ml) collected in serum-free medium [as described in Protocol 3.2.2.1 (g)], 1.5 ml/flask
- Polybrene 1 mg/ml
- HBS (see Section 2.2)
- LCM (see Section 2.2)

Protocol

(a) Incubate replicating liver epithelial cells, 1–2 weeks after isolation, in a 75-cm^2 flask with 1.5 ml of 0.45-μm-filtered high-titer virus collected in serum-free medium [as described in Protocol 3.2.2.1 (g)], in presence of 8 μg/ml polybrene at 37°C, 3.5% CO_2.
(b) After 2 h wash the cultures with HBS and add fresh LCM.

3.3. Subculture of Immortalized Cells

Optimal immortalization efficiency is achieved when the virus-infected cells are cultured at early passages in a modified LCM medium with ornithine replaced with 2mM L-arginine. Use dispase to subculture early passage cells since immediately after transformation the cells are very sensitive to the cell-dissociating enzyme trypsin.

Protocol 3.3

Reagents and Materials

- *Culture medium:* LCM medium with ornithine replaced with 2mM L-arginine
- HBS
- Dispase II (neutral protease) 24 U/ml in PBS
- E-PET (for later-passage cultures)
- Soyabean trypsin inhibitor (SBTI) 30 mg/ml

Protocol

(a) Wash the cells with HBS.
(b) Add dispase II to a final concentration of 2.4 U/ml.
(c) Keep at room temperature for 30 min.
(d) Collect the detached cells in 50 ml HBS.
(e) Centrifuge 5 min at 1000 rpm (200g).

(f) Wash once more with HBS.
(g) Resuspend the cellular pellet in fresh medium.
(h) Reseed at 0.8–1.0$\times 10^6$ cells per 75-cm^2 flask.

To subculture later passage (>10) of immortalized liver epithelial cells use E-PET (see Section 2.1.10).

(a) Wash the cells with HBS and add E-PET solution.
(b) Incubate at room temperature for 5–10 min.
(c) Collect the cells in HBS and centrifuge as described above.
(d) Add 20 μl of 30 mg/ml of SBTI on to the cell pellet before resuspending the cells in fresh medium.
(e) Reseed at 0.8–1.0$\times 10^6$ cells per 75-cm^2 flask.

4. CHARACTERIZATION OF HUMAN LIVER EPITHELIAL CELLS

4.1. Primary Cells

Primary cell cultures of normal liver epithelial cells replicate within 2–5 days after cell isolation in the LCM medium. These cultures usually develop into a confluent monolayer after 10–14 days of incubation. Generally cells average 12 divisions before senescence. The outgrowing cells tend to arise between areas of attached hepatocytes and characteristically are of two different morphologies. One cell type is small and distinctly epithelial-like and has a polygonal shape and distinct nuclear and cytoplasmic characteristics. The second is more spindle-shaped with pale, indistinct cytoplasm and elongated nucleus. The outgrowing cells fill in the areas between the attached hepatocytes, and there is always a progression of single cells to small colonies that evolve into monolayers within 10–14 days of incubation (Fig. 8.1, top left). These cells can undergo as many as four successive subcultures, or an estimated 12 cell divisions, with a cell population doubling of 3days, before division ceases. Immunofluorescence staining of cells after the third passage demonstrates cytokeratin 18 (Fig. 8.1, bottom left) and albumin in 100% and 30–50% of the cell population, respectively. Therefore, the replicating cells could represent "dedifferentiated" hepatocytes or the expansion of biliary epithelial or facultative stem cells as postulated for rodent livers [Miyazaki and Namba, 1991; Thorgeirsson and Evarts, 1992].

4.2. Immortalized Cells

4.2.1. Growth characteristics

Colonies having an altered morphology and high mitotic index arise 6–8 weeks after infection at a frequency of 2×10^{-4}. Virus-exposed

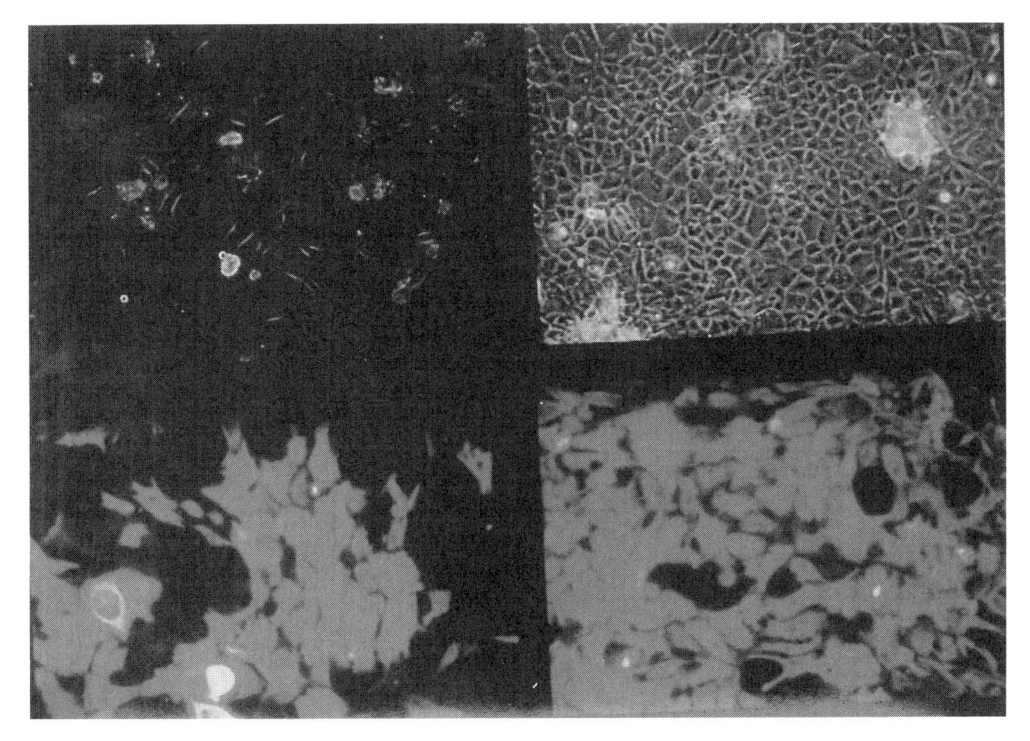

Fig. 8.1. Morphology (top) and cytokeratin 18 immunochemistry (bottom) of normal (left) and SV40-immortalized human liver epithelial cells (right). Indirect immunofluorescence staining for cytokeratin 18 was performed as previously described. [Pfeifer et al., 1993.]

cells divide rapidly in culture, with an average population doubling time of 24–48 h and a colony-transforming efficiency of approximately 15%. Most of the liver epithelial cells do not enter cell crisis, and many of the immortalized lines will have undergone more than 100 cell divisions. Between third and fifth passage, 100% of the cells express SV40 T antigen in their nuclei. After several passages, conditioned medium can be removed from the LCM without modification of the THLE cell growth.

The immortalized liver cells (THLE cells) have a typical epithelial morphology (Fig. 8.1, top right). Karyotype analysis shows that THLE cells are hypodiploid with most karyotypes being near-diploid (Fig. 8.2). Structural alterations such as chromatid breaks, deletions, and acentric fragments may be observed. When previous cell lines were tested for tumor formation by subcutaneous injection of 10^6 cells in athymic nude mice, no tumors were found after 12 months of observation.

4.2.2. Expression of hepatocyte features

Keratin. The epithelial and hepatic origin of the immortalized cells can be demonstrated by specific keratin gene expression. In early-

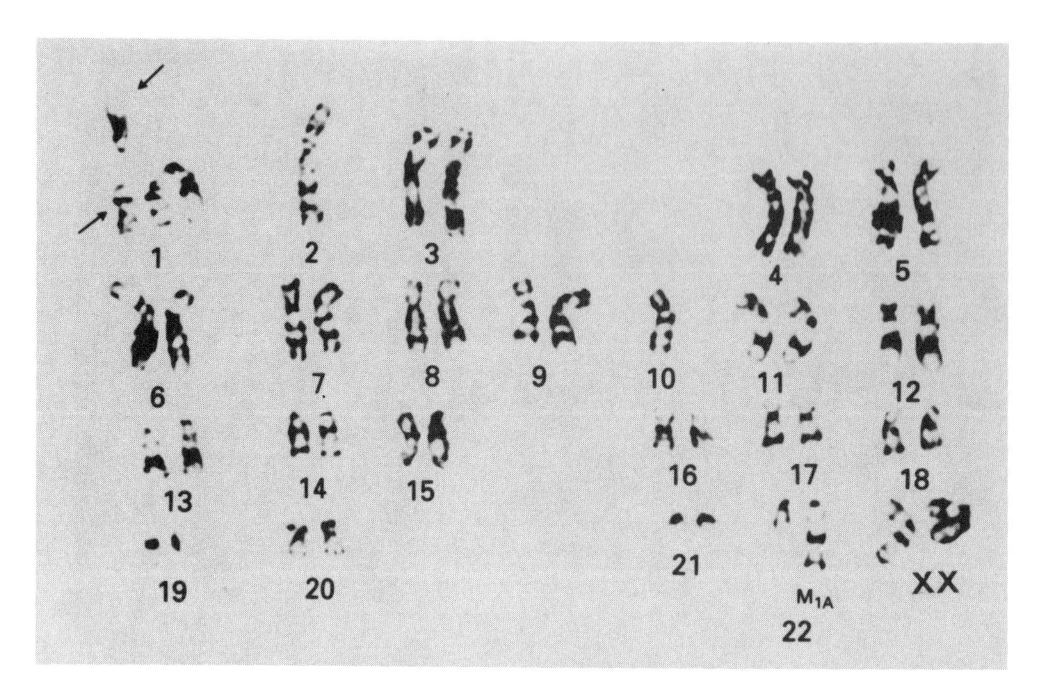

Fig. 8.2. Karyotypes of THLE cells. Monosomy of chromosomes 2 and 10, a break of chromosome 1 (arrow) and a 22q+ translocation leading to the marker chromosome M1A characterize the near-diploid metaphase of THLE cells at passage 18.

passage cells, cytokeratin 18 (Figure 8.1, bottom right) but not cytokeratin 19 is uniformly expressed, whereas at later passages (passage 10–12) all cells also express cytokeratin 19, which is not usually associated with hepatocytes, but present on ductal cells [Germain et al., 1988]. Interestingly, expression of cytokeratin 19 has also been observed in primary cultures of adult liver cells, and has been linked to the presence of high levels of retinoids in the media [Gibson-D'Ambrosio et al., 1993] or the presence of undifferentiated cells.

Albumin. Albumin is readly detectable in the cytoplasm of early passage immortalized liver cells by immunochemistry [Pfeifer et al., 1993]. Islands of albumin-positive cells are surrounded by clusters of less intensely staining cells, indicating different cell clones or types. Although albumin expression tends to decrease with cell passages some cell lines can retain the secretion of albumin (0.07–14.5 ng/ml/day) [Pfeifer et al., 1993]. Similar loss of albumin expression has been observed by Ueno et al. [1993] with several SV40-T-antigen immortalized human epithelial liver cell lines. In agreement with our findings, a subline, previously cloned from one of these immortalized populations, showed relatively stable albumin secretion, indicating the presence of heterogeneous differentiation stages in the culture.

Drug metabolizing enzymes. The metabolism potential of the THLE cells is demonstrated by the detection of DNA-adduct formation of two different chemical classes of carcinogens, benzo[a]pyrene [B(a)P] and aflatoxin B_1 [Pfeifer et al., 1993]. The presence of cytochrome P450 (CYP) 1A1 and 1A2, which are involved in the metabolism of these carcinogens, is confirmed by RNAse mapping and Western blot analysis. CYP1A1 mRNA expression is undetectable in untreated THLE cells but can be induced after B(*a*)P exposure to a similar extent than in Hep G2 cells (Fig. 8.3). As detected by Western blot analysis, the constitutive expression of CYP1A2 in the THLE cell microsomes is approximately 10 times less than in human liver microsomes or in genetically engineered CYP1A2-expressing cells (Fig. 8. 4). Expression of other CYP450, such as CYP3A4, CYP2C has also been detected by RT-PCR, whereas 2E1 and 2A6 has not been detected by Northern and RT-PCR analysis (data not shown).

The THLE cells express mRNA of "phase II" enzymes such as epoxide hydrolase, glutathione S-transferase-π, glutathione peroxidase, NADPH reductase, superoxide dismutase, catalase [Pfeifer et al., 1993], quinone reductase, and aldehyde reductase [E. Offord, unpublished data].

4.2.3. Differentiation stage

Rat oval cells are characterized by the expression of phenotypic markers such as albumin, cytokeratin 18 and 19, γ-glutamyl transpeptidase, α-fetoprotein, and GSTπ [Evarts et al., 1989; Sell, 1990]. The

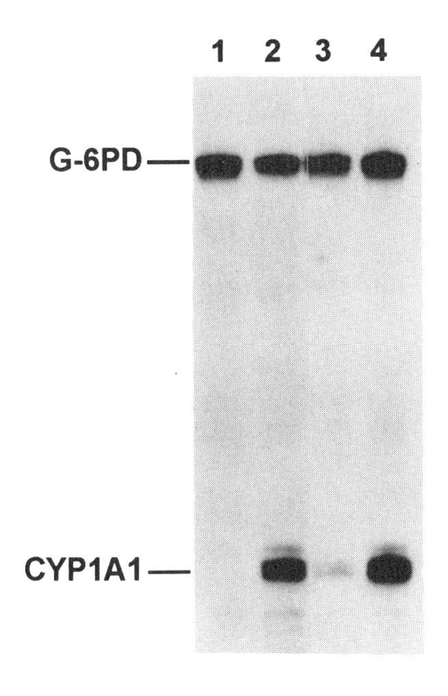

Fig. 8.3. RNase protection analysis of cytochrome P450 1A1 mRNA (CYP1A1) and the control glucose-6-phosphate dehydrogenase mRNA (G-6PD). RNA was from the following: lane 1, untreated THLE cells; lane 2, THLE cells treated with 1.5μM benzo(a)pyrene (B(a)P) for 24 h; lane 3, untreated Hep G2 cells; lane 4, Hep G2 cells treated with 1.5μM B(a)P for 24 h. RNase protection analysis was performed as described previously. Offord EA, Macé K, Rullieux C, Malnoë A, Pfeifer A.M.A. (1995): Rosemary components inhibit benzo(a)-pyrene-induced genotoxicity in human bronchial cells. Carcinogenesis 16:2057–2062.

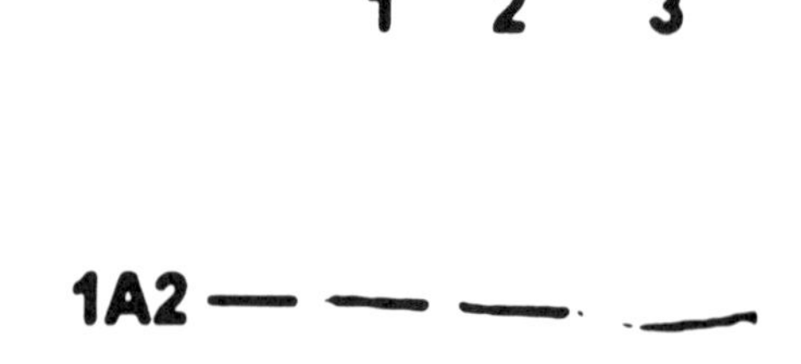

Fig. 8.4. Western blot an-alysis of microsomal extracts. 10 μg of microsomal extract from genetically engineered CYP1A2- expressing AHH-1 TK+/- cells (Gentest Corp., Woburn, MA) (lane 1), from 5 pooled human livers (Human Biologics, Inc., Phoenix, AZ) (lane 2) and 80 μg of microsomes from THLE cells (lane 3) were subjected to electophoresis. The blot was developed with a rabbit antirat CYP1A serum (kindly provided by F. Gonzalez, NIH, Bethesda, MD) as described previously. [Macé et al., 1994.]

immortalized liver cells described here have an epithelial morphology, secrete low amounts of albumin, express cytokeratin 18, GSTπ, and very low γ-glutamyl transpeptidase levels and do not express α-fetoprotein. Moreover, they are uniformly negative for Factor VIII, an endothelial cell marker. Therefore these cells seem to represent a population with a differentiation grade between oval cells and hepatocytes and may be derived from hepatocyte precursors. However, the fact that cytokeratin 18 is expressed and α-fetoprotein is absent in a very early stage of their establishment suggests, more probably, a derivation from differentiated hepatocytes and retrograde differentiation of the cells in culture shown by the appearance of cytokeratin 19 and decreased albumin secretion.

4.3. Conditions for Regulating Growth and Differentiation

Immortalization of liver epithelial cells leads to permanently dividing cells with many cellular and metabolic functions of hepatocytes. Further effort will be necessary for the development of culture techniques and media that will promote the replicating immortalized liver epithelial cells to cease division and, concomitantly, differentiate into cells expressing more differentiated hepatocellular characteristics. This process is well documentated for some cell types such as keratinocytes where an elevated extracellular Ca^{2+} level is generally regarded as a potent inducer of differentiation for primary or immortalized cultures [Boyce and Ham, 1983; Berghard et al., 1990]. For hepatocytes the maintenance of differentiated cultures depends on a variety of conditions, including the nature of the attachment surface, cell–cell interactions, and soluble factors.

4.3.1. Extracellular matrix

Extracellular matrices (ECMs) are insoluble structures composed of three distinct classes of macromolecules: collagens, noncollagenous glycoproteins, (e.g., fibronectin and laminin), and proteoglycans. Several reports indicate that matrix promotes cell differentiation and tissue-specific gene expression [Reid et al., 1986; Lin and Bissell, 1993].

The existence of regional specialization of ECM complicates the choice of an appropriate matrix type for the establishment of highly differentiated hepatocyte cultures. The contrast between interstitial and basement membrane matrix provides the clearest example. Interstitial matrix contains fibrillar collagen types I, II, III, V, and VI associated with fibronectin, while basement membrane is composed of collagen type IV associated with laminin [Schuetz et al., 1988; Bissell and Roll, 1990]. In addition, the matrix types produced, *in vivo*, by liver cells are susceptible to change. Hepatocytes in regenerating liver produce type IV collagen associated with laminin, whereas hepatocytes synthesize collagen types associated with fibronectin, in the quiescent liver [Reid et al., 1988].

Bissell et al., [1986] have shown that only a complex of ECM can support hepatocellular function in culture. Rat hepatocytes plated on a substratum of reconstituted basement membrane, containing collagen IV, laminin and heparan sulfate proteoglycan (hydrated gel matrix from the Engelbreth–Holm–Swarm mouse sarcoma [Orkin et al., 1977] and commercially available as Matrigel®, Collaborative Biochemical Products, Inc.), retain the albumin and CYP450 expression for prolonged periods of culture. In contrast, type I collagen, type IV collagen, laminin, or fibronectin used alone do not maintain a high level of albumin secretion [Bissell et al., 1986]. ECM configuration can also influence the tissue-specific function of cultured hepatocytes. It has been demonstrated that rat hepatocytes sandwiched between two layers of collagen type I gel maintain a level of albumin mRNA similar to that found in the normal liver for at least 6 weeks [Dunn et al., 1992].

Although the mechanism by which ECM induces cell differentiation is not completely elucidated (reviewed in [Piredale and Arthur, 1994]), it appears that ECM components can stimulate cell differentiation by selectively activating transcriptional factors, and that such regulation occurs coordinately with changes in cell shape [DiPersio et al., 1991]. Few data are available concerning interaction of ECM with cultured human hepatocytes. Species differences, especially in hepatic metabolism, restrict the extrapolation from rodent experiments. A recent study indicated that collagen type I gel immobilization supports long-term culture of human hepatocytes with a well-preserved cell morphology, phase II activity (sulfotransferase and glucuronyltransferase), and CYP1A activity over a 2–3-week culture [Knoebe et al., 1994]. The coating that we use for the culture of primary and immortalized liver cells has been previously developed for growth of normal human bronchial epithelial cells [Lechner and La Veck, 1985]. This coating mixture, containing collagen type I, fibronectin, and BSA, allows a high capacity of cell attachment, colony-forming efficiency, and growth rate [Lechner et al., 1982]. With this culture condition, normal human hepatocytes show a faster cell attachment and a higher CYP1A2 and CYP3A4 expression than cells cultured on newborn calf serum-coated plates [M. J. Gòmez-Lechòn, personal communication].

4.3.2. Cell-cell interaction

Cell–cell interactions may provide a stimulus responsible for the long-term maintenance of liver specificity. The need for such cell–cell contacts suggests a key role for plasma membrane interactions and cell–cell signaling. However, the nature of the signal remains unknown but may involve secretion of soluble factors and biomatrix formation by auxiliary cells.

Cell–cell interactions are optimal in a three-dimensional cell architecture. To mimic the three-dimensional cell–cell contact of the parenchymal cells in the liver *in vitro*, a procedure for the culturing of human hepatocytes as multicellular spheroids has been developed. The hepatocyte spheroids were found to retain liver-specific subcellular structures even after prolonged (over 1 month) culturing [Li et al., 1992]. THLE cells have been successfully cultivated in aggregates with or without auxiliary cells and the beneficial effects on the expression of differentiated functions are under evaluation. Preliminary results show an additional increase of albumin expression when THLE cells were cultured in aggregates.

In monolayer cultures, hepatocytes cocultivated with nonparenchymal liver epithelial cells retained the ability to synthesize high levels of liver-proteins for several weeks [Fraslin et al., 1985] and to maintain a reduced but measurable CYP450 expression [Utesch et al., 1991]. It appears that the success of the coculture approach does not require cells derived from liver origin since fibroblastic cell lines, lung, and kidney epithelial cell lines have been successfully used to preserve production of plasma proteins as well as phase I and phase II enzyme activities in rat hepatocytes [Donato et al., 1992b].

In cultures consisting of only one cell type, cell density appears as a critical factor for the induction of differentiation. Human hepatocytes were found to express cytokeratin 18 and albumin more actively in tightly packed cell clusters [Gibson-D'Ambrosio, et al., 1993]. Our own experiments demonstrated that the immortalized human liver epithelial cells showed measurable CYP1A1 and CYP1A2 expression when the cells reached confluence, as detected by RT-PCR and methoxyresorufin *O*-deethylase activity (0.17 ± 0.10 pmol/min/mg total protein).

4.3.3. Soluble factors

Development of hormonally defined media has greatly improved the ability to cultivate many epithelial cell types [Harris, 1987]. Media which support growth of normal human hepatocytes have been developed [Gibson-D'Ambrosio et al., 1993; Pfeifer et al., 1993]. They are supplemented with different additives, such as insulin, transferrin, hydrocortisone, and epidermal growth factor, known to increase the rate of hepatocyte DNA synthesis *in vitro* [Cruise and Michalopoulos, 1985; Luetteke and Michalopoulos, 1987; Lechner et al., 1989; Li et al., 1991]. Additionally some factors present in our medium support the induction of specific liver markers since albumin production of human immortalized liver epithelial cells is 2 times higher in the LCM medium than in RPMI supplemented with 1% FCS. Other soluble factors, improving the expression of differentiated functions of rodent hepatocytes, such as clofibrate, nicotinamide, DMSO, and metyrapone (reviewed in Guguen-Guillouzo [1992]), have been described and will be tested on the immortalized human hepatocytes in culture.

4.3.4. Concluding remarks

Much progress has been made on the isolation and culturing of hepatocytes as well as immortalization of proliferating liver epithelial cells. A second challenge is the induction of a fully differentiated, hepatocyte-like phenotype of these cells. In particular, cells expressing the whole spectra of phase I and II enzymes will require substantial efforts on the improvement of culture conditions. One of our approaches to improve the differentiation of the immortalized cell lines will consist of the reduction, or total removal, of the growth promoting factors present in our medium and in the establishment of organotypic cultures. A second possibility to obtain metabolism-competent hepatocytes constitutes in the reintroduction of the absent or reduced phase I activities by transfection of cDNA-expressing vectors. By this procedure we have already established stable CYP1A2, 2A6, 2B6, 2C9, 2D6, 2E1, and 3A4 expressing-THLE cells with catalytic activities similar to human liver (manuscript in preparation).

REFERENCES

Bayad J, Sabolovic N, Bagrel D, Odoul M, Cassingena R, Siest G (1992): Viral immortalization and phenotypic characterization of a rat hepatocyte cell line. Cell Physiol Biochem 2: 349–358.

Berghard A, Gradin K, Toftgard R (1990): Serum and extracellular calcium modulate induction of cytochrome P450IA1 in human keratinocytes. J Biol Chem 265: 21086–21090.

Berry MN, Friend DS (1969): High yield preparation of isolated rat liver parenchymal cells. J Cell Biol 43: 506–520.

Bissell DM, Arenson DM, Maher JJ, Roll FJ (1986): Support of cultured hepatocytes by a laminin-rich gel. J Clin Invest 79: 801–812.
Bissell DM, Roll J (1990): Connective tissue metabolism and hepatic fibrosis. In Zakin D, Boyer TD (eds): "Hepatology." Philadelphia: Saunders, pp 424–444.
Boyce ST, Ham GH (1983): Calcium-regulated differentiation of normal human epidermal keratinocytes in chemically defined clonal culture and serum-free serial cultures. J Invest Dermatol 81s: 33s–40s.
Butterworth BE, Smith-Oliver T, Earle L, Loury DJ, White RD, Doolittle DJ, Working PK, Cattley RC, Jirtle R, Michalopoulos G, Strom S (1989): Use of primary cultures of human hepatocytes in toxicology studies. Cancer Res 49: 1075–1084.
Cepko CL, Roberts BE, Mulligan RC (1984): Construction and applications of a highly transmissible murine retrovirus shuttle vector. Cell 37: 1053–1062.
Chen C, Okayama H (1987): High efficiency transformation of mammalian cells by plasmid DNA. Mol Cell Biol 7: 2745–2752.
Cruise JL, Michalopoulos G (1985): Norepinephrine and epidermal growth factor: dynamics of their interaction in the stimulation of hepatocyte DNA synthesis. J Cell Physiol 125: 45–50.
DiPaolo JA (1983): Relative difficulties in transforming human and animal cells in vitro. J Natl Cancer Inst 70: 3–7.
DiPersio CM, Jackson DA, Zaret KS (1991): The extracellular matrix coordinately modulates liver transcription factors and hepatocyte morphology. Mol Cell Biol 11: 4405–4414.
Donato MT, Gomez-Lechon MJ, Castell JV (1992a): Biotransformation of drugs by cultured hepatocytes. In Castell JV, Gomez-Lechon MJ (eds): "In Vitro Alternatives to Animal Pharmaco-toxicology." Farmaindustria Madrid: pp 149–178.
Donato MT, Gomez-Lechon MJ, Castell JV (1992b): Co-cultures of hepatocytes: a biological model for long-lasting cultures. In Castell JV, Gomez-Lechon MJ (eds): "In Vitro Alternatives to Animal Pharmaco-toxicology." Farmaindustria Madrid: pp 108–127.
Dunn JCY, Tompkins RG, Yarmush ML (1992): Hepatocytes in collagen sandwich: evidence for transcriptional and translational regulation. J Cell Biol 116: 1043-1053.
Evarts RP, Nagy P, Nakatsukasa H, Marsden E, Thorgeirsson SS (1989): In vivo differentiation of rat liver oval cells into hepatocytes. Cancer Res 49: 1541–1547.
Fraslin J-M, Kneip B, Vaulont S, Glaise D, Munnich A, Guguen-Guillouzo C (1985): Dependence of hepatocyte-specific gene expression on cell-cell interactions in primary culture. EMBO J 4: 2487–2491.
Germain L, Blouin MJ, Marceau N (1988): Biliary epithelial and hepatocytic cell lineage relationships in embryonic rat liver as determined by the differential expression of cytokeratins, alpha-fetoprotein, albumin and cell surface-exposed components. Cancer Res 48: 4909–4918.
Gibson-D'Ambrosio RE, Crowne DL, Shuler CE, D'Ambrosio SM (1993): The establishment and continuous subculturing of normal human adult hepatocytes: expression of differentiated liver functions. Cell Biol Toxicol 9: 385–403.
Guguen-Guillouzo C (1992): Isolation and culture of animal and human hepatocytes. In Freshney RI (ed): "Culture of Epithelial Cells." New York: Wiley-Liss, pp 197–223.
Guillouzo A, Beaune P, Gascoin MN, Begue JM, Campion JP, Guengerich FP (1985): Maintenance of cytochrome P450 in cultured human hepatocytes. Biochem Pharmacol 34: 2995–2997.
Harris CC (1987): Human tissues and cells in carcinogenesis research. Cancer Res 47: 1–10.
Hsu IC Lipsky MM, Cole KE, Su CH, Trump BF (1985): Isolation and culture of hepatocytes from human liver of immediate autopsy. In Vitro Cell Devel Biol 21: 154–160.
Isom HC, Tevethia MJ, Taylor JM (1980): Transformation of isolated rat hepatocytes with simian virus 40. J Cell Biol 85: 651–659.

Kaighn ME, Prince AM (1971): Production of albumin and other serum proteins by clonal cultures of normal human liver. Proc Natl Acad Sci USA 68: 2396–2400.
Koebe HG, Pahernik S, Eyer P, Schildberg FW (1994): Collagen gel immobilization: a useful cell culture technique for long-term metabolic studies on human hepatocytes. Xenobiotica 24: 95–107.
Lafarge-Frayssinet C, Estrade S, Rosa-Loridon B, Frayssinet C, Cassingena R (1984): Expression of gamma-glutamyltranspeptidase in adult rat liver cells after transformation with SV40 virus. Cancer Lett 22: 31–39.
Lechner JF, Babcock MS, Marnell MM, Narayan KS, Kaighn ME (1980): Normal human prostate epithelial cell cultures. In Harris CC, Trump BF, Stoner GD (eds): "Methods in Cell Biology." New York: Academic Press, pp 195–225.
Lechner JF, Haugen A, McClendon IA, Pettis EW (1982): Clonal growth of normal adult human bronchial epithelial cells in a serum-free medium. In Vitro 18: 633–642.
Lechner JF, La Veck MA (1985): A serum-free method for culturing normal human bronchial epithelial cells at clonal density. J Tis Cult Meth 9: 43–48.
Lechner JF, Cole KE, Reddel RR, Anderson L, Harris CC (1989): Replicative cultures of adult human and Rhesus monkey liver epithelial cells. Cancer Detect Prev 14: 239–244.
Li AP, Colburn SM, Beck DJ (1992): A simplified method for the culturing of primary adult rat and human hepatocytes as multicellular spheroids. In Vitro Cell Devel Biol 28A: 673–377.
Li AP, Myers CA, Roque MA, Kaminski DL (1991): Epidermal growth factor, DNA synthesis and human hepatocytes. In Vitro Cell Devel Biol 27A: 831–833.
Lin CQ, Bissell MJ (1993): Multi-faceted regulation of cell differentiation by extracellular matrix. FASEB J 7: 737–743.
Lopez MP, Gomez-Lechon MJ, Castell JV (1991): Glucose: a more powerful modulator of fructose 2,6-biphosphate levels than insulin in human hepatocytes. Biochem Biophys Acta 1094: 200–206.
Luetteke NC, Michalopoulos G (1987): Control of hepatocyte proliferation in vitro. In Rauckman EJ, Padilla GM (eds): "The Isolated Hepatocyte: Use in Toxicology and Xenobiotic Biotransformation." New York: Academic Press, pp 93–118.
Lynch C, Miller D (1991): Production of high titer helper virus-free retroviral vectors by cocultivation of packaging cells with different host ranges. J Virol 65: 3887–3890.
Macé K, Gonzalez FJ, McConnell IR, Garner RC, Avanti O, Harris CC, Pfeifer AMA (1994): Activation of promutagens in a human bronchial epithelial cell line stably expressing human cytochrome P450 1A2. Mol Carcinogen 11: 65–73.
Mann R, Mulligan RC, Baltimore D (1983): Construction of a retrovirus packaging mutant and its use to produce helper-free defective retrovirus. Cell 33: 153-159.
Miyazaki M, Namba M (1991): Liver cell cultures and their utilization for hepatocarcinogenesis. Okayam Igakkai Zasshi 103: 337–348.
Offord EA, Macé K, Ruffieux C, Malnoè A, Pfeifer AMA (1995): Rosemary components inhibit benzo(a)pyrene-induced genotoxicity in human bronchial cells. Carcinogenesis 15: 2057-2062.
Orkin RW, Gehron P, McGoodwin EB, Martin GR, Valentine T, Swarm R (1977): A murine tumor producing a matrix of basement membrane. J Exp Med 145: 204–220.
Pfeifer AMA, Cole KE, Smoot DT, Weston A, Groopman JD, Shields PG, Vignaud JM, Juillerat M, Lipsky MM, Trump BF, Lechner JF, Harris CC (1993): Simian virus 40 large tumor antigen-immortalized normal human liver epithelial cells express hepatocyte characteristics and metabolize chemical carcinogens. Proc Natl Acad Sci USA 90: 5123–5127.
Piredale J, Arthur MJP (1994): Hepatocyte-matrix interactions. Gut 35: 729–732.
Reid LM, Narita M, Fujita M, Murray Z, Liverpool C, Rosenberg L (1986): Matrix and hormonal regulation of differentiation in liver cultures. In Guillouzo A, Guguen-Guillouzo C (eds): "Isolated and Cultured Hepatocytes." Paris: John Libbey Eurotext Ltd., INSERM, pp 225–258.

Reid LM, Abreu SL, Montgomery K (1988): Extracellular matrix and hormonal regulation of synthesis and abundance of messenger RNAs in cultured liver cells. In Arias IM, Jakoby WB, Popper H, Schachter D, Shafritz DA (eds): "The Liver: Biology and Pathology." New York: Raven Press, pp 717–737.

Schuetz EG, Li D, Omiecinski CJ, Muller-Eberhard U, Kleinman KK, Elswick B, Guzelian PS (1988): Regulation of gene expression in adult rat hepatocytes cultured on a basement membrane matrix. J Cell Physiol 134: 309–323.

Sell S (1990): Is there a liver stem cell? Cancer Res 50: 3811–3815.

Thorgeirsson SS, Evarts RP (1992): Growth and differentiation of stem cells in adult rat liver. In Sirica A (ed): "The Role of Cell Types in Hepatocarcinogenesis." Boca Raton, Press, FL: CRC pp 109–120.

Ueno T, Miyamura T, Saito I, Mizuno K (1993): Immortalization of differentiated human hepatocytes by a combination of a viral vector and collagen gel culture. Human Cell 6: 126–136.

Utesch D, Molitor E, Platt K-L, Oesch F (1991): Differential stabilization of cytochrome P450 isoenzymes in primary cultures of adult rat liver parenchymal cells. In Vitro Cell Devel Biol 27A: 858–863.

Woodworth C, Secott T, Isom HC (1986): Transformation of rat hepatocytes by transfection with simian virus 40 DNA to yield proliferating differentiated cells. Cancer Res 46: 4018–4026.

Woodworth CD, Isom HC (1987): Transformation of differentiated rat hepatocytes with adenovirus and adenovirus DNA. J Virol 61: 3770–3779.

APPENDIX: MATERIALS AND SUPPLIERS

Product	Catalog No.	Supplier
Arginine	A3784	Sigma Chemical Co.
Benzo(a)pyrene (B(a)P)	B1760	Sigma Chemical Co.
BES	391334	Calbiochem
Bovine pituitary extract	210	Biofluids Inc.
BSA	343	Biofluids Inc.
Calcium chloride	C7902	Sigma Chemical Co.
Chemically denatured serum	10201	Upstate Biotechnology
Cholera toxin	8052	Sigma Chemical Co.
Collagenase type P	1213-857	Boehringer Mannheim
Collagen type I	PCO-701	Collagen Corp.
Dispase II	165-859	Boehringer Mannheim
DMEM medium	41965-039	Life Technologies
Epidermal growth factor	E4127	Sigma Chemical Co.
E-PET	369	Biofluids Inc.
FBS	3014-32	Paragon Biotechechnology
FCS	10084-077	Life Technologies
Fibronectin	F2006	Sigma Chemical Co.
Gauze, sterile	2317	Johnson & Johnson
Geneticin®	11811-049	Life Technologies
Gentamicin	043-05710 D	Life Technologies
HBS solution	340	Biofluids Inc.
HBSS solution	24221-32	Paragon Biotechechnology
HEPES	305	Biofluids Inc.
Hydrocortisone	346	Biofluids Inc.
Insulin	350	Biofluids Inc.
L15 medium	29150-34	Paragon Biotechechnology

Appendix *(Continued)*

Product	Catalog No.	Supplier
Ornithine	06503	Sigma Chemical Co.
Pasteur pipettes	14672-2000	VWR Scientific
PBS	14040-091	Life Technologies
PC-1 medium	77232	Ventrex Laboratories
P/E353	Biofluids Inc.	
PMFR-4 low calcium	145	Biofluids Inc.
Polybrene	H9268	Sigma Chemical Co.
Polyvinyl pyrrolidone	PVP40	Sigma Chemical Co.
Retinoic acid	348	Biofluids Inc.
Sodium chloride	5886	Sigma Chemical Co.
Sodium phosphate dibasic	S5136	Sigma Chemical Co.
Soybean trypsin inhibitor (SBTI)	367	Biofluids Inc.
Triiodothyronine	354	Biofluids Inc.
Transferrin	352	Biofluids Inc.
Tygon tubing	63012	VWR Scientific

9

Thyroid Epithelium

David Wynford-Thomas

Cancer Research Campaign Thyroid Tumour Biology Research Group, Department of Pathology, University of Wales College of Medicine, Cardiff CF4 4XN, Wales

1. BACKGROUND

The human thyroid gland contains two populations of epithelial cells—the follicular cells, derived from endoderm, which synthesize

Culture of Immortalized Cells, Edited by R. Ian Freshney and Mary G. Freshney.
ISBN 0-471-12134-7

the two main thyroid hormones, thyroxine and triiodothyronine, and the "C" or parafollicular cells, of neuroendocrine differentiation, which secrete calcitonin [Werner and Ingbar, 1978]. The latter, however, amount to a tiny subpopulation (< 0.1% in the human) and can be ignored for the purposes of this chapter.

In the normal adult thyroid gland, mitotic activity is uniformly very low [Smith and Wynford-Thomas, 1989], as the follicular epithelium, like that of liver, behaves, from a cell kinetic standpoint, as a "conditional renewal" population [Wright and Alison, 1984]. Although there are some quantitative variations, the entire follicular cell population is also essentially in a single differentiation state. This striking homogeneity of growth and function contrasts with the complex hierarchies observed in many widely studied stem cell epithelia such as breast or colon, and represents one of the key experimental advantages of thyroid as a model epithelium.

A second advantage derives from its tissue architecture. The follicular cells are organized in discrete structural units (follicles) consisting of hollow spheroids lined by a single layer of epithelium and filled with its secretory product, thyroglobulin. These are embedded in a vascular stroma [Werner and Ingbar, 1978]. Use of a suitable digestion–disaggregation procedure, as described below, allows the epithelium to be released as intact follicles, while digesting the stroma to single cells. A simple differential sedimentation based on the difference in size (and hence sedimentation rate) between follicles and single cells can then be used to achieve a high degree of separation of epithelium from stroma [Williams et al., 1987, 1988].

Unfortunately, these attractions are somewhat offset by the very limited proliferative potential of thyroid follicular cells in tissue culture (rarely more than three population doublings) [Dumont et al., 1992; Wynford-Thomas, 1993], which reflects a similar limitation in the intact gland [Wynford-Thomas et al., 1982]. Several laboratories, including our own, have extensively investigated the effects of manipulating physical and biochemical tissue culture conditions, largely to no avail. Success has been achieved, however, in greatly extending proliferative life-span by the stable expression of viral oncogenes, notably SV40T [Whitley et al., 1987; Lemoine et al., 1989; Wynford-Thomas et al., 1990; Wyllie et al., 1992]. This chapter summarizes our experience with this approach, the optimum protocols for obtaining clones with extended life-span, and the limitations which we have experienced in the long-term culture of such cells.

2. REAGENTS AND MEDIA

RPMI1640 medium. Where antibiotics are used, add 100 U/ml penicillin and 100 μg/ml streptomycin (both from Sigma).

Enzyme mixture for tissue digestion. Prepare fresh on day of use.

Collagenase	50 mg
Dispase	60 mg
RPMI1640, serum-free	60 ml

Filter-sterilize; warm to 37°C immediately before use.

Polybrene. Prepare 0.8 mg/ml solution of polybrene in water. Filter-sterilize. Store at 4°C.

Geneticin (G418). Prepare a 100× stock solution of G418 at 40 mg/ml in 0.1M HEPES.

Plasmid DNA. Closed circular plasmid prepared ideally by alkaline lysis method followed by purification by $CsCl_2$ gradient isopycnic ultracentrifugation [Sambrook et al., 1989], or by ion exchange on commercial resins (e.g., Qiagen or Wizard). Prepare stock solution at 1 mg/ml in TE. Sterilize by centrifugation (12,000*g*, 10 min. in microcentrifuge) in sterile 1.5-ml microtube to pellet microorganisms. Collect supernatant and repeat.

10mM Tris-HCl, 1mM EDTA (TE). Prepare from stock solutions of 1M Tris-HCl (pH8.0) and 0.5M EDTA (pH 8.0) (both sterilized by autoclaving). Store at 4°C.

2M $SrCl_2.6H_2O$

Strontium chloride	10.66 g
H_2O	15 ml

Make up to 20 ml with H_20. Filter-sterlize. Store aliquots at −20°C.

Hepes buffered saline, 2× (2× HBS)

NaCl	1.64 g
$Na_2HPO_4.7H_2O$	0.04 g
H_2O	90 ml

Add 5 ml of 1M HEPES. Adjust pH to 7.8 (accurately) with 0.5M NaOH. Make up to 100 ml with H_2O. Filter sterilize and store at 4°C (up to 3 months).

Glycerol

Glycerol	15 ml
Hanks' BSS	85 ml

Sterilize by autoclaving. Store at 4°C.

Phosphate-buffered saline, without Ca^{2+} and Mg^{2+} (PBSA)

NaCl	45 g
$Na_2HPO_4 \cdot 12H_2O$	13.4 g
$NaH_2PO_4 \cdot 2H_2O$	1.45 g
H_2O	5 liters
Adjust pH to 7.2.	

DAB/H_2O_2 solution

10× DAB stock solution

Diaminobenzidine	0.083 g
PBSA	10 ml

Store at −20°C.
Working solution
Dilute stock 1:10 in PBSA and add 3.4 μl H_2O_2 per 10 ml.

3. PRIMARY CULTURE

The best source of tissue is freshly excised surgical material. We usually use macroscopically normal regions of thyroid lobectomies performed for non malignant disease (ideally isolated "cold" nodules), sampling as far from the lesion as possible. Sections are taken for subsequent histologic confirmation of normality. Success has also been achieved with postmortem material, although viability is, of course, less reproducible.

The sample is transported to the tissue culture laboratory in a universal container containing ice-cold Hanks' balanced salt solution, plus antibiotics (penicillin 100 U/ml; streptomycin 100 μg/ml) (HBSS). Speed is not particularly crucial since viability seems unimpaired up to several hours in this state.

Tissue is processed as follows (see summary in, Fig. 9-1) in a Class II cabinet (see Safety Precautions and Chapter 2) under sterile conditions.

Protocol 3

Reagents and Materials

Sterile

- HBSS
- Ice-cold HBSS
- RPMI medium, serum-free, ice-cold
- " " containing 5% FCS
- " " containing 10% FCS
- Enzyme mixture: collagenase plus dispase in serum-free RPMI (see Section 2)

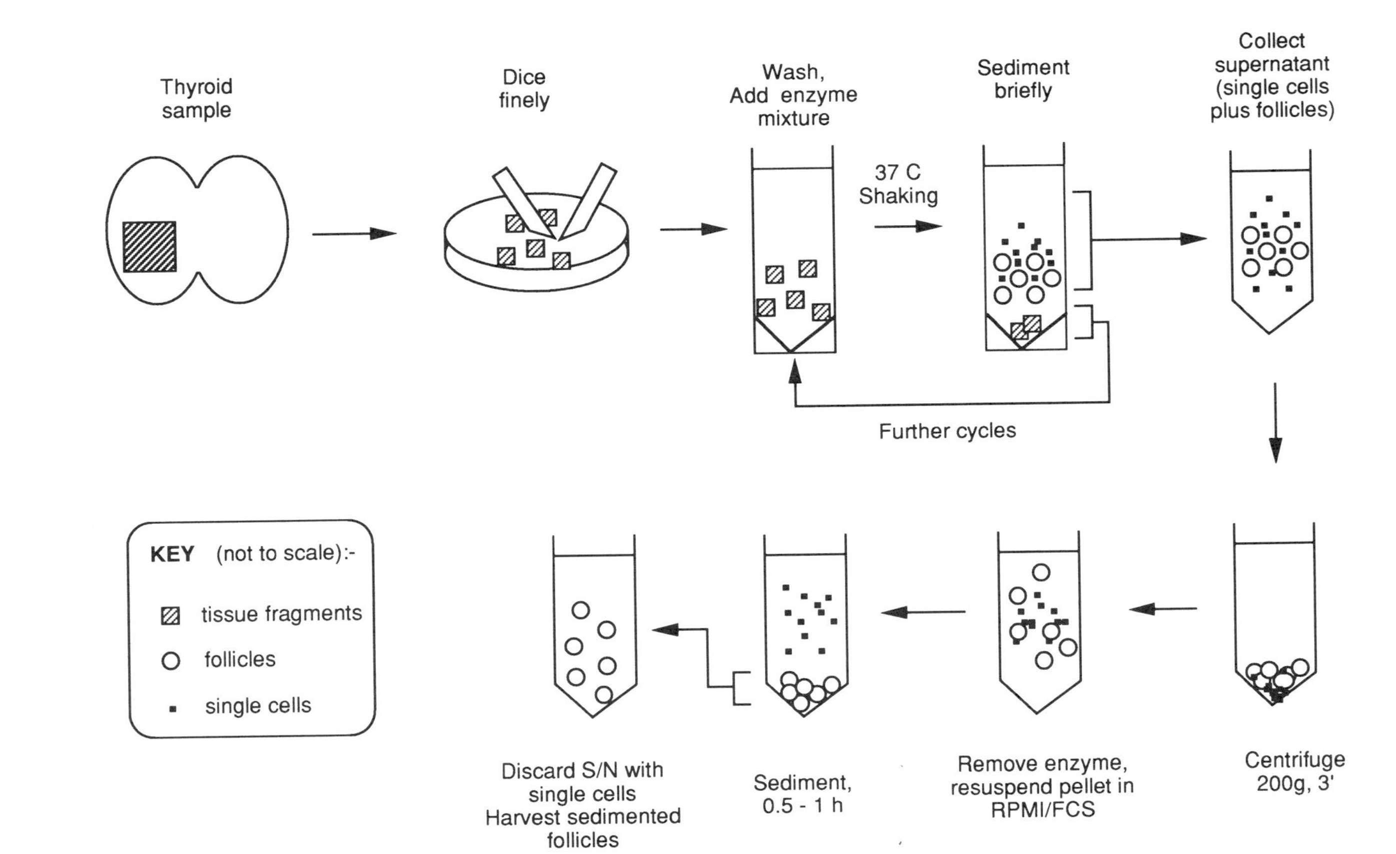

Fig. 9.1. Schematic outline of primary thyroid epithelial culture preparation. (Some steps omitted for clarity).

- Freezing mixture: 1:4 DMSO:FCS at 4°C
- 90-mm plastic petri dishes
- 15-ml centrifuge tubes
- Universal container or 30–50-ml centrifuge tube
- Scalpel and blades, No. 22

Nonsterile

- Ice

Protocol

(a) Rinse twice in HBSS to reduce blood and surface contamination.
(b) Transfer to 90-mm plastic petri dish containing a few milliliters of ice-cold HBSS.
(c) Trim off any connective tissue.
(d) Dice the tissue as finely as possible (~2-mm^3 pieces) using "crossed" scalpel blades. Ensure that tissue remains moist.
(e) Transfer the fragments in HBSS to a suitable container (e.g., a 25-ml universal container for up to 2 g of tissue) using a 25-ml plastic pipette.
(f) Wash 3 times by gentle agitation in ~15-ml ice-cold RPMI medium (serum-free) to remove as much blood as possible, allowing pieces to settle to allow removal of supernatant.
(g) Note: The procedure may be temporarily suspended after this step for up to 18 h, (e.g., overnight). Top up the tube with RPMI, seal, and place on ice until restart.
(h) Wash fragments in a minimal volume of enzyme mixture.
(i) Resuspend in 10 volumes of prewarmed enzyme mixture.
(j) Place in 37°C static waterbath.
(k) Remove tube every 15 min and agitate gently for 20 s.
(l) After 1h harvest the first "fraction" from the digest as follows:
    (i) Remove tube from waterbath, wipe with 70% ethanol, agitate for 20 s, then allow undigested tissue fragments to sediment under gravity.
    (ii) Using a plastic (not glass) 5-ml pipette, carefully remove and save supernatant (containing released single cells and follicles) to a 15-ml centrifuge tube (Falcon).
    (iii) Store on ice.

Note: Strands of sticky connective tissue that occasionally contaminate the supernatant can be conveniently removed by stirring with a glass Pasteur pipette.

(m) Add 10 volumes of fresh, prewarmed enzyme mixture to the remaining tissue fragments and continue the digestion process.
(n) Harvest at 30-min intervals until disaggregation is complete.
(o) While digestion of the next fraction is proceeding, remove the enzyme solution from the supernatant of step (l) by centrifuga-

tion (200g for 2 min in a swinging-bucket rotor) followed by resuspension in ~1ml of RPMI medium containing 5% FCS (to reduce clumping).

(p) Remove a small aliquot (~10 μl) for examination by phase contrast. The content of follicles should reach a maximum from fraction 3 onward. Digestion is normally complete within 3–4 h.

(q) Allow the suspended mixtures of single cells and follicles [step (o)] to sediment on ice for at least 45 min.

(r) Carefully remove most of the supernatant containing single cells and erythrocytes and discard.

(s) Progressively pool the remaining pellets as successive fractions are processed.

(t) Make up to 5 ml with cold RPMI/5% FCS. Rinse tubes with further 5 ml and add to first.

(u) Centrifuge (200g, 3 min) and resuspend in required volume of RPMI containing 10% FCS. Since the preparation is in the form of follicles, cell number cannot be assessed by the usual methods. We have found the simplest guide is to estimate the volume of the follicle pellet: 10^6 cells occupies approximately 8 μl.

(v) Plate out and/or freeze as necessary.

Note: For freezing we resuspend at ~4×10^6 cells per ml in ice-cold RPMI/10% FCS, add an equal volume of freezing mixture and freeze at ~2×10^6 cells per 1-ml vial using standard procedures.

4. IMMORTALIZATION BY SV40 T

Although stable expression of SV40 T can be obtained by conventional plasmid transfection [Whitley et al., 1987; Lemoine et al., 1989; Wynford-Thomas et al., 1990], normal thyroid follicular cells are notoriously refractory to these techniques. Efficiency and reproducibility are enormously improved by the use of a retrovirus-based method of gene delivery, which is now routine in our laboratory and is the principal approach described here. A transfection method is however outlined later for those not wishing to work with retroviruses.

4.1. Retrovirally Mediated Gene Transfer

4.1.1. Principle

The cDNA encoding the large-T region of SV40 is introduced into the target cell in the form of a replication-defective retrovirus, engineered to express only this sequence, and a suitable selectable marker gene, in this case "neo" (permitting selection in G418). In the vector used here, large T and neo are both driven by the retroviral long-terminal repeat promoter/enhancer, the two transcripts being generated by alternative splicing (see Brown and Scott [1987] for principles of vector construction). Since the vector lacks most of the normal structural protein sequences of the virus, these have to be provided in *trans,* by a murine "packaging" line. Depending on the nature of the envelope protein encoded, the resulting virus may be restricted to rodent target cells (ecotropic) or have wider tropism, including human cells (amphotropic).

4.1.2. Source of vector

The SV40 vector used here, SVU19-5 [Jat and Sharp, 1986], was originally obtained in the ecotropic form, in packaging line psi-2, from Dr. P. Jat (Ludwig Institute, London). For human use, we introduced the ecotropic retrovirus into the amphotropic packaging line psi-CRIP [Danos and Mulligan, 1988]. Clones were selected in G418 and screened for titre of amphotropic vector using a human epithelial cell line A431 as target. The highest titer producer clone (1.3×10^4 cfu/ml on A431; $>10^7$cfu/ml on NIH3T3) designated psi-CRIP-SVU19clone8, was used in subsequent experiments. psiCRIPneo [Bond et al., 1994] was used as a negative control.

4.1.3. Gene transfer

Safety note: Procedures (a)–(e) must be performed under ACGM Category II containment guidelines (see below and Chapter 2).

Protocol 4.1.3

Reagents and Materials

- Psi-CRIP-SVU19/8 producer cells
- DMEM medium supplemented with 10% calf serum, G418, 0.4 mg/ml
- 2:1:1 medium: mixture of DMEM:Ham's F-12:MCDB104, 2:1:1, supplemented with 10% FCS with 0.4 mg/ml G418.
- 2:1:1 medium without G418
- 2:1:1 medium containing polybrene at 8 μg/ml. prewarmed
- 25-cm^2 flasks
- 60-mm dishes

Protocol

(a) Several days prior to infection, plate out psi-CRIP-SVU19/8 producer cells at a density (usually ~ 5×10^5 per 25-cm^2 flask) sufficient to give a near-confluent monolayer by the time of infection. These cells, being NIH3T3 derivatives, are cultured in DMEM supplemented with 10% calf serum (rather than FCS). We also routinely maintain them in 0.4 mg/ml G418 to promote retention of the vector.

(b) Two days prior to infection, plate out primary human thyroid follicular cells (derived as above) at a density of ~ 5×10^5 cells per 60-mm dish. For these experiments we have found the optimum medium to be a 2:1:1 mixture of DMEM : Ham's F-12:MCDB104 supplemented with 10% FCS.
(c) Eighteen hours prior to infection, remove the medium from the producer cells, wash, and refeed with 2:1:1 medium plus 10% FCS but *without* G418.
(d) One hour prior to infection, refeed the primary cells with fresh, prewarmed 2:1:1 medium containing polybrene at 8 μg/ml.
(e) To carry out the infection:
    (i) Harvest virus-containing medium from the producer cells (discard cells).
    (ii) Centrifuge (500g for 5 min), and filter through 0.45 μm syringe-end filter to remove debris and any floating cells.
    (iii) Add 0.01 volumes of 0.8 mg/ml polybrene solution and mix well
    (iv) Remove medium from thyroid target cells. Add 1 ml of filtered viral medium from (iii) to each 60-mm dish.
    (v) Return dishes to incubator. Gently rock every 30 min for 2 h. Then add 4 ml of 2:1:1 medium plus 10% FCS (without G418) to each dish (without removing the virus medium).
    (vi) After 24 h, change to fresh growth medium.

4.2. Selection

The initial selection is based purely on the increased proliferative potential of clones expressing large T compared to the very limited capacity of the normal surrounding monolayer. G418 is introduced only once early colonies have formed; we have found empirically that earlier use results in greatly diminished final yield.

Protocol 4.2

Reagents and Materials (Sterile)

- 2:1:1 Medium containing 0.4 mg/ml G418
- 60-mm dishes
- 100-mm dishes

Protocol

(a) Four days after infection, passage each 60-mm dish into a 100-mm dish (standard trypsinization procedure).
(b) Refeed every 3–4days. After 3–4 weeks, check for the appearance of proliferating epithelial colonies against the quiescent background (by phase-contrast examination).
(c) When colonies are evident (3–6 weeks), refeed with medium containing 0.4 mg/ml G418 to kill surrounding cells, which will

take 7–10days. (Note: This results in killing of some epithelial colonies as well, most probably due to occasional failure of co-expression of neo and T through unbalanced transcript splicing.)

(d) Once the colonies have reached ~500 cells, ring-clone (standard trypsin method, see Chapter 7) and transfer to 1.5-cm multiwell plates.

4.3. Characterization

Colonies should be readily recognizable under phase contrast by their "mosaic" or "cobblestone" appearance typical of well-differentiated epithelium (Fig. 9.2b). The morphology closely resembles that of normal thyroid cultures (Fig. 9.2a), except for the presence of frequent mitoses. The only major pitfall is confusion with a second type of colony, which arises from what we believe to be a metaplastic follicular cell variant [Bond et al., 1993]. These, however, are readily distinguishable by their much more scattered cell distribution and fusiform–angular cell shape.

To provide an objective characterisation at the earliest stage possible, we use an immunocytochemical approach. Table 9.1 lists the markers analyzed, the antibodies used, and the expected result.

Cells are subcultured onto coverslips (e.g., Thermanox, ICN Biomedicals) and processed as follows:

Protocol 4.3

Reagents and Materials (Non-sterile)

- Methanol:acetone, 1:1, ice-cold
- 0.6% BSA in PBSA
- PBSA

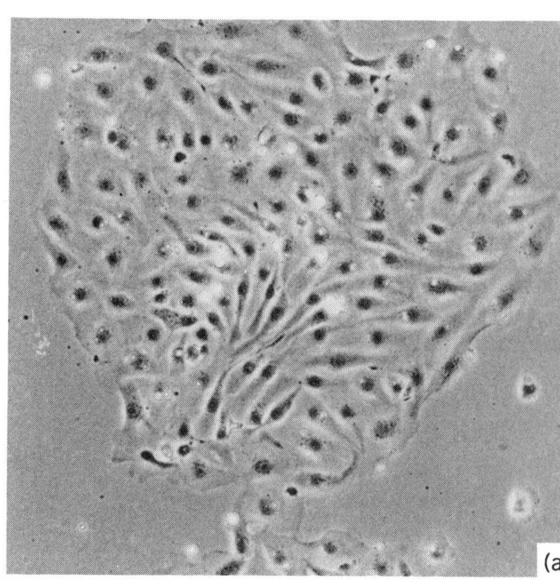

(a)

Fig. 9.2. Typical morphology of (a) normal primary thyroid epithelial culture 3 days after plating; (b) early well-differentiated clone expressing SV40 T (around 10 PD); (c) late dedifferentiated subclone derived from (b) (around 30 PD). (Phase contrast.)

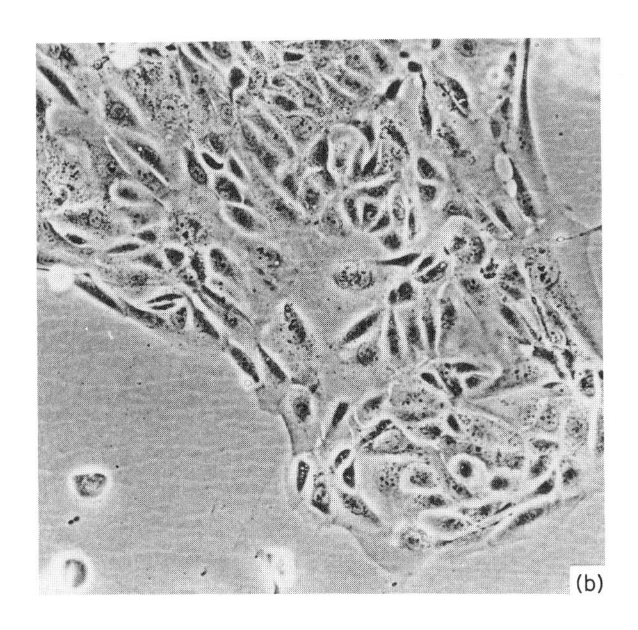

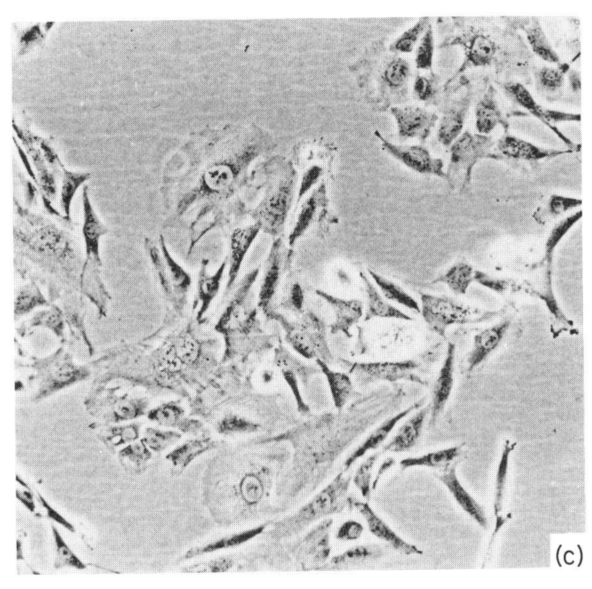

Fig. 9.2. *(Continued).*

Table 9.1. Characterization of SV40-immortalized cell lines

Marker	Antibody	Reference or source	Staining Pattern
SV40 T	PAb419	Harlow et al. [1981]; D. Lane	Nuclear
Cytokeratin 8	LE41	Lane [1982]	Cytoskeleton
Cytokeratin 18	LE61	Lane [1982]	Cytoskeleton
E-Cadherin	HECD-1	R&D Systems	Intercellular
Thyroglobulin	M781	Dako Ltd.	Cytoplasmic (declines rapidly with passage)

- Secondary antibody/peroxidase conjugate, i.e., rabbit antimouse immunoglobulin/HRP

Protocol

(a) Fix in ice-cold methanol:acetone, 1:1, for 10min.
(b) Air-dry. (Coverslips can be stored indefinitely at −20°C in this state.)
(c) Indirect immunoperoxidase procedure (outline):
  (i) Rehydrate in 0.6% BSA in PBSA.
  (ii) Add primary mouse monoclonal antibody, diluted as appropriate in BSA/PBSA.
  (iii) Incubate at room temperature for 45-min.
  (iv) Wash 3 times in PBSA.
  (v) Add secondary antibody/peroxidase conjugate.
  (vi) Incubate, at room temperature for 45-min.
  (vii) Reveal sites of antibody binding by incubation in peroxidase substrate (e.g., diaminobenzidine/hydrogen peroxide solution).

4.4. Clonal Evolution

Typically, infection of each dish of 5×10^5 primary cells with the above vector will generate up to 10 well-differentiated epithelial colonies with the characteristics shown in Table 9.1 and Fig 9.2b. This low yield, even with an optimised vector, is to be expected given the very low proliferative activity of the normal cells (^{3}H-thymidine labelling index usually < 5%) and the dependence of retroviral vectors on existing host cell DNA synthesis for stable integration.

Most colonies will be easily visible by phase contrast by 4–6 weeks after infection. The kinetics of colony development reveals an initial, unexplained delay, perhaps reflecting the requirement for additional genetic events, which is not seen, for example, with other oncogenes such as mutant H-*ras* [Bond et al., 1994].

After 5–15 population doublings, most colonies (whether ring-cloned or still in their original dishes) undergo a marked slowing of growth and eventually senesce. Despite extensive manipulation of media conditions, we have not succeeded in reproducibly overcoming this problem. However, if a sufficiently large number of clones is generated, some will be found that exhibit a much longer lifespan. Out of 29 generated in one experiment, for example, we found 3 that grew well up to at least 25 population doublings. If pooling of colonies is acceptable, enough cells can therefore be generated for most purposes. The only variable that we have found to influence colony lifespan is the age of the donor thyroid. Use of a neonatal gland, for example, generated one clone that reached at least 60 PD without senescence or loss of differentiation [Wyllie et al., 1992].

In some clones (around 50%), a striking evolution occurs before or around the time of senescence, characterized by the emergence of a rapidly growing subclone, which soon becomes the predominant cell type. These cells have a distinct angular–fusiform morphology and form "loose" colonies easily distinguishable from the classic type (Fig. 9.2c). Along with this morphological evidence for loss of differentiation, immunocytochemical analysis shows absence of all the phenotypic markers listed above, except (curiously) cytokeratin 18. These cells always enjoy a greatly extended lifespan, and in many cases have generated apparently immortal lines.

Although a hindrance from the practical standpoint, this dedifferentiation phenomenon has proved interesting biologically. Its tight correlation with escape from senescence, together with its high frequency (at least 1 per 3×10^3 cell divisions) suggests that it represents a spontaneous switch in gene expression regulating both proliferation and differentiation, rather than a second mutational event. The more general implication is that it may be a model for an epigenetic mechanism of progression from well- to undifferentiated thyroid cancer [Bond et al., submitted].

5. ALTERNATIVE PROCEDURES

5.1. DNA Transfection

As stated above, normal thyroid cells are extremely resistant to all transfection protocols. Nevertheless, if efficiency of yield is not important, colonies can be obtained from large-scale experiments using the protocol below. We use the strontium phosphate coprecipitation method [Brash et al., 1987]; the earlier calcium phosphate technique is toxic to some primary cell types. In both cases the principle is to form a fine coprecipitate of DNA with an insoluble inorganic phosphate, which is then phagocytosed by the target cells. The plasmid used can be any mammalian expression vector coding for SV40 T. We have used the plasmid equivalent of the retroviral vector, pZIPSVU19, and a simpler vector lacking neo sequences: SV40ori- [Gluzman et al., 1980] as well as a variety of *ts* mutants (see below).

Protocol 5.1

Reagents and Materials

- Growth medium: 2:1:1/10% FCS (see above, Section 4.1)
- 1:1, DMEM: F12

- 2× HBS
- Plasmid stock in TE
- HBSS (prewarmed to 37°C)
- 15% glycerol dissolved in HBSS
- 60-mm dishes

Protocol

(a) Plate out thyroid monolayers at approximately 5×10^5 cells per 60-mm dish in 2:1:1/10% FCS growth medium (at least 10 replicate dishes). Allow 2–3 days for attachment and onset of DNA synthesis before transfection.

(b) On the day of transfection (>4 h before) remove medium and refeed with fresh prewarmed medium (to ensure optimum pH). Use 1:1, DMEM: F12 to avoid possible effect on pH of the HEPES present in MCDB104.

(c) Transfect monolayers with plasmid coding for SV40T:

   (i) Prepare plasmid solution. For each 60-mm dish, take 5 μg of (presterilized) plasmid solution (usually kept as a stock at, e.g., 1 mg/ml in TE), make up to 219 μl with *dd*H_2O and add 31 μl of 2M strontium chloride. This is "solution A."

   (ii) Take an equal volume (i.e., 250 μl per dish) of 2× HBS–"solution B"–into a sterile 50-ml polypropylene centrifuge tube.

   (iii) Agitating constantly, add solution A *dropwise* to solution B. This can be conveniently accomplished by dispensing it slowly from a syringe through a 23G (23-gauge) needle, while bubbling sterile air into solution B via a plastic 1-ml pipette (precipitate will stick to glass) attached to an automatic pipettor.

   (iv) Allow precipitate to form without further agitation at room temperature until visibly cloudy (usually around 5–10 min). The suspension should appear as a fine dust.

   (v) Mix by gentle pipetting. Draw up suspension into a 2-ml syringe and "sprinkle" over surface of medium in dishes through. a 18G needle. Add 0.5 ml of suspension to each 60-mm dish containing 4.5 ml of medium. Minimise the length of time dishes are out of incubator to avoid loss of CO_2. (The efficiency of transfection is markedly affected by changes in the quality of the precipitate due to pH fluctuations.)

   (vi) Incubate dishes in CO_2 incubator at 37°C for 90 min to allow adsorption of the coprecipitate.

   (vii) Remove the medium and wash off unadsorbed precipitate by rinsing twice with HBSS (prewarmed to 37°C).

(d) Glycerol shock.
   (i) Remove HBSS and to each dish add 2 ml of 15% glycerol dissolved in HBSS.
   (ii) Leave at room temperature for 30–60 s (depending on observed sensitivity to toxic effect of glycerol).
   (iii) Wash off by rinsing 3 times with HBSS.
   (iv) Refeed with appropriate growth medium.
(e) Select colonies as for retroviral gene transfer.

5.2. Conditional Immortalization

Unfortunately most successful immortalizing genes (particularly SV40 T) have pleiotropic actions and may perturb the growth and differentiation states of the cell to such an extent that it ceases to be a valid representation of the normal. The great attraction of conditional expression is that it should, in principle, permit study of a cell line in the *absence* of such interference by its immortalizing gene.

There are a series of well-characterized mutants of the large-T encoding gene of the DNA tumour virus SV40 [Tegtmeyer, 1975] that code for proteins exhibiting temperature-dependent immortalizing function. One such mutant (tsA58) has been widely used for conditional immortalization of fibroblasts [Jat and Sharp, 1989], and more recently we and others have demonstrated its effectiveness on more refractory cell types, including thyroid [Wynford-Thomas et al., 1990]. At a low ("permissive") temperature (e.g., 33°C) the T protein produced by tsA58 early region is fully active, permitting clonal expansion of the conditionally immortalized lines. By switching to a higher "restrictive" temperature (usually around 40°C), a conformational change in the protein can be induced that results in loss of immortalizing function and cessation of growth. Mass cultures can thus be prepared at 33°C and then switched up to 40°C prior to analysis, thereby hopefully avoiding interfering effects from the pleiotropic actions of SV40 large T.

In our experience with thyroid, this approach works well from the point of view of proliferative behavior. Cessation of growth is complete in nearly all clones at the nonpermissive temperature (40.5°C), and the frequency of emergence of constitutively immortalized variants is negligible. Conditionally immortalized lines have been successfully further transfected with additional oncogenes [Wynford-Thomas et al., 1990], again permitting these to be studied in the absence of functional T. The main limitation is that while proliferation is reversible, the loss of differentiation referred to above is *not* reversed by switch off of T at the restrictive temperature.

Nevertheless a useful line B-thy-ts1 was successfully generated from neonatal thyroid which shows tight temperature-dependent proliferation up to at least 60 PD and remains well differentiated [Wyllie et al., 1992].

Those wishing to generate such clones should proceed by either retroviral or transfection approaches, using as vectors one of the following:

For retroviral work: psi-CRIP-tsA58-U19 [Wyllie et al., 1992]
For plasmid transfection: pUCSVtsA58 [Wynford-Thomas et al., 1990]; pZIPSVU19tsa [P. Jat ,- unpublished]

All cultures must be performed at 33°C. Expect a diminished yield compared to wild-type vectors at 37°C.

6. SAFETY PRECAUTIONS

The two principal hazards derive from (1) the handling of human tissues (HIV, hepatitis B) and (2) the use of retrovirus vectors expressing a potentially oncogenic sequence (SV40T). The risk should be minimized as follows.

6.1. Human Tissue Handling

Follow standard operating procedures as detailed in Chapter 2.

The only significant danger is from contaminated blood and tissue fluid, which are present in the initial stages of tissue processing. (Take care to avoid penetrating injuries when working with sharp implements, particularly when "dicing" the tissue with scalpel blades.) Once prepared, the primary epithelial cultures themselves (and lines derived from them) are extremely unlikely to pose any threat, since there is no evidence that they can support replication of viral pathogens such as HIV.

6.2. Retroviral Vectors

Although replication defective, the vectors used, being amphotropic, are potentially capable of a "one-off" delivery of SV40T to the tissues of the operator. Two features reduce the practical risk: (1) the great fragility of these viruses that, as with the HIV retrovirus, restricts the mode of infection to direct inoculation; and (2), the low efficiency of gene transfer expected from accidental exposure, given the absence of agents such as polybrene.

Nevertheless, needle-stick injuries remain a real risk, and use of sharp implements should be avoided wherever possible. Disposal of waste containing retroviruses should also be carefully performed; liquids must be treated with an appropriate concentration of disinfectant (e.g., hypochlorite) at the point of use, and solids sterilized by autoclaving (as close as possible to the laboratory) before disposal as refuse.

A safe system of work must be established in accordance with the regulations set out by ACGM for use of Group II organisms in Category II containment and all laboratory users made aware of the work being performed. (Further recommendations can be found in the ACGM/HSE Guidance Note 5). A dedicated tissue culture facility operating at Category II will be required, equipped with Class II hoods having either external venting or recirculation via suitable HEPA filters (see also Chapter 2).

Finally, regular checks should be performed to ensure the continued replication-defective status of the vector. Although minimised in psi-CRIP by separation of the packaging functions onto different DNA segments, multiple recombinations can theoretically lead to helper virus being produced. This can be tested by first infecting a target cell line and then observing the ability of medium from the resulting clones to transduce G418 resistance in a second round of infection

REFERENCES

Bond JA, Wyllie FS, Ivan M, Dawson T, Wynford-Thomas D (1993): A variant epithelial sub-population in normal thyroid with high proliferative capacity in vitro. Mol Cell Endocrinol 93: 175–183.

Bond JA, Wyllie FS, Rowson J, Radulescu A, Wynford-Thomas D (1994): In vitro reconstruction of tumour initiation in a human epithelium. Oncogene 9: 281–290.

Brash DE, Reddel RR, Quanrud M, Yang K, Farrel MP, Harris CC (1987): Strontium phosphate transfection of human cells in primary culture: stable expression of the simian virus 40 large T-antigen gene in primary human bronchial epithelial cells. Mol Cell Biol 7: 2031–2034.

Brown AMC, Scott MRD (1987): Retroviral vectors. In DM Glover (ed) "DNA Cloning, a Practical Approach," Vol 111. Oxford: IRL Press, pp 189-212.

Danos O, Mulligan RC (1988): Safe and efficient generation of recombinant retroviruses with amphotropic and ecotropic host ranges. Proc Natl Acad Sci USA 85: 6460–6464.

Dumont JE, Lamy F, Roger P, Maenhaut C (1992): Physiological and pathological regulation of thyroid cell proliferation and differentiation by thyrotrophin and other factors. Physiol Rev 72: 667–697.

Gluzman Y, Sambrook J, Frisque RJ (1980): Origin-defective mutants of SV40. Cold Spring Harbor Symp Quant Biol 44: 293–300.

Harlow E, Crawford LV, Pim DC, Williamsen NW (1981): Monoclonal antibodies specific for simian virus 40 tumour antigens. J Virol 39: 861–869.

Jat P, Sharp PA (1986): Large T antigens of simian virus 40 and polyomavirus efficiently establish primary fibroblasts. J Virol 59: 746–750.

Jat P, Sharp PA (1989): Cell lines established by a temperature-sensitive simian virus 40 large T-antigen gene are growth restricted at the non-permissive temperature. Mol Cell Biol 9: 1672–1681.

Lane EB (1982): Monoclonal antibodies provide specific intramolecular markers for the study of epithelial tonofilament organisation. J Cell Biol 92: 665-673.
Lemoine, NR, Mayall ES, Jones T, Sheer D, McDermid S, Kendall-Taylor P, Wynford-Thomas D (1989): Characterisation of human thyroid epithelial cells immortalised in vitro by simian virus 40 DNA transfection. Br J Cancer 60: 897-903.
Sambrook J, Fritsch EF, Maniatis T (1989): "Molecular Cloning," 2nd ed. Cold Spring Harbor Laboratory Press, pp 1.21–1.52.
Smith P, Wynford-Thomas D (1989): Control of thyroid follicular cell proliferation–cellular aspects. "Thyroid Tumours: Molecular Basis of Pathogenesis." Edinburgh: Churchill Livingstone.
Tegtmeyer P (1975): Function of the simian virus 40 gene A in transforming infection. J Virol 15: 613–618.
Werner SC, Ingbar SH (1978): "The Thyroid: A Fundamental and Clinical Text. New York: Harper & Row.
Whitley GS, Nussey SS, Johnstone AP (1987): SGHTL-34, a thyrotrophin-responsive immortalised human thyroid cell line generated by transfection. Mol Cell Endocrinol 52: 279–284.
Williams DW, Williams ED, Wynford-Thomas D (1988). Loss of dependence in IGF-1 for proliferation of human thyroid adenoma cells. Br J Cancer 57: 535–539.
Williams DW, Wynford-Thomas D, Williams ED (1987): Control of human thyroid follicular cell proliferation in suspension and monolayer culture. Mol Cell Endocrinol 51: 33–40.
Wright NA, Alison M (1984): "Biology of Epithelial Cell Populations." Oxford, Clarendon Press.
Wyllie FS, Bond JA, Dawson T, White D, Davies R, Wynford-Thomas D (1992): A phenotypically- and karyotypically-stable human thyroid epithelial line conditionally immortalised by SV40 large T antigen. Cancer Res 52: 2938–2945.
Wynford-Thomas D (1993): In vitro models of thyroid cancer. Cancer Surv 16: 115–133.
Wynford-Thomas D, Bond, JA, Wyllie FS, Burns JS, Williams ED, Jones T, Sheer D, Lemoine NR (1990): Conditional immortalisation of human thyroid epithelial cells: a tool for analysis of oncogene action. Mol Cell Biol 10: 5365–5377.
Wynford-Thomas D, Stringer BMJ, Williams ED (1982): Dissociation of growth and function in the rat thyroid during prolonged goitrogen administration. Acta Endocrinologica 101: 210–216.

APPENDIX: SOURCES OF MATERIALS

Materials	Suppliers
BSA (bovine serum albumin)	Boehringer Mannheim BSA Fraction V
Calf serum	Life Technologies.
Centrifuge tube 15 ml	Falcon
Collagenase	Worthington CLS Type 4 or Boehringer Type A (Cat. 1088793)
Diaminobenzidine	Sigma
Dispase	Boehringer Grade II (Cat. 165859)
DMEM, Ham's F-2 and MCDB104 media	Life Technologies
E-cadherin antibody	R&D Systems
Fetal calf serum (FCS)	Imperial Laboratories (but many sources successfully used
G418	Life Technologies
Glycerol	Merck
Hank's balanced salt solution, calcium- and magnesium-free	Life Technologies
HEPES	ICN/Flow
Penicillin and streptomycin	Sigma
Polybrene	Aldrich
Qiagen	Qiagen
Rabbit antimouse immunoglobulin/HRP	Dako
RPMI1640 medium	ICN/Flow
Strontium chloride (Analar)	Merck
Thermanox	1CN Biomedicals
Thyroglobulin antibody	Dako Ltd.
Wizard	Promega

10

Production of Immortal Human Umbilical Vein Endothelial Cells

Neville A. Punchard, Duncan Watson, Richard Thompson, and Mick Shaw

Department of Applied Sciences, University of Luton, Luton LU1 3JU, England (N.A.P., D.W.), Gastrointestinal Laboratory, The Rayne Institute, St. Thomas' Hospital, London, SE1 7EH, England (R.T.), and Zeneca Pharmaceuticals plc, Alderley Park, Macclesfield SK10 4TG, England (M.S.)

Culture of Immortalized Cells, Edited by R. Ian Freshney and Mary G. Freshney.
ISBN 0-471-12134-7

1. CULTURE AND IMMORTALIZATION OF ENDOTHELIAL CELLS

The endothelial cells lining the blood vessels have many vital functions such as the regulation of blood flow and pressure, thrombosis and coagulation, and inflammation through the control of leukocyte recruitment into tissues. In addition, through the process of angiogenesis, they are also involved in wound healing and cancer. Hence, there is much interest in modelling endothelial cell behavior *in vitro* (for an extensive review of the biology of endothelial cells, refer to Warren [1990]).

The most readily accessible and most easily isolated endothelial cells are those lining the large blood vessels, such as the aorta and pulmonary vein. Early studies concentrated on these endothelial cells, isolating them from animal tissues, mainly pig or cow, obtained from abattoirs. Later studies used human cells derived from the same vessels obtained at surgery. The need to study human endothelial cells has been largely satisfied by using blood vessels of umbilical cords, avoiding problems arising from the relative scarcity of availability of human surgical specimens. Umbilical cords are available in relatively large numbers from local maternity hospitals. Today human umbilical vein endothelial cells (HUVEC) are probably the most commonly used endothelial cells in experimental work.

The fact that the endothelial cells are a homogeneous population lying on the surface of the vessel lumen, attached to a relatively robust matrix, makes them comparatively easy to isolate. This has facilitated their use in the study of large vessel endothelial cells. Once isolated, these cells are relatively resilient and can be grown in normal mammalian tissue culture media, supplemented with fetal calf serum to provide necessary growth factors and nutrients. Although endothelial cells grow and divide in these media the fact that they grow better in more recently developed specialized media suggests that normal medium does not provide all the needs of endothelial cells.

However, isolation of human endothelial cells is time-consuming and requires regular access to human tissues, with the associated risks of infection from viruses such as HIV and hepatitis. An alternative is to use immortal endothelial cells. However, there are few immortal human umbilical vein endothelial cell lines currently available, and those that exist differ significantly from normal endothelial cells or have an extended life only in culture, rather than being immortal.

There is thus still a need for immortal endothelial cells lines. Moreover, the emphasis of research into endothelial cells is moving from large vessel endothelial cells to those of the microvasculature. Microvascular cells are isolated from large pieces of whole tissue and thus are less accessible and part of a heterogeneous mixture of cell types. Hence they require more complex digestion procedures and extensive isolation techniques, which often result in low yields. As for other human cells, the supply of tissue in the form of surgical specimens, even of foreskins from which dermal microvascular endothelium is isolated, is relatively limited. Thus there is a greater need for immortal microvascular cells.

It has already been demonstrated that the methods required to produce immortal large vessel endothelial cell lines are readily applicable to producing immortal cell lines of microvascular endothelial cells [Ades et al., 1992]. We therefore describe two different approaches to immortalization of HUVEC as undertaken in two separate laboratories: (1) the use of radiation to generate mutations that affect the regulation of cell growth and (2) the introduction of foreign DNA in the form of oncogenes that produce proteins that interfere with normal control of cell growth, as explained in more detail elsewhere in this book. A general method is given for the use of oncogenes. It is assumed that those undertaking research in this area will either be proficient in the relative technology for preparation of the appropriate oncogene-containing vectors or collaborate with other laboratories that possess the appropriate expertise. Although it is often possible to obtain such genetic material from other groups working in the field, it is not recommended that inexperienced laboratories should attempt to construct or handle vectors containing oncogenes, due to the safety hazards and special precautions required as detailed in later sections.

The methods of isolation, culture, and characterisation are an amalgamation of those used in both laboratories. Where methods differ significantly both variations are given.

2. PREPARATION OF MEDIA AND REAGENTS

All general reagents used are Analar quality unless stated. Tissue culture reagents, growth factors, and so on are of tissue culture quality where available and solutions are prepared in sterile tissue-culture-grade water [ultrapure water (UPW)], preferably guaranteed endotoxin-free.

2.1. Tissue Culture Reagents and Materials

2.1.1. Buffers

PBSA: 10mM phosphate buffered saline w/o (without) Ca^{2+}/Mg^{2+} (Dulbecco's modified phosphate buffered saline)

HBSS w/o Ca^{2+}/Mg^{2+}: Hanks' balanced salts solution lacking Ca^{2+} $/Mg^{2+}$ ions

2.1.2. Media recipes

Media for transfection

M199SF: medium 199, serum-free

M199S: medium 199 supplemented with 10% fetal calf serum (FCS) plus 10% newborn calf serum (NBCS)

M199SG: medium 199 supplemented with 10% FCS, 10% NBCS, and 15 μg/ml endothelial cell growth supplement (ECGS)

Medium for irradiation and growth

M199Ir: medium 199 with Earle's salts, supplemented with 10% FCS and 10% M1 serum-free concentrate (Imperial Labs.)

Note: 100 U/ml penicillin and 100 μg/ml streptomycin should be added for the first 24 h following isolation (optional thereafter).

Alternative medium: Clonetics endothelial cell growth medium based on MCDB131 (a complete medium ready to use; see Section 8)

Freezing media for transfected cells

10% dimethyl sulfoxide (DMSO)
90% FCS

TrypsinEDTA

2.5% trypsin solution (to give final concentration of 0.125%)	1ml
0.08% (2.7mM) EDTA in PBSA	20 ml

2.1.3. Preparation of gelatin-coated flasks

Reagents and materials required

- Gelatin, 1% in UPW, sterilized by autoclaving
- M199SF
- Flasks

Protocol 2.1.3

(a) Add 2 ml 1% (w/v) gelatin for each 25 cm^2 of flask surface area.
(b) Leave for at least 30 min at 37°C.
(c) Immediately prior to use, aspirate off the gelatin and rinse the flask once with 3 ml M199SF.

2.1.4. Collagenase

Collagenase A—0.5 mg/ml in serum-free medium, or collagenase type I—1 mg/ml in HBSS w/o Ca^{2+}/Mg^{2+}. Sterilize by filtration and use immediately.

2.2. Immortalization Reagents and Equipment

2.2.1. Reagents and materials for radiation exposure

The effect of irradiation is enhanced by having rapidly growing cultures prior to and immediately after receiving the dose of irradiation. This is best achieved by culturing the cells in a nutrient-rich medium. M199 supplemented with 10% FCS and a serum-free concentrate (Imperial Lab.) ensures a good growth rate. Other materials and reagents are those commonly used in routine culture of HUVEC.

The radiation source should be capable of giving a dose of γ- (or x-)irradiation equivalent to 2 Gy over a time period of approximately 30–60 s. (e.g., Isomedix Gammacell 10).

Other than the radiation source, no special equipment is required that would not be required for the routine culture of HUVEC.

2.2.2. Reagents and materials for use of oncogenes

It is assumed that laboratories will either have the expertise to produce recombinant cloning vectors containing oncogenes or will obtain such material from other laboratories in the field. It is recommended that those with no previous expertise in the preparation and use of recombinant DNA initially collaborate with others that do. If the vector does not contain an appropriate selection characteristic, then the endothelial cells can be cotransfected with an additional vector carrying this, for instance, pSVneo carrying resistance to the antibiotic G418.

Protocol 2.2.2

Reagents and Materials

- DNA for transfection
- UPW
- Lipofectin reagent

Protocol

On the day of use prepare a fresh solution for each oncogene as follows:

(a) Prepare 10 μg recombinant DNA in 50 μl UPW. Where co-transfection is necessary, for example, with pSVneo, prepare solutions of each at 10 μg of oncogene-containing vector DNA plus 1 μg of pSVneo in 50 μl UPW.
(b) Prepare the Lipofectin reagent by mixing 30 μl stock with 20 μl UPW.
(c) Add 50 μl of this solution to the 50 μl of DNA from above.
(d) Leave the resulting solution to clear (~5 min) before adding to the cultures.
(e) Proceed as in Section 3.4.

3. ISOLATION, CULTURE, AND IMMORTALIZATION OF HUVEC

3.1. Isolation of Endothelial Cells

The following is adopted from the two different, but similar, procedures used in the authors' laboratories. Thus some of the steps have alternative solutions or processes given, while others in italics are marked as being optional, that is, used by only one author's laboratory.

3.1.1. Preparation of the cord

Protocol 3.1.1

Reagents and Materials

Sterile

- 500-ml plastic containers
- Saline, 0.9% (w/v), or normal growth medium (i.e., M199SF)
- Crocodile clip or artery clamp
- Strong thread or suture
- Washing solution: either PBSA or HBSS w/o Ca^{2+}/Mg^{2+}
- Blunt-nosed or umbilical scissors

Nonsterile

- Waste container

Protocol

(a) Collect human umbilical cords from the delivery suite or maternity ward of a local hospital. Collect the cords in alcohol sterilized 500-ml plastic containers, containing either sterile 0.9% (w/v) saline or M199SF. Prepare cultures within 6 h of delivery for optimum results.

(b) Perform all stages after collection under sterile conditions in a Class II cabinet. Wipe this area down with 70% ethanol (w/v). Prepare a small area of the cabinet as a work area by covering it with aluminum foil or paper towel (taking care not to block area of airflow). This is to prevent the spillage of blood onto the cabinet's surface.

(c) Open the jar and remove the cords. Dry the cord and remove any blood clots from the external surface by wiping with a clean paper towel. Examine the cord for clamp marks, cuts, and any needle puncture holes where cord blood has been removed in the hospital. Severely damaged areas can be cut and either the cord processed as two cords if each piece is of a reasonable length, or the damaged piece discarded. Needle holes in the vein resulting from blood collection can be closed with a small crocodile clip. The ideal length for a piece of cord is approximately 15–30 cm, the longer the length the better the yield.

(d) Cut approximately 1 cm from each end of the cord with blunt-nosed or umbilical scissors to present a clean surface with which to work (this also helps to reduce the risk of contamination by removing exposed surfaces). Locate the three blood vessels. The cord has (normally!) two arteries, recognized by their narrow diameter and thick walls, and a thin-walled, large lumen, vein, which is the one required (Fig. 10.1).

(e) *Optional: squeeze the blood out of the cord, into the waste container, by applying gentle pressure with finger and thumb of one hand, while running the hand down the length of the cord.*

(f) Either

   (i) Fill a 20-ml disposable sterile syringe with sterile washing solution. Attach a cannula made from a 16G, 1.5-in. (40-mm) needle, which has had 0.5 in. (10 mm) cut off the end to give a blunt end, and insert it into the left-hand end of the vein. Secure in place by a thin piece of strong thread or suture tied around the outside of the cord over the area containing the cannula and drawn tight enough to prevent leaks occurring. Some thread can be looped over the end of the cannula once on the syringe to help hold the cannula in the vein.

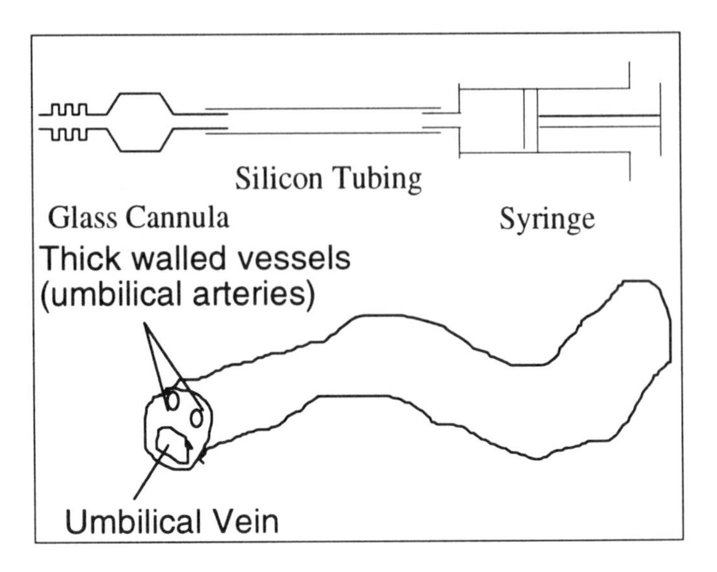

Fig. 10.1. Diagram of umbilical cord, illustrating the appearance of relative position of the two arteries and vein. The shape of the glass cannula and its attachment to the syringe is also shown.

or

(ii) Use glass cannula as shown in Figure 10.1. Insert into the umbilical vein and tie in position with catgut. Tie twice to ensure that the cord does not come loose while perfusing. If a cannula as shown in Figure 10.1 is used, the ridges prevent the cord from sliding off and the glass bulb prevents air bubbles from entering the vein. Fill a 20-ml syringe with sterile washing solution and connect it to the cannula.

(g) Hold the cord over the waste container, containing a suitable disinfectant, and flush out any remaining blood by perfusing the washing solution through the vein by applying gentle pressure on the syringe plunger.

Option: Any blood clots that cause resistance to flow can be sometimes cleared by gently massaging the cord between the finger and thumb of one hand. If necessary the vein can be flushed with a further 20 ml.

3.1.2. Digestion Stage

Each of the two laboratories uses a different method of digestion, one using a 37°C incubator (method A) the other using a waterbath (method B). Both are listed separately below.

Protocol 3.1.2A (*Method A*)

Reagents and Materials

- Collagenase A solution (see Section 2)
- Washing solution: either PBSA or HBSS w/o Ca^{2+}/Mg^{2+}
- 20ml disposable sterile syringe
- 50ml conical centrifuge tube
- Aluminium foil or paper towel

Protocol

(a) Lay the cord flat, from left to right, on the foil or paper.
(b) Fill another 20-ml disposable sterile syringe with washing solution and attach to the right-hand end of the vein as in Protocol 3.1.1, above.
(c) Perfuse the solution through the vein and collect it using the left syringe.
(d) Disconnect the left-hand syringe and discard the solution, draining as much as possible by gently squeezing the cord between finger and thumb. If necessary, air can be flushed through to facilitate this procedure.
(e) Refill the syringe with filter-sterilized collagenase A and reattach to cannula.
(f) Fill the cord with the collagenase, under hand pressure, until the vein is completely distended.
(g) Loosely cover the cord and syringes with aluminium foil or "cling-film" and incubate at 37°C in an Incubator for 5 –10 min (depending on batch of collagenase; see later comment).
(h) Uncover the cord and, taking as little time as possible, push the collagenase back and forth along the vein several times with the aid of the syringes, gently massaging the cord between fingers and thumb at the same time. Finally, collect the enzyme solution in one of the syringes. Disconnect this syringe from the cannula and empty the collagenase (now containing endothelial cells) into a sterile, disposable, 50-ml conical centrifuge tube.
(i) Refill this syringe with washing solution, reconnect it to the vein, and push the washing solution back and forth several times. Collect this washing into the same 50-ml centrifuge tube.

Protocol 3.1.2B (*Method B*)

Reagents and Materials

- 20-ml syringe
- Collagenase Type I

- Waste container
- Artery forceps
- 500-ml beaker
- Centrifuge tube.
- Washing buffer, prewarmed to 37°C: either PBSA or HBSS w/o Ca^{2+}/Mg^{2+}

Protocol

(a) Refill the syringe with 20-ml collagenase Type I solution and reattach to the cannula.

(b) Perfuse it through the vein into the waste container, stopping when there is approximately 5ml left in the syringe.

(c) Clamp the free end with artery forceps and place into a sterile 500-ml beaker.

(d) With the remaining 5 ml collagenase solution, slightly distend the vein, then add 200 ml of prewarmed washing buffer to the beaker, cover with foil, and incubate in a 37°C waterbath for 10-15 min.

(e) Remove the cord from the beaker, fill the syringe with 20 ml washing buffer, and release the artery forceps, collecting the collagenase solution in a sterile 50-ml centrifuge tube.

(f) Flush the vein with the fresh washing buffer, then clamp the free end again.

(g) Refill the syringe with washing buffer and inject around 5 ml into the vein.

(h) *Gently* massage the whole length of the vein between the finger and thumb to roll off the endothelial layer into the lumen of the vein.

(i) Release the artery forceps once again, flush the vein through with washing buffer and collect all washings in a centrifuge tube.

3.1.3. Isolation of HUVEC

Protocol 3.1.3

Reagents and Materials

- 25-cm^2culture flasks (*pregelatinized; optional*)
- *Serum-free medium:* M199SF
- *Growth medium:* M199S, M199SG, or M199Ir

Protocol

(a) Centrifuge at 100g for 5 min, or 150g for 3 min (bench centrifuge).

(b) During centrifugation, aspirate the gelatin from a pregelatinized 25-cm^2 culture flask and wash with M199SF.
(c) Aspirate and discard the supernatant from the centrifugation.
(d) Resuspend the cell pellet in growth medium.
(e) Transfer the cell suspension to 25-cm^2 culture flasks.
(f) Incubate under an atmosphere containing 5% CO_2 at 37°C.

Antibiotics may be added for the first 24 h in culture, but may not be necessary thereafter. Normally the cells from a 10-cm length of cord can be plated into a 25-cm^2 flask in 5 ml growth medium to give a confluent monolayer within 48 h (See Fig. 10.2). If higher yields are attained, for example, from bigger pieces of cord, then multiple 25-cm^2 flasks can be seeded, or 75-cm^2 flasks should be considered. Yields can vary from group to group and depend on a variety of factors. It is not unusual for yields to be quite low initially, but they should improve with experience of the technique.

3.2. Basic Culture of HUVEC

If there is a high degree of contamination of blood, then, 24 h after isolation, aspirate off the medium from the culture flask, wash the cells twice by addition, and removal M199SF or balanced salts solution, and add fresh growth medium. *Option: Replace only half of the medium each time in subsequent changes of medium.* Thereafter endothelial cells can be treated much the same as any other cell line. They should be passaged prior to reaching confluence for continued culture, using trypsin/EDTA solution. Split ratios of 1:3–1:5 should be used. When using trypsin/EDTA the exposure time to trypsin should be kept to a minimum. The cells can be frozen using standard techniques.

3.3. Immortalization of HUVEC Using Low-Level Radiation

Protocol 3.3

Reagents and Materials

- Trypsin/EDTA
- Growth medium: M199S, M199SG, or M199Ir
- Flasks
- γ-Irradiation source

Protocol

(a) Use cells within one passage from isolation.
(b) Harvest prior to reaching confluence, using trypsin/EDTA.

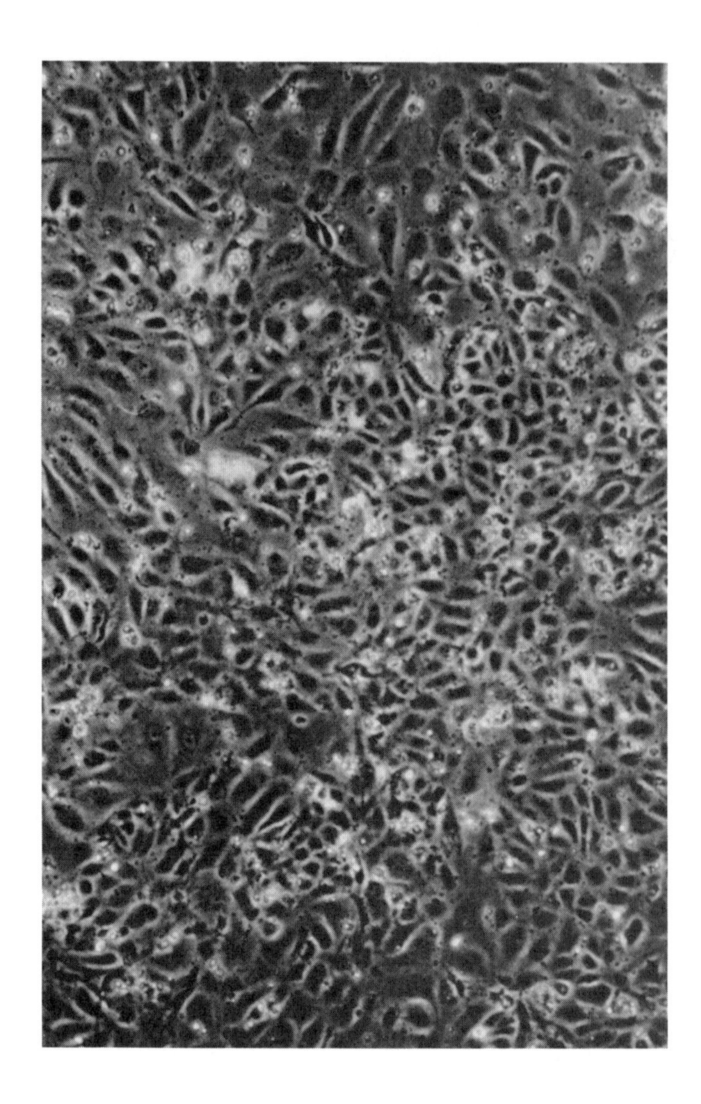

Fig. 10.2. Phase-contrast photomicrograph of a nontransformed HUVEC monolayer demonstrating typical cobblestone morphology of endothelial cells (cells at passage 3, 100× magnification).

(c) Resuspend in growth medium at a density of $0.5–1 \times 10^6$/ml.
(d) Expose immediately to 2-Gy γ-irradiation.
(e) Immediately following irradiation plate the cells at a density of $4 \times 10^4/cm^2$ in tissue culture flasks.
(f) Incubate at 37°C.

Within the first 24 h many of the cells will die (approximately 30%), but the remaining cells will grow as normal such that after 48 h the culture will require splitting. After a further two or three passages there should be sufficient cells available to establish a substantial primary cell bank. Cells should be frozen by harvesting cultures, resuspending at 2×10^6/ml in growth medium supplemented with 8–10% DMSO, aliquoting into 1-ml freezing vials and freezing in the gas phase of a liquid nitrogen refrigerator in a polystyrene box. Cells can be transferred to the liquid phase after 4 h. One of these vials can then be recovered and grown on to produce a secondary bank within a further two or three passages. Further banks can be prepared in this manner to give an indefinite supply of the cells.

Using the above method the normal characteristics that are so readily lost from primary cultures, such as prostacyclin production, responsiveness to growth factors and some cell surface markers, are retained for extended periods. The cells produced from this technique may not be immortal but will grow for extended periods and will retain the normal characteristics of primary HUVEC, making them invaluable tools for research.

3.4. Immortalization of HUVEC Using Oncogenes

Protocol 3.4

Reagents and Materials

- Medium M199SG
- Medium M199SF.
- Liposomal DNA
- 35×10-mm gelatinized culture dishes

Protocol

(a) Grow a sample of HUVEC in M199SG in 35×10-mm gelatinized culture dishes, until approximately 40% confluent.
(b) Remove the medium and replace with 2 ml of M199SF.
(c) Add 100 µl of liposomal DNA (see Section 2.2.2) to the cells in the culture dish, placing the solution as drops evenly over the surface of the dish. Incubate the cells at 37°C for 16–20 h (overnight).
(d) Remove the medium and replace with fresh M199SG.
(e) The cells are then cultured as for normal HUVEC as described previously.

3.5. Selection of Clones

When to select suitable clones can pose a problem. If clones are selected out from colonies that become established soon after immortalization, then, to ensure that some of these will be suitably long-lived and yet maintain differentiated characteristics, a large (unspecified) number of clones will need to be isolated, grown on, subcultured, and then characterized. This will require a large amount of laboratory work, increasing with the number of differentiated characteristics being determined.

Alternatively the cells can be maintained as mixed clones, characterized as a whole, and then only when it is confirmed that basic desirable characteristics, such as staining for von Willebrand factor (vWf) and lectin binding, are present is selection of single clones performed. The disadvantage of this approach is that if grown for too long, fast growing clones that tend to be less differentiated can overgrow and outpopulate the slower growing differentiated clones. Testing for viral infections is probably best carried out on freshly isolated HUVEC rather than clones. It is recommended that testing for mycoplasma contamination should be carried out before immortalization and also confirmed after production of single clones.

As an aid to selection of clones, simple examination of the clones to judge whether they look like normal HUVEC can be performed. However, in our experience all immortalized HUVEC we have examined look different from freshly isolated cells. The other problem is whether the cells are contact-inhibited, or whether they overgrow. This can be important if the cells are being used for permeability studies. All the immortalized cells we have examined or produced ourselves with oncogenes are not contact inhibited, while those produced by irradiation are. Generally we have found that HUVEC transfected with c-*myc* and c-Ha-*ras* tend to produce distinct colonies and, if not plated out correctly, will produce multilayer clusters.

An alternative to physical selection, if it is available, is to initially separate the cells based on membrane proteins. This can be performed by tagging the living cells with fluorescently labeled antibodies to these proteins and then separating the labeled cells out and collecting them using a fluorescence activated cell sorter (FACS), such as Becton and Dickenson FacsStar plus.

3.5.1. Geneticin (G418) selection procedure

Protocol 3.5.1

Reagents and Materials

- Medium M199SG with Geneticin G418, 250 μg/ml

Protocol

(a) Seven days posttransfection, select cells by incorporating 250 μg/ml G418 into the M199SG.
(b) Replace medium every 2 days.
(c) Remove the G418 after 14 days.

This method will still result in the production of a culture containing a mixture of clones, and thus other selection processes will still be required to produce a homogeneous clonal population. There is an argument for keeping the G418 in the culture medium in that the continuous presence of the selection criteria will ensure that the cells continue to express the foreign DNA. However, G418 is expensive and slows down endothelial cell growth. It may be possible to maintain positive selection pressure if other selection methods are used.

3.5.2. Selection of clones

Clones can be selected using standard techniques, e.g., cloning rings and limited dilution. However, normal endothelial cells do not grow well as single cells isolated from others. Either clumps of cells forming a distinct colony should be selected or single clones initially grown with ordinary endothelial cells until they are established.

4. ADVICE ON SAFETY

For advice on the hazards of particular reagents used, please refer to manufacturers' data sheets, reference books, and local COSSH assessments (see also Chapter 2 for further details on safety). It is advisable to produce a list of these and append it to the back of any written protocols.

4.1. Initial Isolation of Cells: Handling of Human Tissue Samples

The local safety officer should be consulted and local regulations followed at all times. All personnel handling human blood should be offered the protection of a hepatitis B vaccination. All abrasions and cuts should be adequately protected before starting tissue culture. All materials should be confirmed as being free of risk of infection by HIV and hepatitis B. All material should be assumed to be potentially harmful with regard to unknown infectious agents and suitable precautions taken, as follows: wearing glasses to protect the eyes, sterile surgical gloves while handling the human tissue, and use of a Class II cabinet as a minimum standard. All blood spillage should be immediately cleaned with 70% alcohol solution (or other sterilizing fluid). All work surfaces should be cleaned with 70% alcohol before and after the isolation procedure. All contaminated waste should be double-bagged and autoclaved or incinerated and glassware autoclaved or washed by immersion and soaking in a suitable sterilizing solution. Liquid waste should be decontaminated by treating with hypochlorite (2500 ppm available chlorine) or 2% Virkon for 2 h prior to washing down the sink with copious amounts of water. The sink should be cleaned with undiluted bleach afterward.

Personal hygiene should be maintained at all times; hands should be washed immediately before leaving the tissue culture laboratory. Any spillage should be immediately washed off the body, and contaminated clothing should be changed. All accidents, especially needle-stick or cuts from broken glass, or serious abrasions, should be immediately reported, after washing of the affected area, and medical advice sought.

For more extensive guidelines and recommended procedures for handling human tissues refer to the paper by Grizzle and Polt [1988] (See also Chapter 2).

4.2. Immortalization Procedures

The irradiation of cell cultures does not present the user with any great safety issues provided local operating procedures are adhered to. The radiation source can be operated following the manufacturers instructions and should require no special protective clothing. The radiation source is normally enclosed in a protective layer of lead and/or concrete. The equipment should be monitored before, during, and after use to ensure that no leakage has occurred. The use of recombinant DNA containing oncogenes does present some problems due to the chance of exposure to potential carcinogenic material. All institutions involved in the use of molecular biology and recombinant DNA technology should have such an appropriate safety committee or officer and local regulations concerning the handling of genetically modified material. The advice of this safety committee or officer should be sought and approval gained before the work is begun. The transfection of the cells and subsequent handling will initially have to be undertaken in a laboratory, approved for such work by the local committee or officer. All precautions to avoid exposure to the recombinant DNA should be taken. The transfection should be undertaken in a Class II cabinet and the operator protected by use of gloves. Any other local regulations for handling, operation, and disposal should be closely followed. After the transfection, and the removal of the recombinant DNA, the risk is the same as for handling any other immortal cell line.

4.3. Culture of Early-Passage and Immortalized Cell Lines

There will always be the risk of infection from human tissues, even exhaustive testing can rule out the risk only from known pathogens. It is thus highly recommended that it is assumed that there is a real risk of infection from early passage cell lines and that they are treated as for fresh tissue and reasonable precautions taken as above. If not previously tested for the presence of HIV and hepatitis-B, then, if it is possible to do so, early passage cells should be tested before proceeding further. Testing for mycoplasma should also be carried out as soon as possible, and it is recommended that this be carried out on the early passage cells. Once shown to be clear, they should then be kept separate from all untested cell lines, that is, in a different incubator. Even if immortal or other cells and cell lines have been tested and found to be free from HIV, hepatitis, and mycoplasma, it is recommended that they still be treated as potentially dangerous. Human cells in culture are still susceptible to infection by bacteria or viruses as are humans, more so as they offer easy access, ideal medium for growth, and a highly homogeneous culture. Single cells in culture do not have the advantages of the physical defences or complete immune system that humans have. Antibiotics are not proof against all pathogens. Thus there is a risk, especially in research labs with close links with hospitals, that cells in culture can become infected and act as a source of infection of the operator(s). For a more extensive guide and recommended procedure for handling immortalized and other cell lines, refer to the paper by Caputo [1988] and Chapter 2 in this volume.

4.4. Biosafety of Products Derived from Immortalized Cell Lines

Where foreign DNA such as oncogenes have been introduced into cells, or mutations induced, it cannot be ruled out that, in the presence of some infective unidentified virus, there will be production of infectious and possibly cancer-inducing virus or DNA. Similarly, an unidentified virus may be passed on in the products of cells, including waste products such as media. The risk, although theoretically small, should therefore always be considered when handling purified DNA or other products or reagents that have come into contact with cell lines.

5. VARIATIONS ON THE METHODOLOGY

5.1. Isolation and Culture

These methods refer to the isolation of HUVEC performed in two different laboratories and are thus a summary of the two slightly different procedures. The isolation of endothelial cells from human umbilical cord veins given above is derived from the method of Jaffe et al. [1973]. Virtually all the other methods published are also variations on the method of Jaffe. A variety of cannula are used and some groups prefer to dissect out the vein and ligate it to hold the cannula. Some modifications use HEPES-buffered serum-free medium to collect the cords, different collagenases for release of cells, and additional growth factors, heparin, or hydrocortisone, or omission of gelatin.

One of the groups contributing to this chapter has found success isolating endothelial cells from HUVEC collected 3–24 h after delivery. The other reports a distinct change in characteristics of the cells isolated from cords more than 6 h old. Although good yields can be obtained from cords over 6 h old, the cells senesce earlier and do not retain normal characteristics for prolonged periods. Overall it is perhaps best to use cords as fresh as possible, depending on local availability. If it is difficult to collect the cords immediately after delivery, then to minimize the growth of contaminating infectious organisms, we recommend using saline or PBSA, rather than media, in the collection vessel, and temporary storage at 4°C.

The digestion time is critical to the isolation of a pure endothelial population. Too short a time or too weak a collagenase mixture results in low yields, too high a strength, or too long a digestion time produces endothelial cell cultures contaminated with fibroblasts or smooth muscle cells. The fact that collagenase digestion times and conditions vary from group to group probably reflects the relative activity and purity of the collagenase used. Some collagenases seem to suffer from a high degree of batch variability. If yields are low or fibroblast contamination is detected, then investigators should experiment with a particular batch of collagenase to determine an appropriate time for the digestion.

With regard to the antibiotics used, we recommend as low a strength as possible. Some groups have added fungizone to their particular cocktail. Although this will reduce fungal infection, we have not found this infection a particular problem, whereas we have found that the presence of fungizone slows endothelial cell growth.

The medium used in the majority of HUVEC cultures is Medium 199, a general tissue culture medium, although some have used Eagle's Minimum Essential Medium [Balconi et al., 1983]. Other factors may be added in addition to endothelial growth factor to improve the basic medium (heparin, selenium, and thrombin). In a study on growth factors, MCDB107 was found to be superior to other media, an effect enhanced by the presence of tissue-derived growth factors [Hoshi and McKeehan, 1984]. Recently an optimized medium has become available from Clonetics, based on MCDB131, that supports better growth of HUVEC in culture with extended passage number. Using this medium we have been able to increase passage levels two- or threefold above that for M199.

It is not even necessary to isolate HUVEC yourself as these, and other endothelial cells, are also commercially available, tested free of HIV and hepatitis, from Clonetics.

5.2. Immortalization

There have been several different approaches to immortalizing endothelial cells in the past. Immortal cell lines arising by spontaneous generation have been found, namely, the ECV304 cell line [Takahashi et al., 1990] and C11STH [Cockerill et al., 1994]. ECV304 has been reported to contain Weibel–Palade bodies but not vWf. They also stain positive for Ulex europaeus agglutinin (UEA)-1 and other endothelial markers and possess angiotensin-converting enzyme (ACE). More importantly, ECV304 has also been reported to produce levels of PGI_2 comparable to HUVEC (results we have not been able to repeat in our laboratory), although possession of other normal endothelial cell characteristics was variable [Takahashi and Sawasaki, 1992]. The cause of this immortalization is not clear, and, although intact viral particles have not been found in these cells, the role of undiscovered viral activity cannot be excluded. However, this is not a recommended way of producing immortal cells lines as over 600 HUVEC isolations were required for ECV304 to arise spontaneously. Although the same group has reported some success in repeating this by isolating cells from HUVEC that grew for up to 7 months [Takahashi and Sawasaki, 1992], it still represents a very low probability of success when compared to that achieved with active induction of immortalization.

One such approach using fusion of HUVEC with the lung carcinoma cell line A549/8 has produced the immortal cell line EA.hy926 [Edgell et al., 1983]. This cell line had undergone more than 60 passages, in excess of 200 cumulative doublings by 1983, and has been reported to produce PGI_2 [Suggs et al., 1986] and stain positive for vWf [Edgell et al., 1983]. It suffers from being polychromosomal, as are many immortal cell lines, and as it is the product of a fusion, may express nonendothelial cell genes.

However, the most common technique has been to use either intact viruses or just the viral oncogenes. The main "trick" with this method is to induce immortalization, while maintaining differentiated characteristics and avoiding transformation.

Amphotropic transforming murine sarcoma viruses containing *v-ras* or *v-mos* oncogenes successfully immortalized HUVEC without morphologically transforming the cells or causing a loss of differentiated characteristics [Faller et al., 1988]. The cell lines produced grew in excess of 300 doublings, possessed a density-dependent inhibition of growth, a diploid karyotype, synthesised vWf measured by ELISA, contained Weibel–Palade bodies and were not themselves tumorigenic. However, on the negative side the cell lines produced grew to 50% of the normal density of HUVEC, were growth factor–independent, and some clones possessed chromosomal abnormalities. Moreover, the resulting pseudo-typed sarcoma virus that is produced has the potential for infecting human cells and use of both the virus and the immortal cells produced requires extreme precautions. Thus this approach is severely limited. However, both intact wild type SV40 and parts of the SV40 DNA were equally successful in transforming HUVEC [Gimbrone and Fareed, 1976]. These cells had an increased growth potential, loss of anchorage dependence, and a reduced serum requirement. However, the cells did not possess Weibel–Palade bodies or vWf, while angiotensin-converting activity was only occasionally detected. These cells maintained HUVEC-like characteristics for 10–15 passages, and then the growth rate slowed down beyond passages 42–50 until it was not possible to subculture the cells further. Length of time in culture was matched by increasing chromosomal aberrations typical of SV40 "crisis".

More success has been achieved by others with SV40, HUVEC lines being produced with a higher growth potential, reduced serum requirement, and positive staining for vWf and *UEA-1* up to passage 45 [Ide et al., 1988]. However, the cells had a modal chromosomal number of 45 and again senesced after passage 50, ceasing to grow after passage 62. Although these transformed cells did not appear to be infectious, the use of the intact viruses always carries this risk.

An origin-defective SV40, has been used to immortalise HUVEC [Iijima et al., 1991]. The resulting cell lines at best stained only weakly positive for vWf, produced more plasminogen activator (tPA) than HUVEC, were growth factor–independent, and had abnormal numbers of chromosomes. After 100 generations the cell lines either died or morphologically transformed, with dramatic increases in chromosome number and loss of differentiated characteristics. Given the risk with intact viruses, a safer method is to use recombinant viral DNA in the form of oncogenes carried in appropriate vectors. The use of such vectors also facilitates selection based on antibiotic resistance or other selection criteria contained in the vectors.

Initially only limited success was achieved just using the SV40 oncogenes, the small-t and large-T antigens, in the pSV3-neo vector to produce the cell line SGHEC-7 [Fickling et al., 1992]. This cell line had a HUVEC-like appearance, released tPA in response to thrombin, and had a normal diploid karyotype up to passage 16. However, after passage 22 (1:1 splits) there was loss of chromosomes, decreased tPA release, reduced cell growth, and a loss of the normal cobblestone appearance. Greater success was achieved with the same oncogene, namely, large-T antigen, in immortalizing human dermal foreskin microvascular cells, but in a different vector, namely the plasmid pBR-322 [Ades et al; 1992]. These cells also have a typical cobblestone appearance, secrete vWF, take up acetylated low-density lipoprotein, form tubes on Matrigel, and have survived in excess of 60 passages.

EA.hy926 and ECV304 have been used and characterized by many other research groups. There is insufficient space here to list all the publications, and the reader is recommended to carry out searches of the literature for these and more recent publications for details of characterization of these cell lines.

One of the authors' groups has had success extending the life of

HUVEC using pSV3neo containing small- and large-T antigens of SV40 (See Fig. 10.3), pSVneo containing the HPV-16 DNA (pSV16), pUC plasmid containing c-Ha-*ras* (pEJ*ras*), and pUC plasmid containing c-*myc* (pLTR*myc*) [Punchard et al., 1994]. All four lines produced, one for each oncogene, maintained staining for vWf and UEA-1 and a general HUVEC-like appearance. However, cells immortalized with SV40 T antigens did not produce PGI_2 in response to stimulation with thrombin or another stimulant, hydrogen peroxide, while their basal production was much lower than that of normal HUVEC, as given later.

Although the strength of staining was maintained above passage 25 for SV40 transfected cells it was much reduced for the others. Although all four oncogenes extended endothelial cell life, cells trans-

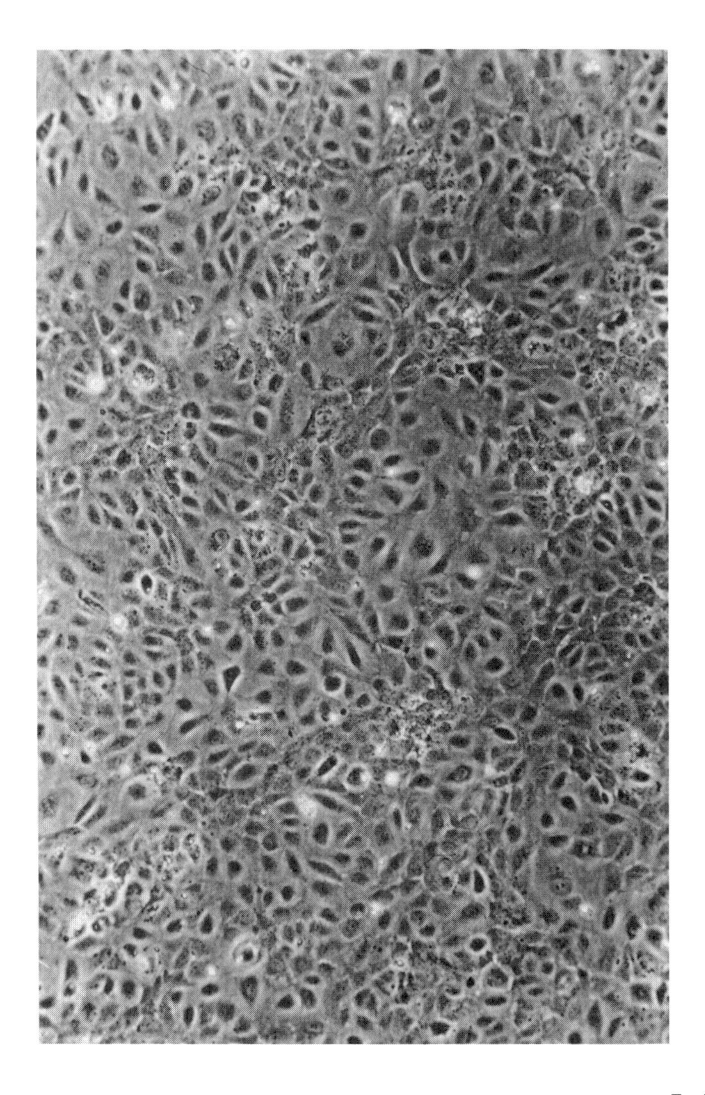

Fig. 10.3. Phase-contrast photomicrograph of a SV40-transformed HUVEC monolayer (cells at passage 9, 100× magnification).

fected with c-*myc,* H-*ras,* and HPV-16 DNA slowed their growth between passages 25–32 (based on $\frac{1}{3}$ splits for each passage), with increasing numbers of giant cells becoming evident (Fig. 10.4). The presence of giant cells, containing 5–10 nuclei and massively enlarged cytoplasm with dendritic processes, is a common sign found in HUVEC cultures as they become senescent and has been reported by others [Gimbrone and Fareed, 1976; Ide et al., 1988].

We have also found that ECV304 and EA.hy926, do not produce PGI_2 in response to stimulation with thrombin or hydrogen peroxide, with basal production being lower than that of normal HUVEC. This confirms our previous findings for EA.hy926 [Boswell et al., 1992]. Nor have we been able to confirm the presence of vWf in EA.hy926

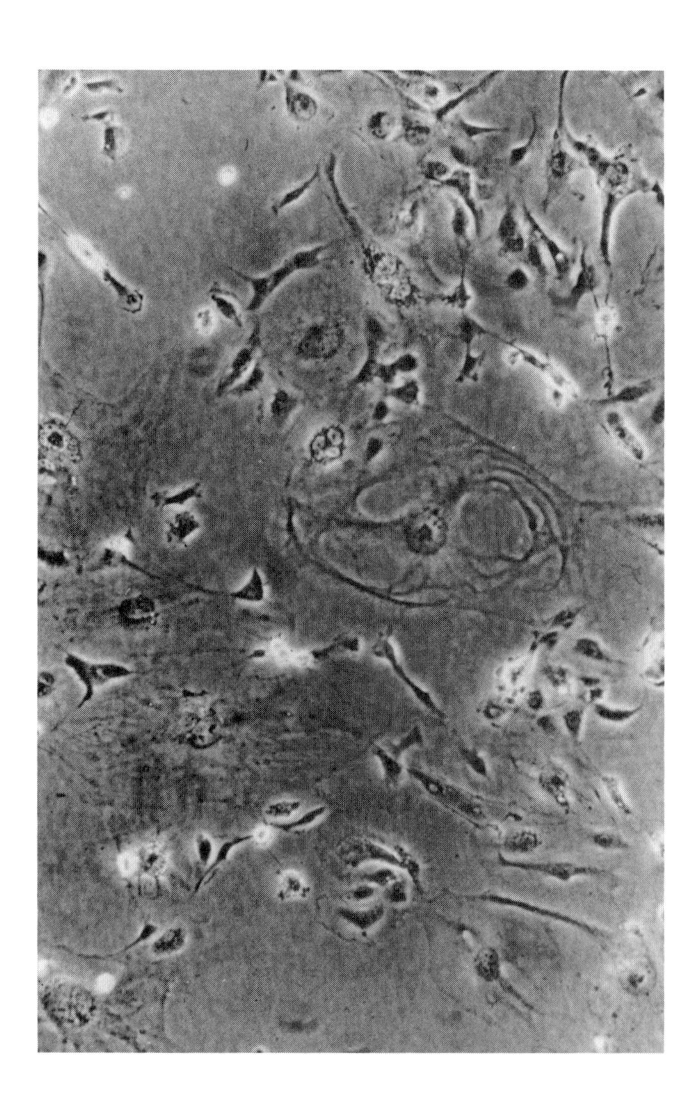

Fig. 10.4. A transformed HUVEC culture demonstrating presence of typical "giant cell" (cells at passage 35, 100× magnification).

cells, measured by either immunohistochemical staining or ELISA. We have found no vWf in ECV304, in agreement with the original findings.

Discussion with others indicates that we are not alone in finding changes in immortal HUVEC cell lines from those original reports. This may be due to genetic drift away from the original characteristics. This is known to happen with all immortal cell lines to some extent. Subsequent to single cell cloning, further genetic alterations can arise that produces a heterogeneous population of cells [Iijima et al, 1991]. Changes in growth medium, plastics, split ratios, or times of splitting and absence of selection antibiotics or nutrient factors may all encourage the growth of a particular clone, leading to changes in the nature and behavior of the immortal cell line in culture. Mycoplasma or other infection may also cause loss of desired characteristics. All immortal cell lines should thus be regularly checked to confirm that they are maintaining the characteristics of their parent lines.

Native or intact viruses come complete with their own mechanisms for infecting cells. With oncogenes a variety of means have been used to get the foreign DNA into the endothelial cells, including electroporation [Fickling et al., 1992], calcium phosphate precipitation [Gimbrone and Fareed, 1976], and dextran [Iijima et al., 1991]. The present authors have used Lipofectin, a 1:1 liposome formulation of the cationic lipid *N*-(1-(2,3-dioleyloxy) propyl)-*n*,*n*,*n*-trimethylammonium chloride and dioloeoyl phosphatidylethanolamine, which is reported to be more efficient than calcium phosphate and less toxic than electroporation.

The use of radiation presents a novel approach to the production of immortal endothelial cell lines. It is much simpler and does not carry the risks of infection to the operator presented by viruses and oncogenes. Certainly the results from the two methods favor the use of radiation, as shown by the ability of the cells to maintain differentiated characteristics, including PGI_2 production, for extended periods and to maintain contact inhibition in culture when grown on collagen.

The general impression from our own studies and those of others above is that the process of producing immortalization is a relatively random process. Factors that could increase the success in producing differentiated clones that are long-lived still need to be identified. The term immortal is misleading as most cell lines produced either senesce and die or transform with loss of differentiated characteristics. It is probably more realistic to aim for a long-lived cell line which maintains differentiated characteristics over a selected number of passages.

6. CHARACTERIZATION OF ENDOTHELIAL CELLS

The commonest criteria used for identification of endothelial cells are those also used by Jaffe et al. [1973]: staining and localisation of von Willebrand factor and binding of the lectin from *Ulex europaeus* (UEA-1). Not only should endothelial cells stain positive for vWf (Fig. 10.5), but the Weibel–Palade bodies should also be clearly identifiable. These are good basic monitors of the endothelial nature of immortalized cells, and methods for these are given below. In addition, the levels of vWf inside cells and released in response to stimuli can be confirmed by ELISA, and gels runs to confirm the polymeric nature of vWf. However, measurement of these parameters will not detect changes in other endothelial characteristics. If the cells are to be used for a specific purpose, it is best that the particular desired characteris-

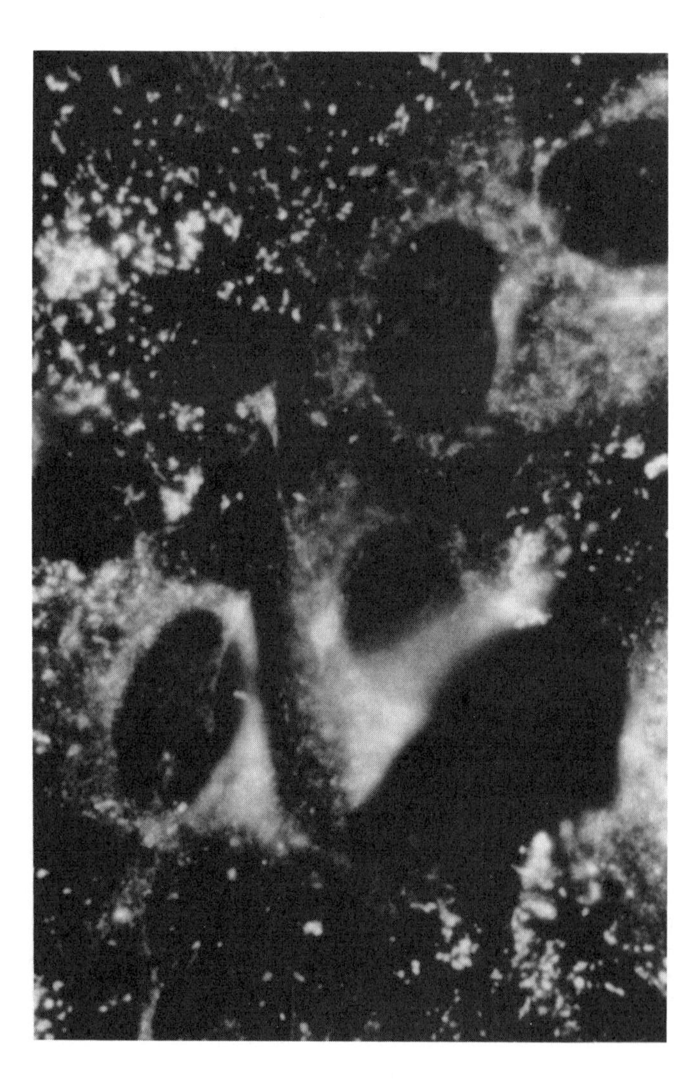

Fig. 10.5. Direct Immunofluorescence of HUVEC for vWf showing positive staining and presence of Weibel–Palade bodies (cells at passage 1, 1000× magnification).

tic be regularly monitored. It is important to monitor regularly the differentiated characteristics of the immortal endothelial cells as continuing genetic rearrangements can result in genetic drift and loss of differentiated characteristics through successive cultures. It is recommended that once suitable clones are produced, they be grown up, stocks kept at different passages, and the cells characterized at intervals over a fixed number of passages. Cells can then be subcultured from those stocks of cells that show differentiated characteristics and used with a degree of confidence over the determined span of passages. In addition to histologic and functional characterization, it is recommended that karyotyping be performed. However, this is extremely difficult to do correctly where chromosomal aberrations can occur and is best done by experts in this area.

6.1. Morphology

HUVEC when confluent and viewed under phase contrast have a polygonal appearance classically referred to as a "cobblestone" appearance (Fig. 10.2). Immediately after exposure to oncogenes the cells will change their physical appearance. They will then gradually revert back to type, although there may be some subtle differences in size and appearance compared to normal HUVEC (Fig. 10.3). Cultures may not always be contact-inhibited.

Irradiated cells will retain their normal morphology and are contact inhibited. The cells will not form multiple layers in culture but will be prone to shedding into the medium if held at confluence for prolonged periods. Growing on collagen-coated dishes will eliminate this problem.

6.2. Immunocytochemistry

The methods given are for fluorescent staining, but if a fluorescent microscope is not available, then one of the many commercial kits, such as Vectastain ABC kit (Vector Laboratories), can be used to detect either rabbit antihuman vWf (Dako), or part of the kit to detect biotinylated antibody to vWf, or even biotinylated lectin. Confirmation that the staining technique works is best undertaken using freshly isolated HUVEC and comparing them to cells that should stain negative, such as fibroblasts.

6.2.1. Preparation of cells for staining

Multichamber slides (e.g., Lab-Tek) are ideal for growing cells for staining as they allow up to 16 different stains to be done on a culture grown under identical conditions. Alternatively, cells can be grown contained in circles of 9-mm radius on a glass slide defined by using a wax pen. The slides are then flame-sterilized. The glass can be pretreated with gelatin, fibronectin, or other attachment factor if desired. The cells should normally be seeded at a density to give confluence within 24 h, although they may be incubated for longer periods to allow the cells to form a mature cell sheet (some surface antigens are only expressed following extended periods at confluence). Cultures may also be examined by FACS to give quantitative measurements. The following methods were devised for preparing cells for examination under a fluorescence microscope, but can also be used for flow cytometry after harvesting a single cell suspension.

6.2.2. Staining for von Willebrand factor (vWf)

Protocol 6.2.2

Reagents and Materials

- PBSA
- Ice-cold methanol
- Antihuman vWf primary antibody, diluted 1:200 in PBSA. These antibodies are available commercially, raised in either goat, rabbit, or sheep. The titer of the antibody may vary, so the manufacturer's recommended concentration should be followed. The dilutions given here are typical values only.
- FITC-conjugated antibody, directed against the species of the primary antibody, 1:200 dilution in PBSA).

Protocol

(a) Wash the cells with PBSA.
(b) Fix the cells, grown on slides, with ice-cold methanol for 15 min at 4°C.
(c) Wash twice with PBSA.
(d) Add 100µl of a 1:200 dilution, in PBSA, of primary antibody, anti human vWf, to the cells.
(e) Leave for 1 h at room temperature.
(f) Wash 3× with PBSA to remove the antibody.
(g) Add 100µl of a 1:200 dilution (in PBSA) of FITC-conjugated secondary antibody directed against the species of the primary antibody.
(h) Incubate for 1 h at 37°C in the dark (CO_2 incubator).
(i) Wash 3× with PBSA to remove the antibody.
(j) Mount the slide in either PBSA or a permanent mounting medium (Polimount).
(k) Examine on a fluorescence microscope using the following filter combinations with UV light: excitation filter, 450–490 nm; barrier filter, 520 nm.

As controls, we recommend running, in parallel to the above, cells treated with normal serum (at the same dilution) instead of the primary antibody. Typical results for fluorescence microscopy are shown in Fig. 10.5 and for flow cytometry in Fig. 10.6.

6.2.3. Staining of cells using *Ulex europaeus* agglutinin-I (UEA-1)

Protocol 6.2.3

Reagents and Materials

- Acetone or 4% paraformaldehyde
- M199 with 10% FCS
- FITC conjugated UEA-1, 1:200 dilution in PBSA
- PBSA
- Normal rabbit serum
- Aqueous mountant
- 0.4M fucose in PBSA

Protocol

(a) Fix in acetone or with 4% paraformaldehyde for 15 min at room temperature.
(b) Add 100 μml 10% FCS in M199 onto the cells and leave for 30 min at room temperature.
(c) Add 100 μl of a 1:200 dilution (in PBSA) of FITC-conjugated UEA-1 onto the cells and incubate in a moist environment for 1 h, at room temperature, in the dark.
(d) Wash in PBSA for 10 min.
(e) Mount using an aqueous mountant and photograph.
(f) Controls:
 (i) Cells treated with normal rabbit serum (at the same dilution) instead of FITC UEA-1
 (ii) Cells treated with FITC UEA-1 preincubated with 0.4M fucose (made up in PBSA).

6.3. Production of Prostacyclin

Prostacyclin (PGI_2) is derived from PGH_2 which is synthesized in the cyclooxygenase pathway. It is a major product of endothelial cells, a potent vasodilator, and is, potentially, a good marker for differentiated endothelial cells. Kits are available that measure 6-keto-prostaglandin $F_{1\alpha}$, which is a stable hydrolyzed product of the unstable PGI_2, used for either ELISA or RIA. It is recommended that commercial kits be used, such as those sold by Cascade Biochem Ltd. Full details of the methods will be given in the kits together with all

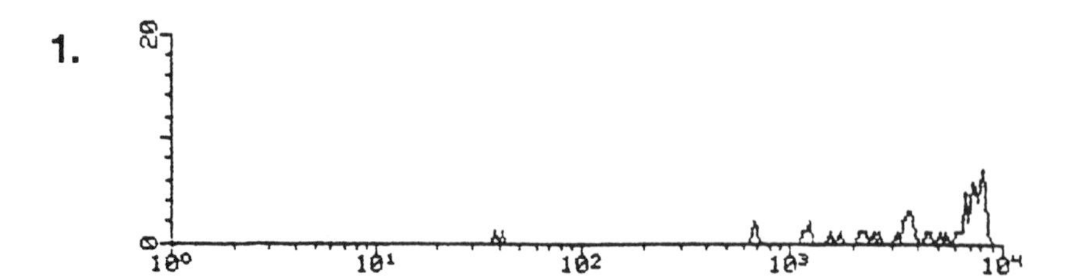
1.
20
0
10^0 10^1 10^2 10^3 10^4

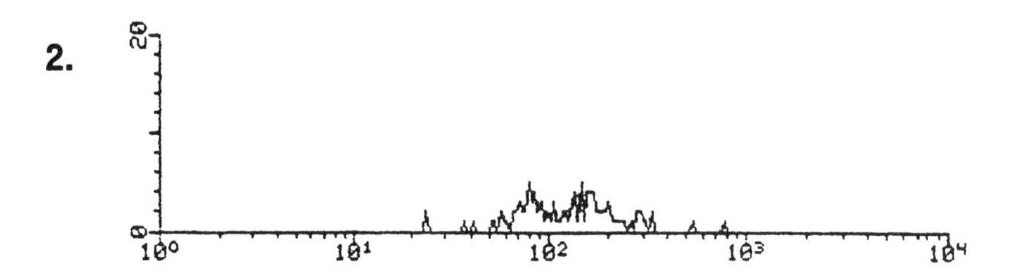
2.
20
0
10^0 10^1 10^2 10^3 10^4

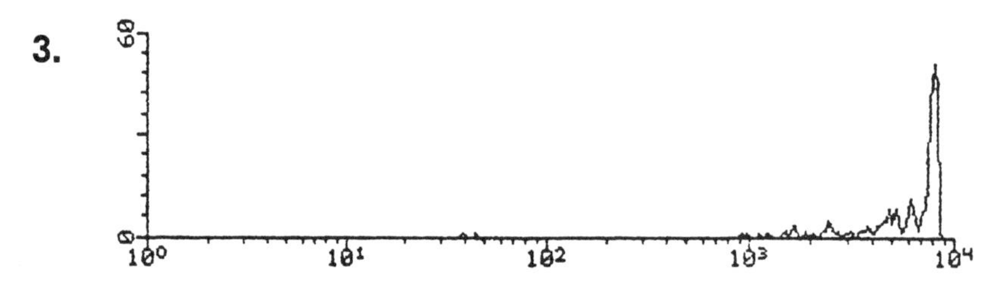
3.
60
0
10^0 10^1 10^2 10^3 10^4

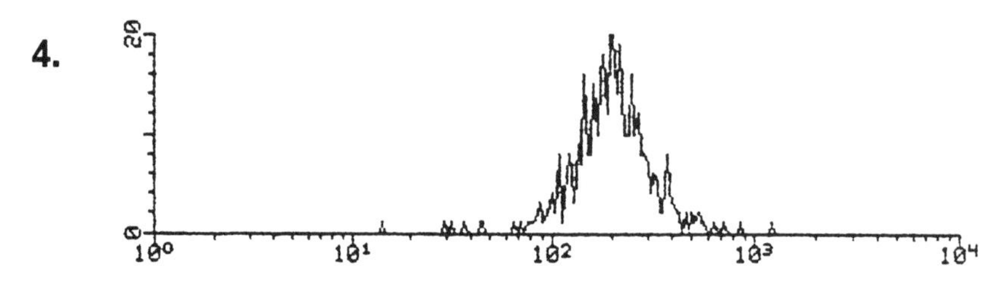
4.
20
0
10^0 10^1 10^2 10^3 10^4

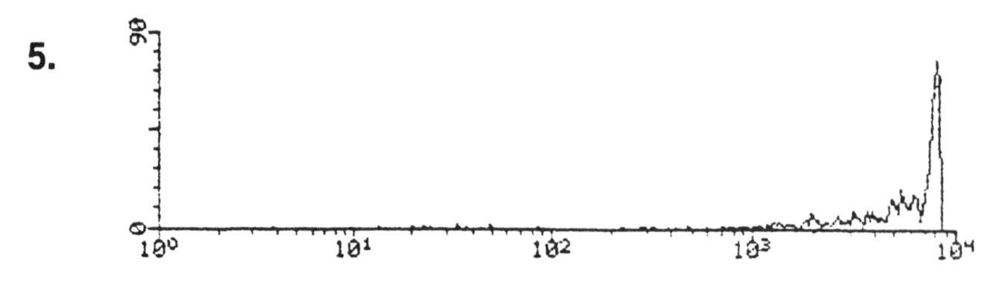
5.
90
0
10^0 10^1 10^2 10^3 10^4

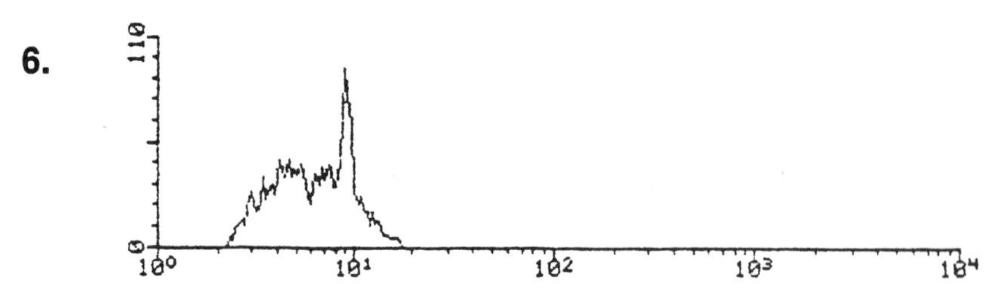
6.
110
0
10^0 10^1 10^2 10^3 10^4

reagents and controls. If large numbers of assays are to be performed, then cheaper assay kits can be constructed using reagents from several different suppliers [Punchard et al., 1989].

Although antibodies and cDNA probes may become available for PGI_2 synthetase, measurement of production will always be advantageous. Production of PGI_2 occurs via a multienzyme system and is dependent on availability of the essential fatty acid arachidonic acid; hence, by measuring production several different characteristics are being monitored as one event. A typical time course of unstimulated, thrombin and arachidonic acid stimulated PGI_2 production by HUVEC measured using Protocol 6.3.2 is given in Fig. 10.7. The methods for measuring PG production given below are the two methods used in the laboratories of the contributing authors. Either is suitable for measuring PGI_2 production.

The ability of normal HUVEC to synthesise and release PGI_2 is rapidly lost, such that by passage 6 the level of stimulation is reduced and, by passage 8–9, has been lost almost completely (Table 10.1). This is therefore a monitor of normal endothelial function. HUVEC cell lines produced by irradiation retain the ability to produce PGI_2 in response to stimulation with arachidonic acid (Table 10.1).

However, HUVEC lines produced by immortalisation with SV40 small and large T did not produce PGI_2 in response to stimulation with thrombin or another stimulant, hydrogen peroxide. Neither did the ECV304 and EA.hy926 cell lines. Constitutive production by the three cell lines was also lower than that of normal HUVEC, (<0.1 ng $PGI_2/10^5$ cells, compared to 0.2 ng/10^5 for HUVEC). Production in response to stimulation with 25μM arachidonic acid was also lower, being <0.2 ng/10^5 for all three cell lines, as compared to 25 ± 3.8 ng/10^5 cells for HUVEC, measured after 20 min of incubation.

Fig. 10.6. FACS analysis of von Willebrand factor expression in irradiated and primary HUVEC. **1.** Irradiated HUVEC 5/94 p 10 stained for vWFactor using FITC conjugated α-Rabbit second antibody. Median fluorescence 6264. **2.** Control for 1. above i.e. second antibody only. Median fluorescence 126. **3.** Irradiated HUVEC 8/93 p20 stained for vWFactor as described in 1. above. Median fluorescence 7234. **4.** Control for 3. above i.e. second antibody only. Median fluorescence 191. **5.** Primary HUVEC p1, not irradiated stained for vWFactor as described in 1. above. Median fluorescence 6264. **6.** Control for 5. above, i.e. second antibody only. Median fluorescence 21. Cells were stained as described in the protocol (section 6.2.2) except that they were first harvested using brief exposure to trypsin/EDTA solution and were centrifuged for each wash. They were examined on a Becton Dickinson FACScan using excitation at 488nm and emmission at 525nm. Results clearly show expression of vWFactor in both normal and irradiated cells although the control levels are slightly elevated in irradiated cultures.

Fig. 10.7. A time course comparing HUVEC PGI_2 production in response to arachidonic acid and thrombin stimulation ($n = 6$).

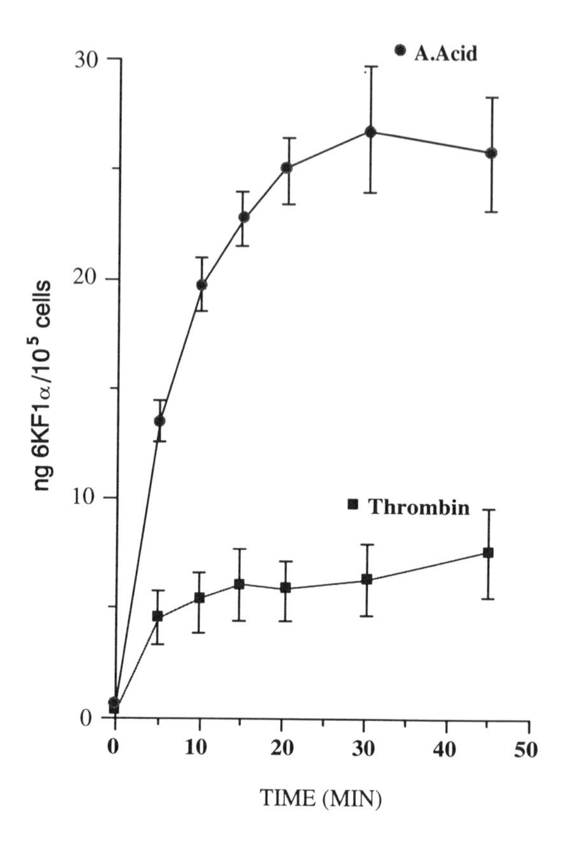

6.3.1. Single time point incubation for prostacyclin production (as used for irradiated cells)

Protocol 6.3.1

Reagents and Materials

- Arachidonic acid is received as a 10-mg vial. This is diluted in 3.28 ml 100mM Na_2CO_3 to give a 10mM stock solution. This is then diluted 1:1000 in M199SF for cell stimulation to give a final concentration of 10μM (for alternative preparation, see Protocol 6.3.2)
- Indomethacin is dissolved in DMSO at a concentration of 0.1M (36 mg/ml). Then 10μl is added to tubes prior to freezing samples at −20°C (for alternative preparation, see Protocol 6.3.2)
- 24-well plates
- Serum-free M199: M199SF
- Sample tubes, 1 ml (Eppendorf)

Table 10.1. Prostacyclin Production in Primary and Irradiated HUVEC[a]

Cells	Passage No.	Prostacyclin Production	
		ng/ml	As %^{age} of primary HUVEC at p1
Primary HUVEC	1	26.7	100
Primary HUVEC	9	2.5	9
Irradiated HUVEC 8/93	10	21.6	81
Irradiated HUVEC 5/94	10	40.5	152
ECV304	Unknown	-0.2	0

[a] This table shows loss of prostacyclin production in normal aged HUVEC between passage 1 and 9, and retention of production in irradiated HUVEC cells at passage 10. Loss of production in spontaneously transformed cells is also included.

Protocol

(a) Grow cells to confluence in 24-well plates in normal growth medium over a period of 3 days. Samples should be in duplicate at least.
(b) Remove growth medium and replace with 1 ml pre-warmed M199SF.
(c) Incubate at 37°C for 15 min.
(d) Remove medium and replace with 1 ml fresh M199SF at 37°C.
(e) Incubate for a further 15 min.
(f) Remove medium and add 1 ml fresh M199SF containing 10 μM arachidonic acid or medium alone (control) at 37°C.
(g) Incubate 2 h.
(h) Remove all of the 1 ml of each sample and add to tubes containing 10 μl indomethacin.
(i) Freeze these tubes at −20°C until ready to assay.

6.3.2. Time course incubation (as used for transfected cells)

Protocol 6.3.2

Reagents and Materials

- *Arachidonic acid:* dilute 10-mg vial, to 12.5mM stock in 2.62 ml ethanol, containing 0.05% butylated hydroxytoluene as an antioxidant (for alternative preparation, see Protocol 6.3.1).
- Store at −70°C.
- On the day of assay dilute the stock to 25 μM in M199SF, 25mM HEPES-buffered, pH 7.4, for stimulation of the HUVEC.
- *Indomethacin:* Dissolve in ethanol at 1 mg/ml and dilute to a final concentration of 1 μg/ml in serum-free M199SF, 25mM HEPES buffered, pH 7.4 (for alternative preparation, see Protocol 6.3.1).

- *Thrombin:* On the day of assay, dilute in same medium as above to a concentration of 0.5 U/ml (final). Store stock commercial supplies at −20°C.

Protocol

(a) Seed cells at 75,000/ml in 0.2 ml M199SG in flat-bottomed 96-well microtitration plates and grow until confluent.
(b) Remove medium.
(c) Wash cells twice with M199SF.
(d) Add 200-μl aliquots of either M199SF or M199 containing thrombin or arachidonic acid (4 wells/time point, 6 time points).
(e) Prepare an additional 15-min time point containing indomethacin as a negative control.
(f) Obtain time zero (T_0) values by removing fresh media immediately and transferring to storage tubes.
(g) Incubate cells at 37°C, 5% CO_2 and at intervals 5, 10, 15, 20, 30, and 45 min remove supernatants
(h) Freeze at −70°C until ready to assay.

6.4. Mitogenic Responses to Growth Factors

Many of the immortalized endothelial cells available show no mitogenic response to growth factors such as fibroblast growth factor (FGF) and epidermal growth factor (EGF), whereas normal primary HUVEC will show good responses. This is probably due to the immortal cells growing at their maximal rate already, but this prevents these cells being used in the study of growth factors. Cells that are irradiated retain their ability to respond to growth factors in a normal fashion. The methods for testing the mitogenic response of endothelial cells are described below.

6.4.1. Growth factor assay

Protocol 6.4.1

Reagents and Materials

- 24-well plates
- Growth medium: M199Ir
- Double-strength growth factor under test
- [^{3}H]-Thymidine [5–10 μCi(0.2–0.4 MBq)/well, 50 Ci/mmol (2 GBq/μmol)]
- or trypsin/EDTA for cell counting

Protocol

(a) Seed 24-well plates with 1.2×10^3 cells per well in 1 ml growth medium without any growth factors added.

(b) Add 1 ml double-strength growth factor in growth medium at concentrations of 0, 2, 20, and 200 ng/ml to give final concentrations of 0, 1, 10, and 100 ng/ml in the wells. These concentrations will cover most growth factors if they are presented in a pure form. To test unknowns a broader range of concentrations may be required.
(c) Incubate the plates for a period of 3–7 days at 37°C.
(d) If the shorter incubation time is used, the cells are pulsed with [^{3}H]-thymidine for 4 h prior to determining incorporation. If the longer incubation time is used, the cells may be counted on a hemocytometer or Coulter counter, following disaggregation in trypsin/EDTA (see Section 2.1.2).
(e) Results are plotted as increase in growth relative to that of control cells grown in the absence of growth factor.

ACKNOWLEDGMENTS

The assistance of Judith Leigh is greatly appreciated in the culture and prostacyclin determinations of the irradiated cultures.

The authors (N.A.P., D.W. and R.P.H.T.) would like to thank Richard Jewers, of The Richard Dimbleby Laboratory of Cancer Virology, The Rayne Institute, St. Thomas' Hospital, London, for the kind donation of the pSVneo containing the full HPV-16 DNA (pSV16), pSV3neo containing small- and large-T antigens of SV40; the pUC plasmid containing c-Ha-*ras*; the pUC plasmid containing c-*myc*, and use of appropriate facilities. They would like to thank the following for contributing to support for this research: The Peel Medical Research Trust; University of London, Lambeth Endowed Charities, The Astra Foundation, the Paterson Charitable Trust, and the Special Trustees for St. Thomas' Hospital.

REFERENCES

Ades EW, Candal F, Swerlick RA, George VG, Summers S, Bosse DC, Lawley TJ (1992): HMEC-1: Establishment of an immortalised human microvascular endothelial cell line. J Invest Dermatol 99: 683–690.

Balconi G, Pietra A, Busacca M, de Gaetano G, Dejana E (1983): Success rate of primary human endothelial cell culture from umbilical cords is influenced by maternal and fetal factors and interval from delivery. In Vitro 19: 807–810.

Boswell DJ, Biggerstaff JP, Wilbourn BR, Thompson RPH, Punchard NA (1992): Prostacyclin production in normal and immortalised umbilical endothelial cells. Clin Sci 82: 19–20.

Caputo JL (1988): Biosafety procedures in cell culture. J Tiss Cult Meth 11: 223–227.

Cockerill GW, Meyer G, Noack L, Vadas MA, Gamble, JR (1994): Characterization of a spontaneously transformed human endothelial cell line. Lab Invest 71: 497–509.

Edgell CJS, Macdonald CC, Graham JB (1983): Permanent cell line expressing human factor VIII- related antigen established by hybridization. Proc Natl Acad Sci 80: 3734–3737.
Faller DV, Kourembanas S, Ginsberg D, Hannan R, Collins T, Ewenstein BM, Pober JS, Tantravahi R (1988): Immortalization of human endothelial cells by murine sarcoma viruses, without morphologic transformation. J Cell Physiol 134: 47–56.
Fickling SA, Tooze JA, Whitely GSt J (1992): Characterization of human umbilical vein endothelial cell lines produced by transfection with the early region of SV40. Exp Cell Res 201: 517–521.
Gimbrone MA, Fareed C (1976): Transformation of cultured human vascular endothelium by SV40 DNA. Cell 9: 685–693.
Grizzle WE, Polt SS (1988): Guidelines to avoid personal contamination by infective agents in Research Laboratories. J Tiss Cult Meth 11: 191–199.
Hoshi H, McKeehan WL (1984): Brain- and liver cell-derived factors are required for growth of human endothelial cells in serum-free culture. Proc Natl Acad Sci 81: 6413–6417.
Ide H, Minamishima Y, Eizuru Y, Okada M, Sakihama K, Katsuki T (1988): Transformation of human endothelial cells by SV40 virions. Microbiol Immunol 32: 45–55.
Iijima S, Ishida M, Nakajima-Iijima S, Hishida T, Watanabe H, Kobayashi T (1991): Immortalisation of human endothelial cells by origin defective Simian Virus 40 DNA. Agric Biol Chem 55: 2847–2853.
Jaffe EA, Nachman RL, Becker CG, Ninick RC (1973): Culture of human endothelial cells derived from umbilical veins. J Clin Invest 52: 2745–2756.
Punchard NA, Murphy GM, Thompson, RPH (1989): Improved radioimmunoassay of prostaglandins with commercial antisera. Biochem Soc Trans 17: 1097–1098.
Punchard NA, Jewers RJ, Sankey MD, Watson DJ, Kell B, Cason J, Best JM, Thompson RPH (1994): Use of viral and cellular oncogenes to immortalise endothelial cells. Biochem Soc Trans 22: 197S.
Suggs JE, Madden MC, Friedman M, Edgell C-JS (1986): Prostacyclin expression by a continuous human cell line derived from vascular endothelium. Blood 4: 825–829.
Takahashi K, Sawasaki Y, Hata J, Mukai K, Goto T (1990): Spontaneous transformation and immortalization of human endothelial cells. In Vitro Cell Devel Biol 25: 265–274.
Takahashi K, Sawasaki Y (1992): Rare spontaneously transformed human endothelial cell line provides useful research tool. In Vitro Cell Devel Biol 25: 380–382.
Warren JB (ed) (1990): "The Endothelium. An Introduction to Current Research." New York: Wiley-Liss.

APPENDIX: MATERIALS AND SUPPLIERS

Materials	Suppliers
Acetone	BDH/Merck
Acetonitrile	BDH/Merck
Arachidonic acid	Sigma
Centrifuge tubes, 50 ml	Bibby Sterilin, Corning
Collagenase A	Boehringer Mannheim
Collagenase Type I	Worthington Biochemical
Cryovials, 2 ml	Sterilin, Corning
Dimethyl sulfoxide	Sigma
Endothelial cell growth medium (CC-3024)	Clonetics Corp. Tissue Culture Services, Sigma
Endothelial cell growth supplement (ECGS) from bovine neural tissue	Sigma
Endothelial cells including HUVECs, human pulmonary artery, human aortic, human dermal microvascular	Clonetics Corp.
FITC swine antirabbit immunoglobulins	Dako; Sigma
Flasks, dishes	Sterilin, Corning
Fetal calf serum	Life Technologies (Paisley, Scotland
Fucose	Sigma
G418	Sigma
Gelatin solution, 2%	Sigma
L-Glutamine	Sigma
Hanks' balanced salt solution (HBSS) w/o Ca^{2+}/Mg^{2+}	Life Technologies, Gibco BRL
HEPES	Sigma
Indomethacin	Sigma
Lipofectin reagent	Life Technologies, Gibco BRL
Earle's 199	Life Technologies, Gibco BRL
M1 serum-free concentrate	Imperial Laboratories (Andover, Hants, England)
M199 (10×)	Sigma
M199 Earle's Salts (1×)	Life Technologies, Gibco BRL
Multichamber slides (Lab-Tec)	Life Technologies, Gibco BRL
Needles, 16G × $1^1/_2$"(cannulae)	Sigma
Newborn calf serum	Life Technologies, Paisley, Scotland
Normal rabbit serum	Dako / Sigma
Paraformaldehyde	BDH/Merck
Penicillin/streptomycin	Sigma; Life Technologies, Paisley, Scotland
Phosphate buffered saline w/o Ca^{2+}/Mg^{2+} (PBSA)	ICN Flow
Plates, 96-well	Sterilin, Corning
Polimount	Schuco International, London, Ltd
6-Keto-prostaglandin F_{1a} kit	Cascade Biochem Ltd, Whiteknights
6-Keto-[5,8,9,11,12,14,15 (*n*)-^{3}H]prostaglandin-F_{1a} ([3H]-6KPGF_{1a})	Amersham
Rabbit anti-6KPGF_{1a}	Advanced Magnetics Inc
Screw-top container, 500ml	Azlon
Sodium bicarbonate 7.5% solution	Sigma
Thrombin	Sigma
Triton X-100	BDH/Merck
Trypan blue	Sigma
Trypsin Solution, 2.5%	Life Technologies (Paisley, Scotland)
Trypsin/EDTA	Sigma
Universals, 30 ml	Sterilin
Vectastain ABC kit, FITC-UEA-1	Vector Laboratories
vWf, rabbit antihuman	Dako / Sigma

Immortalization of Human Mesothelial Cells

Emma L. Duncan, Ruwani De Silva, John F. Lechner[1] and Roger R. Reddel

Children's Medical Research Institute, 214 Hawkesbury Road, Westmead, Sydney 2145, Australia

[1]Inhalation Toxicology Research Institute, P.O. Box 5890, Albuqerque NM 87185

Culture of Immortalized Cells, Edited by R. Ian Freshney and Mary G. Freshney.
ISBN 0-471-12134-7

1. INTRODUCTION

Mesothelial cells are of mesodermal origin and form a simple squamous epithelial layer that lines the serosal surfaces. The culture of mesothelial cells is of considerable interest, as pleural, peritoneal and pericardial mesothelial cells are believed to be the progenitor cells of malignant mesotheliomas. These tumors are associated with exposure to asbestos and other fibrous carcinogens; however, normal human mesothelial cells treated with asbestos fibers *in vitro* do not become tumorigenic [Lechner et al., 1985]. This is in keeping with the observation that normal human cells are highly resistant to neoplastic transformation *in vitro,* while immortal cells are less so. In many experimental systems immortalization appears to be an obligatory prerequisite for tumorigenic transformation [Newbold and Overell, 1983; O'Brien et al., 1986; Reddel et al., 1988]. Therefore, in order to investigate the mechanisms of mesothelial oncogenesis, it is often useful to study immortal mesothelial cells.

Normal mammalian cells undergo only a limited number of divisions in culture, and then senesce [Hayflick, 1965]. Spontaneous immortalization of normal human cells is extremely rare, but genes from DNA tumor viruses such as simian virus 40 (SV40), papillomaviruses and adenoviruses may induce immortalization at low frequency. SV40 is the most commonly used agent for immortalizing nonlymphoid human cells, and provides an extremely useful system for generating stable cell lines that generally retain some differentiated features of normal cells [reviewed in Bryan and Reddel, 1994]. In particular, it has previously been demonstrated that an SV40-immortalized human mesothelial cell line retained many features of normal mesothelial cells [Ke et al., 1989].

After receiving SV40 genes, normal human cells become transformed, and their lifespan is increased by 20–60 population doublings

(PD) beyond the point at which they normally senesce. The cells usually then enter a stage known as crisis, during which no net population increase occurs. During crisis, foci of dividing cells may appear at a frequency of between 10^{-5} [Shay et al., 1993)] and 10^{-9} [E. Duncan et al., unpublished data]; these cells are immortal (Fig. 11.1). SV40 genes are directly responsible for the lifespan extension of SV40-transformed cells, but are not responsible for immortalization since not all SV40-transformed cells become immortal. Escape from crisis presumably involves further, as yet uncharacterized, genetic changes; these changes are probably facilitated by karyotypic instability induced by SV40. Some SV40-transformed cultures become immortal without a period of crisis; presumably these cultures contain a subpopulation of cells in which the genetic changes responsible for immortalization have occurred early, and these already immortal cells overgrow the nonimmortal cells during crisis.

The early region of SV40, which encodes the large-T, small-t, and 17kT antigens, is responsible for the transforming ability of SV40 (reviewed in Bryan and Reddel [1994]). The late region of SV40, which encodes the viral coat proteins, is unnecessary for transformation; human cells become transformed following transfection with a plasmid containing only the SV40 early region. The origin of replication (ori) of the SV40 genome is located with the promoters/enhancers between the coding sequences of the early and late regions. As the ori may cause genetic instability in human cells, it is desirable to use a construct in which the ori function has been deleted, either by mutation of the SV40 promoter/enhancer or by substitution of a heterologous promoter (reviewed in Bryan and Reddel [1994]).

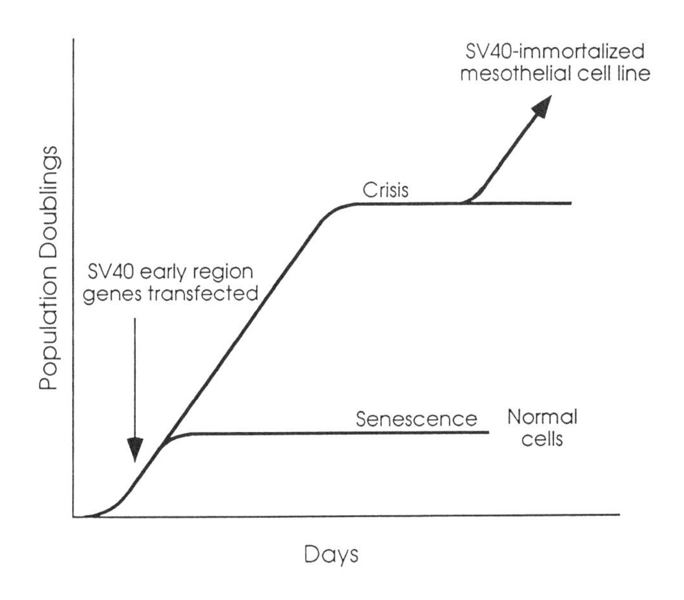

Fig. 11.1. Schematic diagram of the immortalization of human cells by SV40. Cells expressing SV40 early region genes exhibit an extended lifespan compared with normal cells before entering crisis, during which no net population increase occurs. In some, but not all, cultures a subpopulation of cells recommences proliferation and becomes immortal.

This chapter describes the generation of immortal human mesothelial cell lines from normal cells. Normal cells are transfected with the ori-minus plasmid, pRSV-T. This plasmid contains the SV40 early region driven by the Rous sarcoma virus 3' long terminal repeat (RSV 3'LTR). Foci of morphologically transformed cells appear in a background of normal cells 3–4 weeks following transfection. These foci are individually isolated and subcultured until crisis occurs. Several foci must be subcultured since not all SV40-transformed foci give rise to immortal clones. Cells in crisis are maintained by weekly feeding until dividing (immortal) cells appear.

2. PREPARATION OF MEDIA AND REAGENTS REQUIRED FOR CELL CULTURE

The growth medium and growth conditions for human mesothelial cells described below are based on those described previously by Lechner et al. [1985] and Lechner and LaVeck [1985].

2.1. Preparation of LHC Basal Medium

Laboratory of Human Carcinogenesis (LHC) powdered basal medium	154.3 g
Sodium bicarbonate	10 g
Water to	8.0 liters

Adjust the pH to 7.4±0.05 with either 1M hydrochloric acid or 1M sodium hydroxide. Make up to 10 l and then 0.22-μm filter-sterilize. Store at 4°C in the dark.

2.2. Preparation of LHC-MM medium

LHC-MM [Lechner et al., 1985] may be used as the growth medium for normal and immortal mesothelial cells.

Trace-element stock

$MnCl_2$	0.1μM
$(NH4)_6Mo_7O_{24}$	0.1μM
$NiCl_2$	50nM
Na_2SeO_3	3.0μM
Na_2SiO_3	50μM
$SnCl_2$	50nM
NH_4VO_3	0.5μM

Adjust pH to 7.3.

LHC-MM medium

LHC basal medium	500 ml
Insulin, 2mg/ml stock	1.25 ml

Epidermal growth factor (EGF), 5μg/ml	0.5 ml
Gentamicin, 50mg/ml	0.5 ml
Transferrin, 5mg/ml	1.0 ml
Hydrocortisone, 3.6mg/ml in 95% ethanol	0.1 ml
Trace elements stock	5.0 ml
$CaCl_2$, 0.12M	8.0 ml

Filter-sterilize, 0.22 μm. Store at 4°C in the dark.

Just prior to use, add 15 ml fetal bovine serum (FBS) to the medium.

2.3. Preparation of Collagen/Fibronectin-Coated Flasks

The method described below is based on that described previously by Lechner and LaVeck [1985].

Protocol 2.3

Reagents and Materials

- Fibronectin
- LHC basal medium
- Vitrogen 100
- Bovine serum albumin

Dissolve 5 mg fibronectin in 20 ml LHC basal medium at 37°C for 2 h. Add 5 ml Vitrogen 100 (bovine collagen), 50 ml bovine serum albumin (1 mg/ml stock), and 480 ml LHC basal medium. Filter-sterilize, 0.22 μm, and store at 4°C in the dark.

Protocol

(a) At least 2 h before the flasks are required, add 1.5 ml of the collagen/fibronectin solution per 75-cm^2 surface area. Spread the solution over the entire growth surface of the flask. Incubate at 37°C until required.

(b) Just prior to adding the cells to the flask, remove the excess solution by aspiration. If the solution has dried, rinse the flask with 5 ml phosphate buffered saline (PBSA) before adding the cells.

2.4. Preparation of Freezing Medium

L15 medium

Liebowitz 15 (L15) powdered medium (one packet)	14.8 g
Water to	9 liters

Adjust pH to 7.6 ± 0.05 with either 1M hydrochloric acid or 1M sodium hydroxide. Make the volume up to 10 liters and then filter-sterilize, 0.22 μm. Store at 4°C in the dark.

HEPES buffered saline (HBS)

HEPES	4.76 g
NaCl	7.07 g

KCl	0.2 g
Glucose	1.70 g
Na_2HPO_4	1.022 g
Phenol red, 0.5%	0.25 ml
Water to	800 ml.

Adjust the pH to 7.6. Make the volume up, with water to 1 liter. Filter-sterilize, 0.22 μm. Store at 4°C.

2x antibiotic freezing medium

FBS	40 ml
Gentamicin sulfate, 50 mg/ml	0.4 ml
1M HEPES pH 7.6	4 ml
L15 medium	155.6 ml

Filter-sterilize, 0.22 μm. Store at −20°C.

2x DMSO freezing medium

10% w/v polyvinylpyrrolidone, in HBS	40 ml
DMSO	30 ml
1M HEPES pH 7.6	4 ml
L15 medium	126 ml

Filter-sterilize, 0.22 μm. Store at −20°C.

3. IMMORTALIZATION OF HUMAN MESOTHELIAL CELLS

Appropriate biological containment procedures should be observed at all times (see Section 4 and Chapter 2). All equipment and reagents used for cell culture must be sterile and of cell culture grade (see Appendix for suppliers). Except where specified, all solutions should be warmed to 37°C prior to use.

3.1. Initiation of a Normal Human Mesothelial Cell Culture

Mesothelial cells migrate into fluids in the pleural, pericardial, or peritoneal cavities, and may therefore be cultured from such fluids removed for medical indications.

Protocol 3.1

Reagents and Materials

- PBSA
- Medium: LHC-MM
- 150-cm^2 coated cell culture flasks (see Section 2.3)

Protocol

(a) Mesothelial cells are acquired from noncancerous pleural effusions or ascites fluids.

(b) To initiate a normal human mesothelial cell culture, centrifuge the pleural effusion or ascites at 125*g* for 5 min.
(c) Wash the pelleted cells twice in PBSA.
(d) After the final centrifugation, resuspend the cells in 10 ml LHC-MM per 50 ml pleural effusion.
(e) Plate 20 ml resuspended cells per 150-cm^2 coated cell culture flask.
(f) Incubate the cells at 37°C and 3.5% CO_2 in a humidified incubator.
(g) The pleural effusion contains mesothelial cells and blood cells. The mesothelial cells adhere to the coated surface of the flask, while any erythrocytes remain in suspension. The day after the cells are plated, remove the erythrocytes by vacuum aspiration of the medium.
(h) Rinse the mesothelial cells, which remain attached to the growth surface of the flask, with 20 ml PBSA.
(i) Refeed with 20ml fresh medium.
(j) Feed the cells twice weekly until confluent.

3.2. Subculture

The cells are subcultured whenever they become confluent. Cells are generally subcultured at a ratio that allows them to become confluent again in 7 days. Avoid subculturing cells less than 36–48 h after the previous subculture. The protocol is given for a 75-cm^2 flask.

Protocol 3.2

Reagents and Materials

- PBSA.
- Trypsin-EDTA: trypsin, 0.5% (w/v), EDTA, 0.02% w/v
- PBSA/10% FBS
- Medium: LHC-MM
- Centrifuge tubes, 15-ml, or 30-ml Universal containers
- Coated flasks (see Protocol 2.3)

Protocol

(a) Remove the medium and wash the cells twice with 10 ml PBSA.
(b) Add 2 ml trypsin-EDTA and incubate at 37°C for 5 min or until the cells have detached from the flask. Do not leave the cells in the trypsin solution for more than 10 min.
(c) Resuspend the cells in 4 ml PBSA/10% FBS, and transfer to a centrifuge tube.

(d) Wash the flask with a further 4 ml PBSA/10% FBS and transfer this wash to the centrifuge tube.
(e) Centrifuge the cells at 200g for 5 min.
(f) Resuspend the cells in 8 ml LHC-MM, and transfer an appropriate volume to a new coated flask containing 10 ml LHC-MM. We generally transfer at a 1:4–1:16 ratio, depending on the cell growth rate.
(g) Feed the cells twice weekly until confluent.

Normal human mesothelial cells cultured in this way will undergo about 15 population doublings (PD) before senescing, depending on the age of the donor.

3.3. Cryopreservation of Cell Cultures

Protocol 3.3

Reagents and Materials

- 2X antibiotic freezing medium
- 2X DMSO freezing medium
- Medium: LHC-MM
- Coated flasks (see Protocol 2.3)
- Cryopreservation vials
- Insulated container
- −80°C freezer
- Liquid nitrogen freezer

Protocol

(a) Harvest cells by trypsinization (see Section 3.2).
(b) Centrifuge the cells at 200g for 5 min. There should be at least 2×10^6 cells present.
(c) Resuspend the cells in 0.5 ml 2× antibiotic freezing medium.
(d) Add 0.5 ml cold 2× DMSO freezing medium, quickly mix, and transfer to a cryopreservation vial. The cells can be stored on ice for up to 30 min before being transferred to a −80°C freezer in an insulated container, to freeze at ~1°C/min.
(e) The following day, transfer the cells to a liquid nitrogen container for long-term storage.
(f) To thaw cells, warm the vial quickly to 37°C in a waterbath, and transfer the cells to a centrifuge tube containing 10 ml LHC-MM.
(g) Centrifuge at 200g for 5 min.
(h) Resuspend the cell pellet in 10 ml fresh LHC-MM, and transfer the cells to a coated flask.
(i) Culture the cells as usual.

3.4. Transfection of Human Mesothelial Cells with Plasmid pRSV-T

The plasmid pRSV-T [Sakamoto et al., 1993] is illustrated in Figure 11.2. Qiagen® column purification of plasmid DNA produces transfection-quality DNA. The method of transfection described here is that of Lipofectamine™ liposome-mediated plasmid transfer, which is performed essentially as described by the manufacturer. Normal human mesothelial cells can be transfected with pRSV-T at any stage except when they are senescent. The protocol is given for 100-mm dishes.

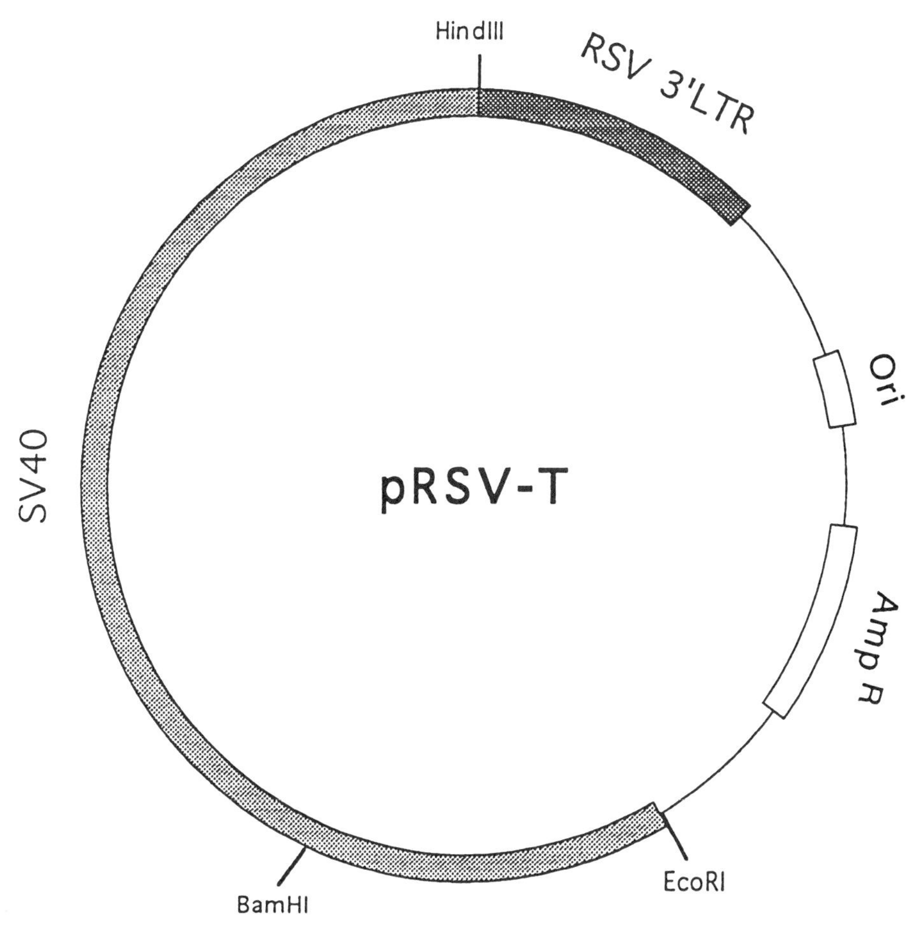

Fig. 11.2. Schematic diagram of plasmid pRSV-T (Sakamoto et al., 1993). The SV40 early region is linked to the Rous sarcoma virus long terminal repeat (RSV 3'LTR) with the plasmid origin of replication (ori) and ampicillin resistance gene (amp^R) from pBR322. The size of the plasmid is 6033 basepairs.

Protocol 3.4

Reagents and Materials

- Coated 100-mm dishes (see Protocol 2.3)
- 17×100-mm polypropylene tubes
- pRSV-T DNA
- Medium: serum-free LHC-MM
- Lipofectamine™

Protocol

(a) Harvest the cells by trypsinization and plate 5×10^5 cells per coated 100-mm dish.

(b) Incubate overnight in a 37°C, 3.5% CO_2 humidified incubator.

(c) At room temperature in a 17×100-mm polypropylene tube, dilute 5µg pRSV-T DNA in serum-free LHC-MM to a total volume of 500µl for each dish to be transfected.

(d) At room temperature in a separate 17×100-mm polypropylene tube, dilute 30 µl Lipofectamine™ reagent in 470 µl serum-free LHC-MM for each dish to be transfected. Mix by drawing the solution once through a sterile plastic pipette.

(e) Mix these two solutions, drawing once through a sterile plastic pipette.

(f) Incubate the final mixture at room temperature for 30 min to allow the DNA-liposome complexes to form. Do not leave the mixture for more than 45 min before adding it to the cells.

(g) Wash the cells twice with serum-free LHC-MM.

(h) Add 5 ml serum-free LHC-MM to each DNA-Lipofectamine™ mixture, to give a final volume of 6 ml per dish to be transfected.

(i) Overlay the cells with 6 ml of the final DNA-Lipofectamine™ mixture.

(j) Incubate the cells for 2–3 h in a 37°C 3.5% CO_2 humidified incubator. Longer incubations may result in extensive cytotoxicity.

(k) Remove the DNA-Lipofectamine™ mixture, and feed the cells with 10 ml LHC-MM.

(l) Feed the cells twice weekly.

3.5. Isolation of SV40-Transformed Colonies

Cells expressing the SV40 early region genes become morphologically transformed and proliferate more rapidly than normal, untransfected cells. These transformed cells form clonal foci which are readily detected on the background of normal cells between 2 and 6 weeks following transfection (Fig. 11.3). At least 10 such foci should be indi-

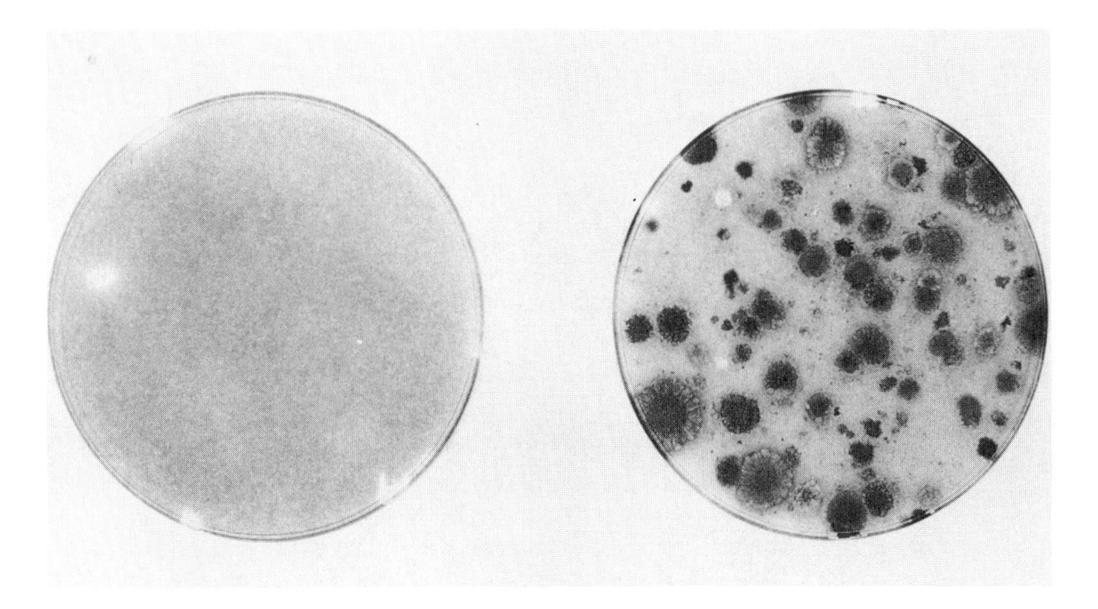

Fig. 11.3. Macroscopic appearance of SV40-transformed foci following transfection of cells with SV40 genes. The plate on the left contains normal human mesothelial cells transfected with a control plasmid; that on the right shows the same cells transfected with pRSV-T. The cells were fixed and stained with Giemsa 6 weeks following transfection.

vidually isolated and subcultured to allow a sufficient probability that an immortal clone will eventually be obtained.

Protocol 3.5

Reagents and Materials

- PBSA
- Trypsin-EDTA: trypsin, 0.5% (w/v), EDTA, 0.02% (w/v)
- Growth medium: LHC-MM
- Glass cloning cylinders
- Coated 6-well cell culture plates
- Coated 25-cm^2 cell culture flasks
- Coated 75-cm^2 cell culture flasks
- Silicone high-vacuum grease
- Forceps
- Glass petri dish

Cover the floor of a glass petri dish with a 2-mm layer of vacuum grease, and embed glass cloning cylinders into the vacuum grease. Minimize the amount of vacuum grease adhering to the inside of the cloning cylinders. Place the lid on the petri dish, and sterilize by autoclaving.

Protocol

(a) To isolate individual colonies, remove the medium from the cell culture dish, and wash the cells twice with PBSA. Place a sterile cloning cylinder around each focus using sterile forceps. Be sure that the cloning cylinder attaches firmly to the dish with the vacuum grease.

(b) Trypsinize the cells within each cloning cylinder with 0.2 ml trypsin-EDTA.

(c) Resuspend the cells in 0.8 ml growth medium.

(d) Transfer the cells from each cloning cylinder into one well of a coated 6-well cell culture plate.

(e) Add 2 ml growth medium. We do not recommend centrifuging the cells before plating them at this step.

(f) Once the cells are confluent, subculture them into a 25-cm^2 cell culture flask, and when confluent again, into a 75-cm^2 cell culture flask.

(g) Once in a 75-cm^2 flask, subculture the cells at an appropriate split ratio as for normal cells.

(h) Cryopreserve the cells as soon as possible (ie when they are confluent in a 75-cm^2 cell culture vessel) and then every second subculture.

3.6. Expansion of Single Clones until Culture Crisis

The isolated foci are subcultured as individual clonal SV40-transformed cell strains. These cells will have an extended lifespan of 20–60 PD compared with their nontransformed counterparts before entering culture crisis. Cells in crisis are generally easy to recognize, often becoming larger and flatter with bizarre shapes. The onset of crisis is usually indicated by a slowing in proliferation rate, while cells in crisis have no net population increase. Cells in crisis may remain viable for up to 9 months, but eventually the number of surviving cells progressively declines.

Immortalization of SV40-transformed human cells occurs at a rate of between 10^{-5} [Shay et al., 1993] and 10^{-9} [E. Duncan et al., unpublished data]. Therefore, each clone should be expanded to at least 5×10^7 cells by the time of crisis. The easiest way to do this is to determine the number of population doublings at which crisis begins by subculturing a single 75-cm^2 flask for each isolated clone until crisis occurs. Cells cryopreserved approximately four subcultures prior to crisis are then thawed and expanded to 5×10^7 cells (5–10 150-cm^2 flasks) per clone by the onset of crisis. SV40-transformed clones in crisis are fed twice weekly. The cultures should be maintained at >50% confluency; thus if the surviving cell population decreases, the cells should be subcultured into fewer or smaller flasks.

3.7. Identification and Subculture of Immortal Clones

In some flasks a small colony of dividing cells may appear in the background of cells in crisis. Such a colony may take up to 9 months to appear. Once a colony appears it is serially subcultured and will usually give rise to an immortal cell line. Cells are often considered immortal when they have undergone 100 PD. It is advisable to cryopreserve cells as soon as possible after escape from crisis, and to regularly cryopreserve newly immortalized cells. This is in order to maintain a stock of immortal cells as close as possible to normal since genetic changes accumulate with prolonged subculturing.

Sometimes dividing cells do not appear as a single colony, but rather appear throughout the flask. This is presumably due to a small subpopulation of cells in which the genetic changes responsible for immortalization have occurred just before crisis, so that in the last subculture before crisis these cells are spread throughout the flask and grow as multiple "colonies."

4. BIOSAFETY

We perform all manipulations at the equivalent of U.S. Biosafety Level 2 (see Chapter 2). Extra caution should be taken with human cells infected with wild-type SV40 virus or when retroviruses containing the SV40 early region genes are used (see Section 5).

5. VARIATIONS

Other growth media may be also used for mesothelial cells. For example, LaVeck et al. [1988] have described a defined serum-free medium capable of sustaining human mesothelial cell multiplication. Briefly, this medium is LHC basal medium supplemented with hydrocortisone, insulin, transferrin, high-density lipoproteins, and one of the following growth factors: epidermal growth factor (EGF), transforming growth factor (TGF)-α, TGF-β, acidic fibroblast growth factor (aFGF), basic FGF (bFGF), platelet-derived growth factor (PDGF), interleukin-1, interleukin-2, interferon-γ, interferon-β, or cholera toxin.

There are two major variations from that described here for the immortalization of human mesothelial cells. The first is the particular plasmid construct and the type of DNA tumor virus DNA used to transform the cells. With regard to SV40, there are several different plasmid constructs expressing SV40 early region genes that will transform human cells as effectively as pRSV-T. It may be desirable to use a temperature sensitive mutant of the SV40 large-T antigen [Radna et al., 1989] or to have the SV40 early region genes under the control of an inducible promoter such as the mouse mammary tumour virus LTR [Wright et al., 1989]; removal of functional large-T antigen by growth at the nonpermissive temperature or in the absence of the inducer respectively will result in growth cessation of immortal cells. It is possible to use a construct that contains the entire SV40 genome, but the presence of late region genes may lower the immortalization frequency [De Silva and Reddel, 1993]. It is also possible to use a selectable marker to isolate transfected cells, but there is no advantage in doing this as cells expressing SV40 early region genes are easily identified by their transformed morphology.

With regard to DNA tumor viruses, we also have immortalized human mesothelial cells following transfection with a plasmid construct expressing the human papillomavirus 16 E6/E7 genes [De Silva et al., 1994].

The second major variation for the immortalization of human mesothelial cells is the method of gene transfer, whether it is the SV40 early region or the transforming genes of other DNA tumor viruses. The method described here for transfer of plasmid DNA into human cells, that of lipofection, is reliable and simple but the reagents used are expensive. Other techniques, such as DNA-phosphate coprecipitation, electroporation, microinjection, or retroviral infection also result in stable expression of transferred genes. It is possible to directly infect human cells with the SV40 virus, but this is not recommended as human cells are semipermissive for SV40 replication, so the cells may give rise to infectious virions, and the SV40 genome may be present at high and variable copy number (compared with one to three copies of transfected DNA).

6. CELL IDENTIFICATION

Nonproliferating cell cultures, such as senescent normal cells and SV40-transformed cells in crisis, are vulnerable to cross-contamination by immortal cell lines. It is therefore extremely important not only to confirm that the cells initiated from pleural effusions are mesothelial cells but also that immortal cells are in fact derived from these normal cells.

6.1. Cell Morphology

Normal human mesothelial cells mostly have a polygonal morphology. Often there are binucleate or polyploid cells within the population (Fig. 11.4A).

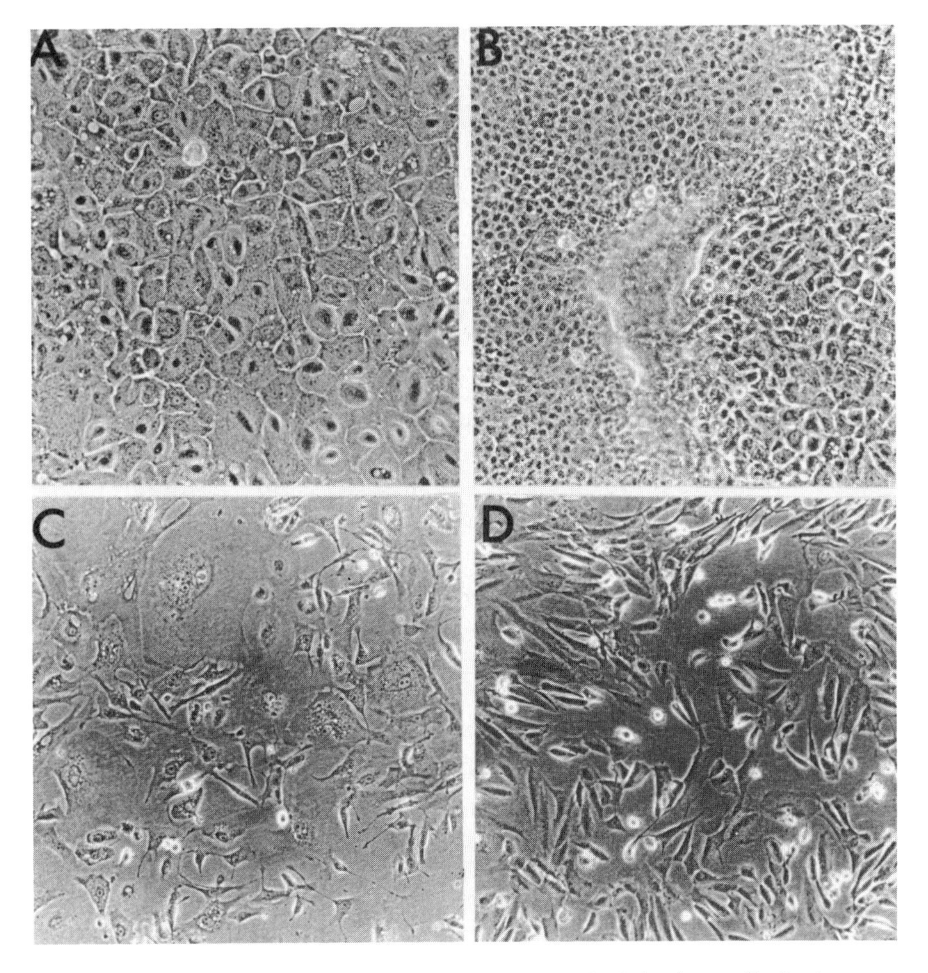

Fig. 11.4. Phase-contrast morphology of human mesothelial cells during immortalization by SV40. (A) Normal cells; (B) SV40-transformed cells (left) overgrowing their normal counterparts (right); (C) SV40-transformed cells in crisis; (D) SV40-immortalized cells.

SV40-transformed human mesothelial cells have an altered morphology compared with that of normal cells, and are capable of forming tightly packed foci of polygonal cells. Figure 11.4B shows SV40-transformed human mesothelial cells on the left overgrowing normal mesothelial cells on the right. When SV40-transformed cells enter crisis, they become larger and flatter, and often exhibit bizarre morphology (Fig. 11.4C). SV40-immortalized human mesothelial cells resemble precrisis SV40-transformed cells (Fig. 11.4D).

6.2. Markers for Mesothelial Cells

There is no one marker that can unequivocally identify mesothelial cells; however, there are several markers, which together give a high probability of cells being mesothelial [Ke et al., 1989]. Mesothelial cells stain positively for keratin and vimentin intermediate filaments. The pattern of keratin staining changes with growth conditions in both normal and SV40-immortalized mesothelial cells [Connell and Rheinwald, 1983; Ke et al., 1989]; confluent cells, or cells lacking hydrocortisone and EGF, display a filamentous keratin structure, while exponentially growing cells display a punctate keratin structure. The two monoclonal antibodies ME1 and ME2 [Stahel et al., 1988] react with the surface membrane of mesothelial cells. Mesodermal cells characteristically express plasminogen activator inhibitor (PAI)-1 [Rheinwald et al., 1987]. Mesothelial cells respond mitogenically to a range of growth factors such as EGF, TGF-α, acidic and basic FGF, PDGF, interleukins-1 and 2, interferon-γ, and TGF-β [LaVeck et al., 1988]. Expression of the Wilms' tumor suppressor gene WT-1 may also be a marker for cells of mesodermal origin [Walker et al., 1994].

6.3. Immunostaining of SV40 T Antigens

SV40-transformed mesothelial cells should be checked for SV40 T-antigen expression. The simplest method for this is immunostaining using the commercially available antibody PAb 419, which recognizes all three SV40 T-antigens. SV40-transformed cells characteristically stain T-antigen positive in the nucleus with nucleolar sparing.

6.4. Confirmation of Immortal Cell Line Origin

It is necessary to confirm that all immortalized cell lines are derived from the parental cell strain. We routinely perform DNA fingerprinting, which relies on differences in repetitive DNA regions between different individuals. Our protocol is single locus DNA fingerprinting using alkaline-phosphatase labeled probes [Duncan et al., 1993]; however, any fingerprinting technique can be used. It is also possible to use isoenzyme or polymerase chain reaction–based microsatellite analysis for this purpose.

In some cases, such as where more than one immortal clone is generated from the one cell strain, it may be useful to determine the integration pattern of SV40 DNA by Southern analysis. This will confirm that the different immortalized cell lines are independent from one another.

ACKNOWLEDGMENTS

We thank Elsa Moy for critical review of the manuscript. This work was commenced in the laboratory of Dr. Curtis Harris at the National

Cancer Institute. Work at the Children's Medical Research Institute was supported by the New South Wales Cancer Council.

REFERENCES

Bryan TM, Reddel RR (1994): SV40-induced immortalization of human cells. Crit Rev Oncogen. 5: 331–357.
Connell ND, Rheinwald JG (1983): Regulation of the cytoskeleton in mesothelial cells: reversible loss of keratin and increase in vimentin during rapid growth in culture. Cell 34: 245–253.
De Silva R, Reddel RR (1993): Similar simian virus 40-induced immortalization frequency of fibroblasts and epithelial cells from human large airways. Cell Mol Biol Res 39: 101–110
De Silva R, Whitaker NJ, Rogan EM, Reddel RR (1994): HPV-16 E6 and E7 genes, like SV40 early region genes, are insufficient for immortalization of human mesothelial and bronchial epithelial cells. Exp Cell Res 213: 418–427.
Duncan EL, Whitaker NJ, Moy EL, Reddel RR (1993): Assignment of SV40-immortalized cells to more than one complementation group for immortalization. Exp Cell Res 205: 337–344.
Hayflick L (1965): The limited in vitro lifetime of human diploid cell strains. Exp Cell Res 37: 614–636.
Ke Y, Reddel RR, Gerwin BI, Reddel HK, Somers ANA, McMenamin MG, LaVeck MA, Stahel RA, Lechner JF, Harris CC (1989): Establishment of a human in vitro mesothelial cell model system for investigating mechanisms of asbestos-induced mesothelioma. Am J Pathol 134: 979–991.
LaVeck MA, Somers ANA, Moore LL, Gerwin BI, Lechner JF (1988): Dissimilar peptide growth factors can induce normal human mesothelial cell multiplication. In Vitro Cell Devel Biol 24: 1077–1084.
Lechner JF, LaVeck MA (1985): A serum-free method for culturing normal human bronchial epithelial cells at clonal density. J Tiss Cult Meth 9: 43–48.
Lechner JF, Tokiwa T, LaVeck M, Benedict WF, Banks-Schlegel S, Yeager H Jr, Banerjee A, Harris CC (1985): Asbestos-associated chromosomal changes in human mesothelial cells. Proc Natl Acad Sci USA 82: 3884–3888.
Newbold RF, Overell RW (1983): Fibroblast immortality is a prerequisite for transformation by EJ c-Ha-*ras* oncogene. Nature 304: 648–651.
O'Brien W, Stenman G, Sager R (1986): Suppression of tumor growth by senescence in virally transformed human fibroblasts. Proc Natl Acad Sci USA 83: 8659–8663.
Radna RL, Caton Y, Jha KK, Kaplan P, Li G, Traganos F, Ozer HL (1989): Growth of immortal simian virus 40 *tsA*-transformed human fibroblasts is temperature dependent. Mol Cell Biol 9: 3093–3096.
Reddel RR, Ke Y, Kaighn ME, Malan-Shibley L, Lechner JF, Rhim JS, Harris CC (1988): Human bronchial epithelial cells neoplastically transformed by v-Ki-ras: altered response to inducers of terminal squamous differentiation. Oncogene Res 3: 401–408.
Rheinwald JG, Jorgensen JL, Hahn WC, Terpstra AJ, O'Connell TM, Plummer KK (1987): Mesosecrin: a secreted glycoprotein produced in abundance by human mesothelial, endothelial, and kidney epithelial cells in culture. J Cell Biol 104: 263–275.
Sakamoto K, Howard T, Ogryzko V, Xu N-Z, Corsico CC, Jones DH, Howard B. (1993): Relative mitogenic activities of wild-type and retinoblastoma binding-defective SV40 T antigens in serum-deprived and senescent human diploid fibroblasts. Oncogene 8: 1887–1893.
Shay JW, Van Der Haegen BA, Ying Y, Wright WE (1993): The frequency of immortalization of human fibroblasts and mammary epithelial cells transfected with SV40 large T-antigen. Exp Cell Res 209: 45–52.

Stahel RA, O'Hara CJ, Waibel R, Martin A (1988): Monoclonal antibodies against mesothelial membrane antigen discriminate between malignant mesothelioma and lung adenocarcinoma. Int J Cancer 41: 218–223.
Walker C, Rutten F, Yuan X, Pass H, Mew DM, Everitt J (1994): Wilm's tumor suppressor gene expression in rat and human mesothelioma. Cancer Res 54: 3101–3106.
Wright WE, Pereira-Smith OM, Shay JW (1989): Reversible cellular senescence: implications for immortalization of normal human diploid fibroblasts. Mol Cell Biol 9: 3088–3092.

APPENDIX: MATERIALS AND SUPPLIERS

Cell Culture Equipment

Equipment	Suppliers
Cell culture dishes and flasks	Corning
Cryopreservation vials (25702)	Corning
Filter unit, 0.22 μm (25942)	Corning
Culture plates, 6-well (25810)	Becton Dickinson
Polypropylene tubes (2059)	Becton Dickinson
Plastic pipettes, sterile	Becton Dickinson
Petri dish, glass (265524)	Bibby Sterilin. Ltd.
Cloning cylinders, 10 mm glass (2090-01010)	Bellco Biotechnology

Media, Chemicals and Reagents for Cell Culture

Materials	Suppliers
LHC basal medium with stocks 4,11, calcium, and glutamine (441)	Biofluids Inc.
Sodium bicarbonate (S5761)	Sigma Chemical Co.
Hydrochloric acid (10125)	BDH/Merck
Sodium hydroxide (10252)	BDH/Merck
Insulin (I6634)	Sigma Chemical Co.
Epidermal growth factor (E4127)	Sigma Chemical Co.
Gentamicin sulfate (G1397)	Sigma Chemical Co.
Transferrin (T8027)	Sigma Chemical Co.
Hydrocortisone (H0888)	Sigma Chemical Co.
Ethanol (10107)	British Drug House
Manganese chloride (M1144)	Sigma Chemical Co.
Ammonium metavanadate (A2286)	Sigma Chemical Co.
Nickel chloride (N6136)	Sigma Chemical Co.
Sodium selenite (S5261)	Sigma Chemical Co.
Sodium metasilicate(S4392)	Sigma Chemical Co.
Stannous chloride (S9262)	Sigma Chemical Co.
Molybdic acid (M1019)	Sigma Chemical Co.
Calcium chloride (127)	Ajax Chemicals
Fetal bovine serum (15-010-0500V)	Trace Biosciences
Fibronectin (F4759)	Sigma Chemical Co.
Vitrogen 100 (PC0701)	Collagen Corporation
Bovine serum albumin (735 086)	Boehringer Mannheim
Phosphate buffered saline, Ca^{2+}/Mg^{2+} free 11-075-0500V)	Trace Bioscience
Liebowitz 15 medium (L4386)	Sigma Chemical Co.
HEPES (H9136)	Sigma Chemical Co.
Sodium chloride (10241)	British Drug House

APPENDIX (*Continued*)

Materials	Suppliers
Potassium chloride (383)	Ajax Chemicals
Glucose (G8270)	Sigma Chemical Co.
Disodium hydrogen phosphate (10249)	British Drug House
Phenol red (20090)	British Drug House
Polyvinylpyrrolidon (20611)	United States Biochemical Corp.
Dimethylsulfoxide (D-2650)	Sigma Chemical Co.
Trypsin-EDTA (1 : 250) (21-160-0100VG)	Trace Biosciences
Vacuum grease (D1400)	Corning
Lipofectamine reagent (18324-012)	Gibco-BRL

Other Reagents

Reagents	Suppliers
Qiagen Plasmid Mega Kit (12181)	Qiagen GmbH
Monoclonal antibody PAb 419 (DP01)	Oncogene Science Inc.
Keratin polyclonal antibody (657922)	ICN/Immunobiologicals
Vimentin polyclonal antibody (657942)	ICN/Immunobiologicals
Alkaline phosphatase labeled single-locus DNA fingerprinting probes	CellMark Diagnostics

12

Chondrocytes

Bénédicte Benoit, Sophie Thenet-Gauci, and
Monique Adolphe

Laboratoire de Pharmacologie Cellulaire, Ecole Pratique des Hautes Etudes, 15 rue de l'Ecole de Médecine, Paris 75006, France

Culture of Immortalized Cells, Edited by R. Ian Freshney and Mary G. Freshney.
ISBN 0-471-12134-7

1. INTRODUCTION

Chondrocytes are highly specialized cells of mesenchymal origin that are responsible for synthesis, maintenance, and degradation of cartilage matrix. This matrix consists of complexes of large proteoglycan aggregates (aggrecan, hyaluronan, and link protein), collagen fibrils (mostly Type II, with smaller amounts of Type XI and Type IX) and other matrix proteins [Petit et al., 1992; Handley and Ng, 1992].

Attempts of immortalization of this cell type have been justified by the high phenotypic instability of chondrocytes in culture, which rapidly lose their polygonal morphology for a fibroblastoid one, with a concomitant switch from predominant synthesis of Type II to Type I collagen and loss of aggrecan synthesis [Von der Mark, 1986].

This chapter deals with the generation of immortalized cells from normal well-differentiated chondrocytes. However, establishment of cell lines from embryonal carcinoma or teratocarcinoma cells [Kellermann et al., 1987, Atsumi et al., 1990], fetal bone [Ohlsson et al., 1993], or chondrosarcoma [Fujisawa et al., 1991] that are able to express a cartilage phenotype, has also been described.

Chondrocyte immortalization seems to be relatively easily obtained using classical methods and immortalizing genes, and some cell lines that retain, at least transiently, the expression of a differentiated phenotype, have been described. Table 12.1 presents the characteristics of some immortalized chondrocyte lines from animal origin, with details concerning their generation and characterization. From most studies, it seems that SV40 large-T antigen and the *myc* oncogene are able to immortalize chondrocytes and are not incompatible with the expression

Table 12.1. Origin and Characteristics of Chondrocyte Cell Lines

Species	Chondrocyte	Oncogene	Phenotype	Possibility of Reexpression	Reference
Quail	Embryo tibiae	v-*myc* (MC29 retrovirus)	Expression of Collagen II Alcian blue staining Expression of proteoglycan core protein	Not studied	Gionti et al. [1985]
Rat	Fetal costal	v-*myc* (recombinant retrovirus)	Expression of proteoglycan core and link proteins Low level of Collagen II (1% mRNA, no protein detected) Expression of proα 2(I) mRNA	Not studied	Horton et al. [1988]
Mouse	Embryo limb	SV40 large T (recombinant retrovirus)	Expression of Collagen II, IX, XI, aggrecan, link protein Expression of Collagen I	Not studied	Mallein-Gerin and Olsen [1993]
Rabbit	Articular	SV40 large T + small t (plasmid transfection)	No Collagen II expression Expression of Collagen I No alcian blue staining	No phenotype modification after Tridim. culture (agarose, collagen) DHCB or staurosporin treatment	Thenet et al. [1992]
Rabbit	Articular	SV40 large T or *myc* (plasmid transfection)	Expression of Collagen II mRNA at early passages From passage 10: loss of Collagen II and expression of Collagen I mRNA	Same as above	Thenet-Gauci, unpublished data
Human	Fetus epiphyseal	SV40 large T (plasmid transfection)	No Collagen II mRNA expression Expression of α 1(I) mRNA	No phenotype modification after tridim. culture in alginate beads	Bonaventure et al. [1994]
Human	Juvenile costal	SV40 large T (plasmid and recombinant retrovirus)	Expression of Collagen II, IX, XI, aggrecan, link protein in presence of an insulin-containing serum substitute	—	Goldring et al. [1994a]
Human	Fetus epiphyseal	SV40 large T (plasmid transfection)	No Collagen II expression Expression of Collagen I and Collagen III	—	Benoit et al. [1995]

[a]This table indicates species and type of cartilages chondrocytes are isolated from, immortalizing oncogenes and vectors that have been employed, phenotypes of immortalized chondrocytes concerning the main components of the cartilaginous matrix, the possibility of reexpression of a more differentiated phenotype under conditions that have been shown to trigger redifferentiation in normal dedifferentiated chondrocyte, and bibliographic references.

of the differentiated phenotype. We recently confirmed, using transient transfection experiments, the compatibility of SV40 large-T antigen and *myc* expression with the activity of Collagen II regulatory regions (unpublished data).

Oncogene-mediated immortalization seems to delay, but does not seem to totally prevent, the process of culture-dependent loss of differentiated functions that occurs more or less rapidly depending on the cell line. Moreover, it may block reexpression signals as it has been suggested for rabbit or human articular chondrocytes [Thenet et al., 1992; Bonaventure et al., 1994].

Data from the literature concerning human chondrocyte immortalization are also shown in Table 12.1. In accordance with general data on human cell immortalization, only SV40 large T was used. In two studies reported [Bonaventure et al., 1994; Benoit et al., 1995], the differentiated phenotype was lost on immortalization. However, Goldring et al. [1994a] described two SV40-immortalized cell lines from human juvenile costal chondrocytes that expressed most chondrocyte-specific markers. This expression was observed only after the replacement of serum by a serum substitute (see Section 7).

Considering that dedifferentiation seems often to be, in the long term, inevitable, more specific strategies to immortalize chondrocytes would be worth testing. The expression of the immortalizing gene could be driven by the promoter and the enhancer of the collagen II gene [Kohno et al., 1985; Horton et al., 1987], namely by regulatory regions activated only in differentiated chondrocytes. We are currently investigating this strategy that could permit a better maintainance of the differentiated phenotype.

Methodologies used in our laboratory to immortalize rabbit or human articular chondrocytes are described in details in the following sections. Other methods concerning different types of chondrocytes will also be mentioned. Criteria for characterization of immortalized chondrocytes will then be described.

2. MEDIA AND REAGENTS

2.1. Culture Media

2.1.1. Medium for cartilage dissociation (F12G)

Ham's F12 medium supplemented with 20 μg/ml gentamycin is used for dissection of the joints and removal of cartilage as well as for enzymic digestion (Section 2.2.1.). For dissection of human fetal samples, this medium is supplemented with 5 μg/ml metronidazole.

2.1.2. Complete culture medium

Ham's F12, 5 μg/ml gentamicin (from aliquots stored at 5 mg/ml at −20°C) supplemented with 10% v/v fetal calf serum (FCS).

Note: Different batches of serum have to be compared for their ability to support chondrocyte growth as well as expression of the differentiated phenotype (assessed, for example, by immunolabelling of Type II collagen). When one lot is selected, several months supply should be purchased to avoid frequent change of serum batch.

2.2. Enzymes

All solutions are sterilized using 0.22-μm filters.

2.2.1. Enzyme solutions for cartilage dissociation

0.05% w/v porcine pancreatic trypsin (Biosys) in Ham's F12
0.3 % w/v collagenase Type 1 CLS1 (Worthington Biochemical Corp.) in Ham's F12
0.06% w/v collagenase in Ham's F12 + 10% FCS

Note: New batches of collagenase should be tested for efficient cartilage dissociation by comparison with previous batch.

2.2.2. Trypsin-EDTA solution

Used for subculture

0.1% w/v porcine pancreatic trypsin-0.02% w/v EDTA in PBSA

2.2.3. Hyaluronidase solution

Used before and during transfection (Section 3.2.)
4 U/ml hyaluronidase (Type I-S, Sigma) in complete medium

Prepare 4 U/μl aliquots in Hanks' BSS (HBSS) and store at −20°C.

2.3. Plasmids

2.3.1. Choice of plasmid

Two plasmids must be chosen. One will carry the immortalizing oncogene and the second one, a selection gene that will allow elimination of nontransfected cells from the culture. Ideally, the two genes should be carried on the same vector (as is often the case with recombinant retrovirus; see Section 4.3.), but cotransfection of both plasmids gives very good results if one respects the following proportions: 1 copy of "selection" plasmid for 5 or 10 copies of "immortalizing" plasmid.

We used the plasmid pSV2tkNeoβ [Nicolas and Berg, 1983], which carries the Neo^r gene encoding aminoglycoside phosphotransferase. This confers to eukaryotic cells resistance to geneticin (G418; see Section 3.3).

As discussed in the introductory section, SV40 large T or the *myc* oncogene seem to be the most suitable for chondrocyte immortalization. We used the following plasmids: pSVmyc [Land et al., 1983], encoding the *myc* oncogene; pAS [Benoist and Chambon, 1981] encoding SV40 early region without replication origin; pS701 encoding for SV40 large-T antigen without replication origin (generous gift of Dr. J. Feunteun, Villejuif, France). However, many other vectors carrying these genes have been published and could be used as well.

2.3.2. Preparation

Plasmids are purified from bacterial cultures using a plasmid-DNA-isolation kit (Plasmid Maxi Kit, Qiagen) following the manufacturer's instructions. At the final step, air-dry DNA, dissolve in sterile milliQ H2O, adjust the DNA concentration to 0.5 or 1 mg/ml and store at −20°C. Plasmid solutions can be thawed and refrozen several times without any problem.

2.4. Calcium Phosphate–DNA Coprecipitate

2.4.1. Stock reagents

All solutions are sterilized using 0.22-µm filters. Do not heat sterilize. Prepare 1-ml aliquots and store at −20°C.

2.0M $CaCl_2.2H_2O$

2X HBS:

HEPES	50mM
NaCl	280mM
$Na_2HPO_4 \cdot 12H_2O$	1.5mM

Adjust to pH 7.1 with 1N NaOH.

5X TE:

Tris	50mM
EDTA	5mM

Adjust to pH 7.6 with concentrated HCl.

2.4.2. Coprecipitate preparation

Coprecipitate must be prepared just before transfection (see Section 3.2.). For each 100-mm dish of chondrocytes to be transfected prepare 1 ml of coprecipitate. Prepare in sterile 5-ml plastic tubes as follows.

Protocol 2.4.2

Reagents and Materials

- *Tube A*

Plasmid DNA: immortalizing plasmid	20 μg
pSV2tkNeoβ	4.0 μg
2M $CaCl_2$	62.5 μl
1X TE to	500 μl

- *Tube B*

2X HBS	500 μl

- Pasteur pipette drawn to <0.25 mm

Protocol

(a) Add A dropwise into B while bubbling air through the solution using an electric pipetting device and a very thin Pasteur pipette.
(b) Incubate the mixture for 30 min at room temperature.
(c) The solution becomes opalescent on formation of the precipitate.

3. ARTICULAR CHONDROCYTE ISOLATION AND IMMORTALIZATION

3.1. Isolation

3.1.1. Dissection

Protocol 3.1.1

Reagents and Materials

- Primary cultures can be prepared from knee, shoulder, and hip joints. Whenever possible, fetal or young donors should be selected in preference to adult donors in order to obtain higher quantities of cells and to be free from cellular alterations that can accompany cellular aging. One-month-old "Fauve de Bourgogne" rabbits or over 12-week-old human fetus are suitable donors.
- Dissection medium: F12G.

Sterile instruments:
- Scalpels
- Blades: 10,15, 24.
- Forceps
- Scissors: 115 mm, 160 mm

Protocol

(a) Wash tissue pieces thoroughly with F12G (see Section 2.1) before dissection. Dissection should begin without delay.

(b) Remove skin, muscle, and tendons from joints.

(c) Using a scalpel, carefully cut cartilage fragments from articulations, free of adherent connective tissue.

3.1.2. Dissociation and primary culture [Green, 1971]

Avoid cartilage drying by carrying out all dissociation steps in 100-mm non-tissue-culture-treated dishes containing Ham's F12 medium supplemented with antibiotics (see Section 2.1). Quantities of enzyme solutions indicated below are suitable for tibiofemoral and scapulohumeral articulations from one rabbit or about two articulations from one human fetus; they must be adapted to the quantity of cartilage to be dissociated.

Protocol 3.1.2

Reagents and Materials

Sterile
- Complete culture medium (see Section 2.1)
- Trypsin solution, 0.05% (see Section 2.2.1)
- Collagenase solution, 0.3% (see Section 2.2.1)
- Collagenase solution, 0.6% (see Section 2.2.1)
- Scalpels
- 30-ml flat-bottomed vial
- Magnetic follower ("flea")
- Non-tissue-culture-treated petri dishes, 100 mm
- Tissue-culture grade petri dishes, 100 mm
- 50-ml centrifuge tubes
- Nylon filter, 70 μm (Falcon "cell strainer" 2350)

Nonsterile
- Vortex mixer
- Hemocytometer
- Magnetic stirrer

(a) Mince cartilage slices into 1-mm^3 pieces using crossed scalpels.
(b) Transfer cartilage fragments into a 30-ml flat bottom vial with small magnetic follower.
(c) Add 10 ml of 0.05% trypsin solution and incubate under moderate magnetic agitation for 25 min in sealed vial at room temperature.
(d) Remove trypsin solution after fragment sedimentation. Add 10 ml of 0.3% collagenase solution and incubate under moderate magnetic agitation for 30 min in sealed vial at room temperature.
(e) Remove collagenase solution after fragment sedimentation.
(f) Add 10 ml of 0.06% collagenase solution.
(g) Transfer into a 100-mm dish (non-tissue-culture-treated).
(h) Rinse vial with another 10 ml of 0.06% collagenase solution, transfer into dish and incubate cartilage fragments overnight at 37°C in a 5% CO_2 incubator.
(i) Transfer to 50-ml centrifuge tube and vortex for a few seconds.
(j) Remove residual materials left after digestion by passing the digested material through a 70-μm nylon filter (Falcon).
(k) Centrifuge at 400g for 10 min.
(l) Resuspend the cell pellet in a volume of complete medium and quantify the cell concentration using a hemocytometer.
(m) Distribute the cell suspension, 10^7 cells/100-mm plastic culture dish.
(n) Incubate at 37°C in a 5% CO_2 incubator.

Cells in primary culture adhere in about 2 days and display a polygonal morphology at confluency (Fig. 12.1).

3.2. Transfection in Primary Culture

Transfect when cells reach approximately 60–80% confluence.

Protocol 3.2

Reagents and Materials

- Hyaluronidase, 4 U/ml in complete medium, prepared from stock solution (see Section 2.2.3.)
- Calcium phosphate-DNA coprecipitate as described in Section 2.4.2
- PBSA
- Glycerol, 15% in 1X HBS
- Complete culture medium (see Section 2.1)

Protocol

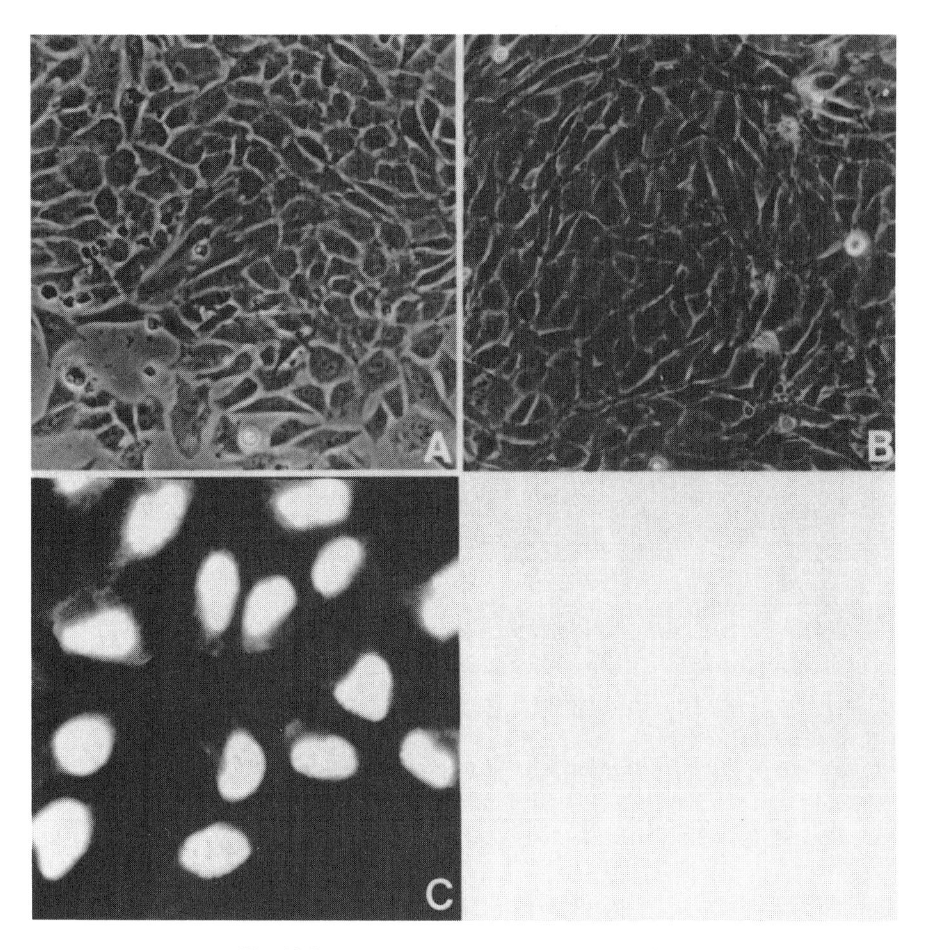

Fig. 12.1. (A,B) Phase-contrast micrographs of rabbit articular chondrocytes, in primary culture (A), or after immortalization by the SV40 large-T antigen (B). Magnification 80×. (C) Indirect immunofluorescent staining for SV40 large-T antigen within the nuclei of immortalized chondrocytes. Magnification: ×630. Reproduced from Thenet et al. [1992], with permission from the publisher.

Protocol

(a) Hyaluronidase pretreatment

One day before transfection, replace the medium by fresh complete medium containing 4 U/ml hyaluronidase.

Note: The presence of hyaluronidase in the culture medium before and during the transfection was shown to enhance by several times the rate of transfection by the calcium phosphate–DNA coprecipitation method in rabbit articular chondrocytes, probably by partially digesting the surrounding matrix and enhancing the accessibility of the coprecipitate to the cell membrane.

(b) Transfection
   (i) Prepare the calcium phosphate-DNA coprecipitate (see Section 2.4.2).
   (ii) While the precipitate is forming at room temperature, replace the medium by fresh complete medium containing 4 U/ml hyaluronidase (9 ml medium/100-mm dish).
   (iii) Add dropwise 1 ml precipitate/dish. Incubate at 37°C for 5 h in a 5% CO_2 incubator.
   (iv) The calcium phosphate–DNA coprecipitate can be visualized under the microscope as small black grains. These should not form large aggregates.
(c) Glycerol shock
   (i) Remove the medium by aspiration and wash the monolayer twice with PBSA.
   (ii) Add 2 ml of 15% glycerol in 1X HBS for exactly 30 s.
   (iii) Add 10 ml PBSA and remove immediately by aspiration.
   (iv) Wash twice with PBSA.
   (v) Add 10 ml of complete culture medium and incubate the cells at 37°C.
   (vi) Change medium every 2 days during 5 days.

3.3. G418 Selection

Protocol 3.3

Reagents and Materials

- Geneticin (G418 sulfate) 400 μg/ml, freshly prepared in Ham's F-12 medium + 10% FCS

Protocol

(a) Remove medium and add 10 ml per 100-mm dish of 400 μg/ml geneticin.
(b) Repeat treatment every 2 or 3 days for 3 weeks. Cell number dramatically decreases on geneticin treatment. At the end of selection, the remaining cells are usually very sparse.
(c) Allow immortalized clones to grow for a few days until they reached ~2-mm diameter before cloning.

According to the literature and our experience, the processes of immortalization of cells from nonhuman or human origin are very different. After the 3-week selection period, immortalized rabbit chondrocytes appeared as small clones of dividing cells. For human cells, the very rare remaining cells after geneticin selection entered a latency period of 7–8 weeks. At the end of this period, clones appeared and grew regularly for approximately 35 population doublings and then reentered a second latency period called "crisis." Crisis has been described as a period of balanced cell growth and cell death [Shay et al., 1991]. This "crisis" period, which has not always been reported for immortalized human chondrocytes [Bonaventure et al., 1994], lasted for approximately 2 months in our experiments (Fig. 12.2) [Benoit et al., 1995].

3.4. Cloning Procedure

Transfected clones are individually picked with cloning rings.

Protocol 3.4

Reagents and Materials

- PBSA
- Trypsin-EDTA (Section 2.2.2)
- Complete culture medium (see Section 2.1)
- Heat sterilized cloning rings and silicone grease in glass petri dish
- Micropipette tips (e.g., Eppendorf)

Protocol

(a) Remove medium.
(b) Rapidly select and circle clones to be isolated by viewing against the light.
(c) Check chosen clones under microscope.
(d) With sterile forceps, press ring onto a dish containing sterilized silicone grease, then gently press ring onto an empty dish to make sure that all the circumference is covered and to avoid grease excess on the ring.
(e) Place rings delicately over the selected clones.
(f) Trypsinize cells within the rings using sterile Eppendorf tips to deliver the solutions to the rings:
    (i) Add trypsin-EDTA to fill the ring, leave 15 s, and remove.
    (ii) Incubate for approximately 10 min (check trypsinization under microscope).
    (iii) Add medium to fill the ring and disperse cells by pipetting up and down.
(g) Seed individually picked clones in 35-mm dishes.

3.5. Cryopreservation

Chondrocytes can be cryopreserved in complete medium supplemented with 10% DMSO. However, this procedure is to be avoided

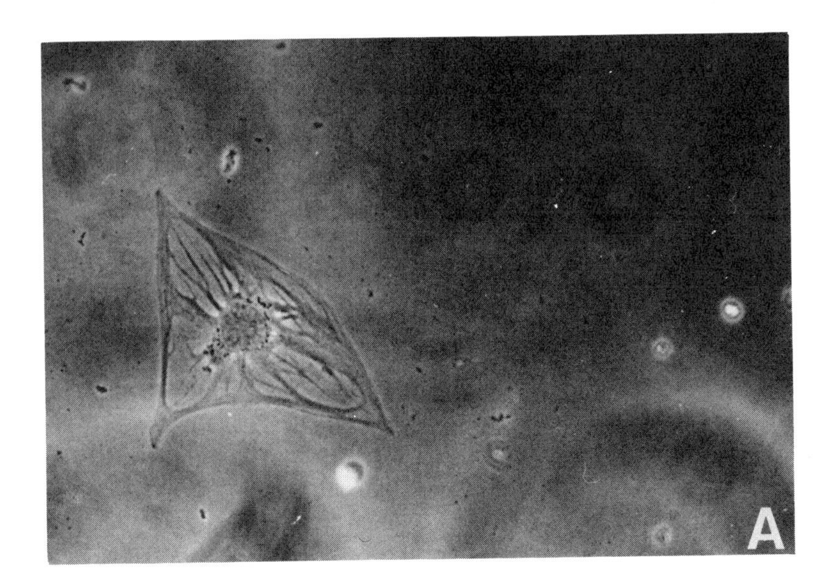

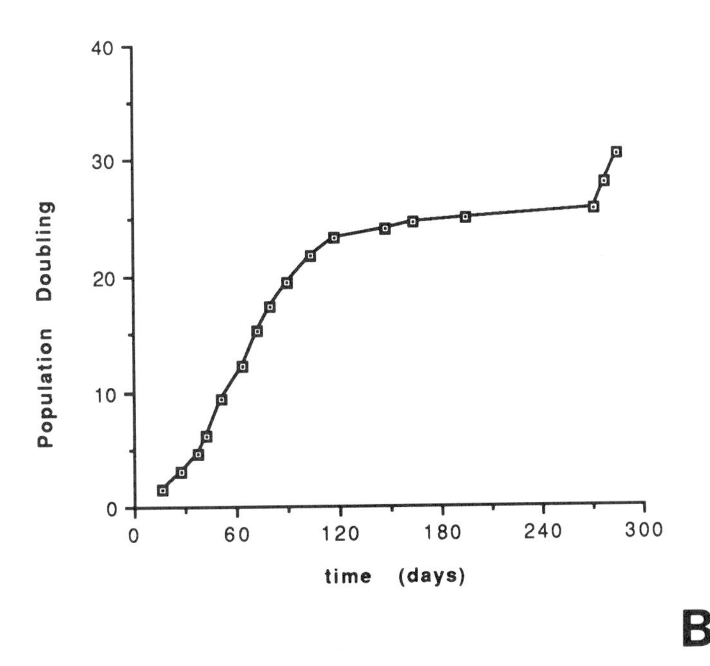

Fig. 12.2. (A) Micrograph of giant human chondrocyte as seen in the crisis period. Example of change in cell size and morphology associated with the crisis. Magnification: 250×. (B) Growth curve of SV40-transfected human chondrocytes after emergence from geneticin treatment. Cumulative population doublings were calculated for each passage; reproduced from Benoit et al., [1995] with permission of the copyright owner.

when working on cellular differentiation as, from our experience, freezing–thawing cycles are responsible for an important loss of differentiated properties.

4. ALTERNATIVE METHODS FOR CHONDROCYTE IMMORTALIZATION

4.1. Early Transfection

It may be more appropriate to transfect chondrocytes earlier than after a few days in primary culture. Indeed, chondrocytes can be transfected either after rapid attachment on a mussel protein extract called "Cell Tak" (Collaborative Biomedical Products), or in suspension immediately after their release from cartilage, as transfection efficiency is weaker in the latter case.

4.1.1. Transfection after rapid cell attachment on Cell Tak

Protocol 4.1.1

Reagents and Materials

- 35-mm culture dishes
- Cell Tak
- Ham's F12G medium without serum
- Complete medium
- Calcium phosphate–DNA coprecipitate (see Section 2.4.2)
- PBSA
- Glycerol, 15% in 1X HBS
- 100-mm dishes

Protocol

(a) Prepare 35-mm culture dishes coated with Cell Tak according to the manufacturer's instructions. These dishes can be used immediately or kept for a few days at 4°C.
(b) Seed per 35-mm dish 5×10^6 freshly released chondrocytes in medium without serum.
(c) Incubate at 37°C; chondrocytes adhere on Cell Tak in about 30–40 min.
(d) Replace medium by complete medium.
(e) Add 0.5 ml coprecipitate per dish and transfect for 5 h at 37°C in a CO_2 incubator.
(f) Perform glycerol shock (see Section 3.2., step (c)).
(g) Incubate overnight at 37°C.

(h) Detach chondrocytes from Cell Tak:
   (i) Remove medium and add 0.3% collagenase (Section 2.2.1) for 30 min at 37°C.
   (ii) Add $\frac{1}{5}$ volume of trypsin-EDTA solution (Section 2.2.2) and incubate at 37°C until complete cell dispersion (10–15 min, check under microscope).
   (iii) Centrifuge at 400g for 10 min.
   (iv) Resuspend the pellet in complete medium and seed in 100-mm dishes for G418 selection.

4.1.2. Transfection in suspension

Protocol 4.1.2

Reagents and Materials

- Calcium phosphate–DNA coprecipitate
- Complete medium

Protocol

(a) Centrifuge 10^7 chondrocytes freshly released from cartilage at 400g for 10 min.
(b) Suspend the cell pellet in 1-ml calcium phosphate–DNA coprecipitate and incubate at room temperature for 30 min.
(c) Add 9 ml of complete medium and incubate at 37°C for 4–5 h (preferentially in a non-tissue-culture-grade culture dish in order to avoid cell attachment).
(d) Wash the cells by centrifugation and seed in standard conditions for G418 selection.

4.2. SELECTION OF IMMORTALIZED CHONDROCYTES BY THEIR ABILITY TO ESCAPE FROM SENESCENCE

Normal rabbit articular chondrocytes usually cease to proliferate after 8–9 passages [Dominice et al., 1986]. Therefore, successfully transfected cells can be selected by their capacity to escape from senescence after the number of passages at which the nontransfected cells usually senesce and die. In that case, cells are transfected according to the protocol described in Section 3.2., the selection plasmid being omitted. When cells reach confluency, subculture weekly using trypsin-EDTA solution (Section 2.2.2) until complete proliferation arrest. Immortalized cells may emerge at that time (Fig. 12.3).

This type of selection is less appropriate for human cells since human chondrocytes are able to accomplish more than 20 passages before dying [Benoit, unpublished data]. Moreover, immortalization oc-

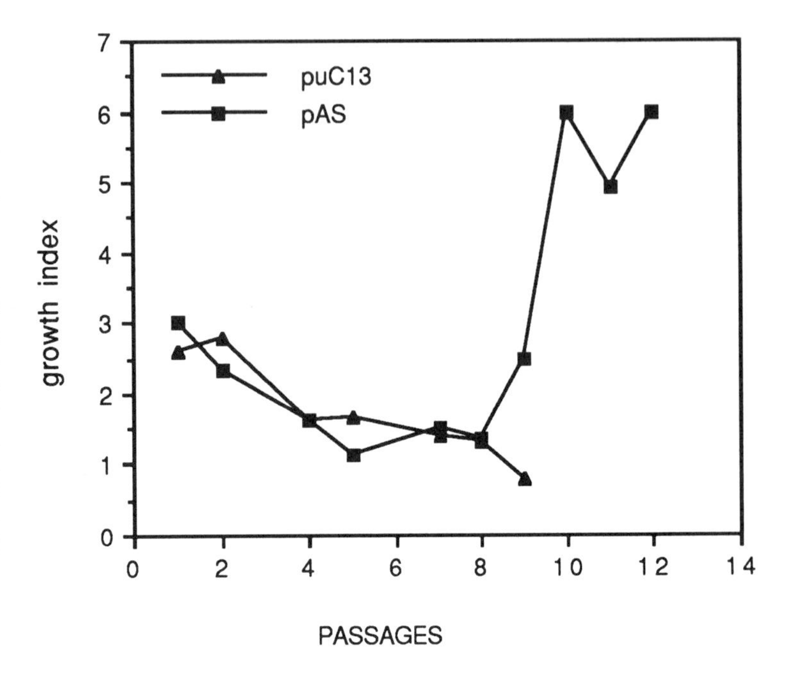

Fig. 12.3. Selection of immortalized chondrocytes by their ability to escape from senescence. Chondrocytes were transfected in primary culture either by the plasmid pAS encoding SV40 early functions or by a control plasmid (puC13) as described in Section 3.2. Then they were subcultured weekly and reseeded at 1.5×10^6 cells per 100-mm dish. Growth index was assessed at each passage by the number of cells per dish after 7 days/number of seeded cells. Mock-transfected chondrocytes ceased to proliferate after 9 passages, whereas pAS-transfected chondrocytes resumed growth from the 8th passage.

curs much less frequently for human than for animal cells and may be hindered by the influence of normal cells that are eliminated in the case of G418 selection.

4.3. Use of Recombinant Retrovirus

The use of recombinant retrovirus carrying either the *myc* oncogene or SV40 large-T antigen to immortalize rodent chondrocytes has been described by several authors [Horton et al., 1988; Mallein-Gérin and Olsen, 1993] (see also Table 12.1). The advantages of using recombinant retrovirus are, in particular, the efficiency of cell infection (almost 100% infected cells in comparison to about 1% transfected cells using the calcium phosphate–DNA coprecipitate method), and the low number and the stability of integrated oncogene copies. Moreover, most recombinant retrovirus contain the Neo^r gene.

Ecotropic helper retroviruses limit their use to rodent cells. This may be a disadvantage when working on articular cartilage, because excision from mouse or rat is much more tricky than on young rabbits or human donors. However, other types of cartilage (costal, sternal or embryo limb cartilage) are easily available on rodent animals [Hunziker, 1992]. Argentin et al. [1993] underlined the greater phenotypic stability of mouse chondrocytes in monolayer culture in comparison to other species. On the other hand, rabbit or human chondrocytes may also be immortalized by recombinant retrovirus, but this requires the use of amphotropic helper retrovirus and this may be more hazardous in terms of biosafety. High-security cell culture installations are then necessary.

5. SAFETY PRECAUTIONS

5.1. Handling Biopsy Material

For human tissue handling, refer to the rules and regulations applicable in the country (see also Chapter 2). Anonymity and consent of donor are usually required. Whenever possible, it is wise to ask for biological testing before taking organs. *Postmortem* samples may be hazardous. Make sure that operators are vaccinated against hepatitis B and that they are aware of the risks of infectious pathogens, especially retroviruses.

5.2. Immortalization and Culture of Immortalized Cell Lines

The immortalization process and the culturing of the immortalized cells should be carried out in a Class II cabinet. Plastic and biological wastes should be incinerated.

6. CHARACTERIZATION OF IMMORTALIZED CHONDROCYTES

6.1. Growth and Transformation

Apart from their unlimited growth potential, immortalized chondrocytes usually exhibit growth properties that differ only slightly from those of the normal chondrocytes from which they were derived. Population doubling times may be shorter (although this is not always the case) and the saturation density higher. Verify regularly that growth properties do not change drastically with passaging.

In regard to human chondrocytes, a crisis period may appear after a period of regular growth (see Section 3.3). Deciding that the emergence of cells eventually escaping from this crisis are definitively immortalized can require several weeks; it is wise to keep dishes and change their medium once a week over a period of a few months to avoid throwing away potentially immortalized clones.

The transformation state is not easy to evaluate since normal chondrocytes share some properties with transformed cells. For example, normal chondrocytes form colonies in 0.5% agarose [Benya and Shaffer, 1982], although *src*-transformed chondrocytes form colonies of increased size and density compared to normal or *myc*-immortalized chondrocytes [Alema et al., 1985].

Tumorigenicity can be evaluated by injecting immortalized chondrocytes into nude mice. This must be performed in parallel with injection of normal chondrocytes since the latter were shown to form slowly growing nodules on grafting on nude mice [Lipman et al., 1983; Takigawa et al., 1987; Bravard et al., 1992]. In our hands, SV40-immortalized rabbit chondrocytes form slowly growing tumors when injected into nude mice [Thenet et al., 1992; Bravard et al., 1992], and this weak tumorigenicity does not increase with passaging.

6.2. Oncogene Integration into Cellular Genome

The presence of the immortalizing oncogene in the cellular genome can be shown using conventional Southern transfer and hybridization techniques [Sambrook et al., 1989].

For example, the integration profile may be analyzed as follows: use the whole immortalizing plasmid as probe and digest the cellular DNA with one restriction enzyme that does not cut into the plasmid (this will give information about the number of integration sites) and one restriction enzyme that has one single restriction site within the plasmid (this will give information about the eventual presence of integrated "tandem" copies).

6.3. Oncogene Expression

The expression of the immortalizing oncoprotein can be very easily shown by indirect immunofluorescence. Specific antibodies for SV40 large-T antigen or c-*myc* oncoprotein are commercially available (e.g., Oncogene Science, Inc.). The use of a fluorescent secondary antibody allows detection of nuclear labelling in 100% immortalized cells (Fig. 12.1). The presence of the oncoprotein can also be analyzed by Western blot; this allows verification of the size of the expressed oncoprotein. In the case of SV40 large-T-immortalized cells, immunoprecipitation followed by Western blot analysis shows the complex formed between the tumor suppressor protein p53 and SV40 T antigen.

6.4. Chondrocyte Phenotype

The chondrocyte phenotype should be analyzed at the earliest possible passages since it has been shown that the differentiated properties may be unstable with passaging.

6.4.1. Collagen

Type II collagen represents 85–90% of total collagen synthesized by chondrocytes in cartilage or primary culture. It is very specific for this cell type and is considered as its main differentiation marker.

The expression of Type II collagen can be analyzed by indirect immunofluorescence using specific antibodies. Antibodies must be very carefully checked for absence of cross-reactivity with Type I collagen, which unfortunately often occurs. Polyclonal antibodies against Type II collagen can be purchased from Dr. D. Hartmann (Institut Pasteur, Lyon, France) or from Southern Biotechnology Associates, Inc.

However, for a precise analysis of the collagen phenotype, SDS-PAGE and two-dimensional cyanogen bromide (CNBr) peptide maps of purified collagen after ^{3}H-proline incorporation have to be performed [Benya, 1981]. Concerning the major collagens, SDS-PAGE analysis can reveal the presence of Type III collagen and of the α2(I) chain of collagen I. However, it cannot separate the α1 chain of Type I and Type II, that is, distinguish between Type II and Type I trimer collagen. That is why two-dimensional CNBr peptide mapping is necessary to assert Type II collagen expression.

Concerning minor collagens, Type IX, Type X, and Type XI can be separated in SDS-PAGE [Ricard-Blum et al., 1988; Petit et al., 1992], although Type XI chains are not separated from Type V chains that are not cartilage-specific. Type X collagen is specific for hypertrophic chondrocytes, and its expression was shown to be inhibited by c-*myc* and v-*myc* overexpression [Quarto et al., 1992; Iwamoto et al., 1993].

The expression of these genetically distinct collagens can also be analyzed at the transcriptional level (Northern blot) with conventional transfer and hybridization techniques [Sambrook et al., 1989]. Probes for pro1(II) [Elima et al., 1985; Metsäranta et al., 1991], proα1(I) [Mäkelä et al., 1988; Metsäranta et al., 1991], α1(IX) [Metsäranta et al., 1991], and proα1(X) [Apte et al., 1992] have been published. Because of sequence homology between different collagen genes, hybridization and washing conditions must be tested and be stringent enough to avoid cross-hybridizations.

6.4.2. Proteoglycans

Alcian blue staining at acidic pH is widely used as an indicator of the presence of sulfated aggrecan. [Ahrens et al., 1977; Lev and Spicer, 1964]. A more precise analysis of the different types of proteoglycans synthesized can be performed after $^{35}SO_4$ incorporation [Verbruggen et al., 1990]. cDNA probes for aggrecan [Doege et al., 1986a] and link protein [Doege et al., 1986b] allow an analysis at the transcriptional level of the expression of these specific proteins.

6.5. Karyotype

As chromosomal alterations often accompany SV40 immortalization, karyotypic analysis is worth performing, using classical techniques. Sometimes, this permits visualization of the karyotypic alterations with time in culture. Moreover, it is recommended to perform such analysis to check for the origin of chondrocytes especially when working with human samples.

7. CONDITIONS FOR REGULATING GROWTH AND DIFFERENTIATION

7.1. Culture Medium

Over the last 20 years, important progress has been made in defining the requirements for cellular growth. The identification and purification of growth and adhesion factors have led to the development of reduced-serum and serum-free media formulations. With regard to chondrocytes, a medium supplemented with insulin, transferrin, selenium, FGF, hydrocortisone, fibronectin, and BSA was shown to support proliferation and synthesis of matrix components [Adolphe et al., 1984]. Medium supplement ITS+ (Collaborative Biomedical Products), which contains some of these products, can be recommended for chondrocyte culture. Recently, Goldring et al. [1994a] showed that human juvenile chondrocytes, immortalized by SV40 large-T antigen, expressed the differentiated phenotype only after removal of serum and addition of an insulin-containing serum substitute (Nutridoma-SP, Boehringer Mannheim). This serum substitute was previously shown to be permissive for expression of endogenous Type II collagen or the transfected collagen II gene regulatory sequences [Goldring et al. 1994b].

7.2. Three-Dimensional Culture

Articular chondrocytes are readily grown in monolayer culture. However, when cultured in monolayer, chondrocytes rapidly lose their differentiated phenotype and changes in the types of collagens, and proteoglycans synthesized can occur [Von der Mark, 1986]. A major part of this dedifferentiation is, however, reversible, if cellular senescence is not well established.

Three-dimensional culture, such as entrapment of chondrocytes in agarose gel, has been reported to permit the restoration of the original properties. Benya and Shaffer [1982] have established that chondrocyte culture in agarose helped to promote the synthesis of collagens and proteoglycans typical of those found in the matrix of the cells from which they were isolated. More recent studies have shown that chondrocytes could also be entrapped in alginate [Guo et al., 1989; Ramdi et al., 1993; Häuselmann et al., 1994]. Articular chondrocytes embedded in alginate gel produce a matrix rich in collagens and proteoglycans. This type of culture offers all the advantages of the agarose system. However, unlike agarose, alginate is readily depolymerized by the addition of a calcium chelator that otherwise maintains the alginate in its gel state. By this means, the cells can be easily recovered and good quality RNAs obtained.

Although it was shown in most cases (Table 12.1) that three-dimensional culture was not able to restore differentiated functions of immortalized articular chondrocytes, in our hands, culture in alginate beads has permitted demonstration of redifferentiation of the collagen phenotype in one cell line of immortalized chondrocytes. Hence we would recommend testing three-dimensional culture in comparison with monolayer culture, when studying the phenotype of immortalized chondrocytes and its modulation. Similarly, intermittent short periods of suspension culture were shown to allow the maintenance of an homogeneous polygonal morphology of immortalized chondrocytes when replated and subcultured in monolayer [Goldring et al., 1994a].

ACKNOWLEDGMENTS

The authors are very grateful to Dr Jean Feunteun (Institut Gustave-Roussy, Villejuif, France) for the gift of the pS701 plasmid and to Dr Frédéric Mallein-Gérin (Institut de Biologie et de Chimie des Protéines, Lyon, France) for interesting discussions and good advices. Part of this work was supported by grants (BMH*-92-0013-FR and BRIDGE programme BIOT 0196) from the European Communities.

REFERENCES

Adolphe M, Froger B, Ronot X, Corvol MT, Forest N (1984): Cell multiplication and type II collagen production by rabbit articular chondrocytes cultivated in a defined medium. Exp Cell Res 155: 527–536.

Ahrens PB, Solursh M, Reiter RS (1977): Stage-related capacity for limb chondrogenesis in cell culture. Devel Biol 60: 69–82.

Alema S, Tato F, Boettiger D (1985): Myc and Src oncogenes have complementary effects on cell proliferation and expression of specific extracellular matrix components in definitive chondroblasts. Mol Cell Biol 5: 538–544.

Apte SS, Seldin MF, Hayashi M, Olsen BR (1992): Cloning of the human and mouse type X collagen genes and mapping of the mouse type X collagen gene to chromosome 10. Eur J Biochem 206: 217–224.

Argentin G, Cicchetti R, Nicoletti B (1993): Mouse chondrocytes in monolayer culture. In Vitro 29A: 603–606.

Atsumi T, Miwa Y, Kimata K, Ikawa Y (1990): A chondrogenic cell line derived from a differentiating culture of AT805 teratocarcinoma cells. Cell Differ Devel 30: 109–116.

Benoist C, Chambon P (1981): In vivo sequence requirements of the SV40 early promoter region. Nature 290: 304–310.

Benoit B, Thenet-Gauci S, Hoffschir, F, Penfornis P, Demignot S, Adolphe M (1995): SV40 large T antigen immortalization of human articular chondrocytes. In Vitro 31: 174–177.

Benya PD (1981): Two dimensional CNBr peptide patterns of collagen types I, II and III. Coll Relat Res 1: 17–26.

Benya PD, Shaffer JD (1982): Dedifferentiated chondrocytes reexpress the differentiated phenotype when cultured in agarose gels. Cell 30: 215–224.

Bonaventure J, Kadhom N, Cohen-Solal L, Ng KH, Bourguignon J, Lasselin, C, Freisinger P (1994): Reexpression of cartilage-specific genes by dedifferentiated human articular chondrocytes cultured in alginate beads. Exp Cell Res 212: 97–104.
Bravard A, Beaumatin J, Luccioni C, Fritsch P, Lefrançois D, Thenet S, Adolphe M, Dutrillaux B (1992): Chromosomal, mitochondrial and metabolic alterations in SV40-transformed rabbit chondrocytes. Carcinogenesis 13: 767–772.
Doege K, Fernandez P, Hassel JR, Sazaki M, Yamada Y (1986a): Partial cDNA sequence encoding a globular domain at the C terminus of the rat cartilage proteoglycan. J Biol Chem 261: 8108–8111.
Doege K, Hassel JR, Caterson B, Yamada Y (1986b): Link protein cDNA sequence reveals a tandemly repeated protein structure. Proc Natl Acad Sci USA 83: 3761–3765.
Dominice J, Levasseur C, Larno S, Ronot X, Adolphe M (1986): Age-related changes in rabbit articular chondrocytes. Mech Ageing Devel 37: 231–240.
Elima K, Mäkelä JK, Vuorio T, Kauppinen S, Knowles J, Vuorio E (1985): Construction and identification of a cDNA clone for human type II procollagen mRNA. Biochem J 229: 183–188.
Fujisawa N, Sato NL, Motoyama TI (1991): Establishment of clonal cell lines, with or without cartilage phenotypes, from a hamster mesenchymal chondrosarcoma. Lab Anim Sci 41: 590–595.
Gionti E, Pontarelli G, Cancedda R (1985): Avian myelocytomatosis virus immortalizes differentiated quail chondrocytes. Proc Natl Acad Sci USA 82: 2756–2760.
Goldring MB, Birkhead JR, Suen L-F, Yamin R, Mizuno S, Glowacki J, Arbiser JL, Apperley JF (1994a): Interleukin-1-β-modulated gene expression in immortalized human chondrocytes. J Clin Invest 94: 2307–2316.
Goldring MB, Fukuo K, Birkhead JR, Dudek E, Sandell LJ (1994b): Transcriptional suppression by interleukin-1 and interferon-γ of type II collagen expression in human chondrocytes. J Cell Biochem 54: 85–99.
Green WT (1971): Behavior of articular chondrocytes in culture. Clin Orthop 75: 248–260.
Guo J, Jourdan GW, MacCallum DK (1989): Culture and growth characteristics of chondrocytes encapsulated in alginate beads. Connect Tis Res 19: 277–297.
Handley CJ, Ng CK (1992): Proteoglycan, hyaluronan, and noncollagenous matrix protein metabolism by chondrocytes. In Adolphe M (ed): "Biological Regulation of the Chondrocytes." Boca Raton, FL: CRC Press, pp 85–104.
Häuselmann HJ, Fernandes RJ, Mok SS, Schmid TM, Block JA, Aydelotte MB, Kuettner KE, Thonar EJMA (1994): Phenotypic stability of bovine articular chondrocytes after long-term culture in alginate beads. J Cell Sci 107: 17–27.
Horton W, Miyashita T, Kohno K, Hassell JR, Yamada Y (1987): Identification of a phenotype-specific enhancer in the first intron of the rat collagen II gene. Proc Natl Acad Sci USA 84: 8864–8868.
Horton WE, Cleveland J, Rapp U, Nemuth G, Bolander M, Doege K, Yamada Y, Hassel JR (1988): An established rat cell line expressing chondrocyte properties. Exp Cell Res 178: 457–468.
Hunziker EB (1992): The different types of chondrocytes and their function in vivo. In Adolphe M (ed): "Biological Regulation of the Chondrocytes." Boca Raton, FL: CRC Press, pp 1–31.
Iwamoto M, Yagami K, LuValle P, Olsen BR, Petropoulos CJ, Ewert DL, Pacifici M (1993): Expression and role of c-myc in chondrocytes undergoing endochondral ossification. J Biol Chem 268: 9645–9652.
Kellermann O, Buc-Caron MH, Gaillard J (1987): Immortalization of precursors of endodermal, neuroectodermal and mesodermal lineages, following the introduction of the simian virus (SV40) early region into F9 cells. Differentiation 35: 197–205.
Kohno K, Sullivan M, Yamada Y (1985): Structure of the promoter of the rat type II procollagen gene. J Biol Chem 260: 4441–4447.

Land H, Parada LF, Weinberg RA (1983): Tumorigenic conversion of primary embryo fibroblasts requires at least two cooperating oncogenes. Nature 304: 596–602.
Lev R, Spicer SS (1964): Specific staining of sulfate groups with Alcian blue at low pH. J Histochem Cytochem 12: 309.
Lipman JM, McDevitt CA, Sokoloff L (1983): Xenografts of articular chondrocytes in the nude mice. Calcif Tissue Int 35: 767–772.
Mäkelä JK, Raassina M, Virta A, Vuorio E (1988): Human proα1 (I) collagen: cDNA sequence for the C-propeptide domain. Nucleic Acids Res 16: 349.
Mallein-Gérin F, Olsen BR (1993): Expression of simian virus 40 large T (tumor) oncogene in mouse chondrocytes induces cell proliferation without loss of the differentiated phenotype. Proc Natl Acad Sci USA 90: 3289–3293.
Metsäranta M, Toman D, De Combrugghe B, Vuorio E (1991): Specific hybridization probes for mouse type I, II, III and IX collagen mRNAs. Biochim Biophys Acta 1089: 241–243.
Nicolas JF, Berg P (1983): Regulation of expression of genes transduced into embryonal carcinoma cells. Cold Spring Harbor Conf Cell Prolif 10:469–485.
Ohlsson C, Nilsson A, Swolin D, Isaksson OGP, Lindahl A (1993): Establishment of a growth hormone responsive chondrogenic cell line from fetal tibia. Mol Cell Endocrinol 91: 167–175.
Petit B, Freyria AM, van der Rest M, Herbage D (1992): Cartilage collagens. In Adolphe M (ed): "Biological Regulation of the Chondrocytes." Boca Raton, FL: CRC Press, pp 33–84.
Quarto R, Dozin B, Tacchetti C, Robino G, Zenke M, Campanile G, Cancedda R (1992): Constitutive myc expression impairs hypertrophy and calcification in cartilage. Devel Biol 149: 168–176.
Ramdi H, Legay C, Lièvremont M (1993): Influence of matricial molecules on growth and differentiation of entrapped chondrocytes. Exp Cell Res 207: 449–454.
Ricard-Blum S, Ville G, Hartmann DJ (1988): Use of the Pharmacia Phast System for sodium dodecyl sulphate polyacrylamide gel electrophoresis of non-globular proteins: application to collagens. J Chromatogr 431: 474–476.
Sambrook J, Fritsch EF, Maniatis T (1989): "Molecular Cloning, A Laboratory Manual," 2nd ed. Cold Spring Harbor Laboratory Press.
Shay JW, Wright, WE, Werbin H (1991): Defining the molecular mechanisms of human cell immortalization. Biochim Biophys Acta 1072: 1–7.
Takigawa M, Shirai E, Fukuo K, Tajima K, Mori Y, Susuki F (1987): Chondrocytes dedifferentiated by serial monolayer culture form cartilage nodules in nude mice. Bone Mineral 2: 449–462.
Thenet S, Benya PD, Demignot, S, Feunteun, J, Adolphe M (1992): SV40-immortalization of rabbit articular chondrocytes: alteration of differentiated functions. J Cell Physiol 150: 158–167.
Verbruggen G, Veys EM, Wieme N, Malfait AM, Gijselbrecht L, Nimmegeers J, Almquist KF, Broddelez C (1990): The synthesis and immobilization of cartilage-specific proteoglycan by human chondrocytes in different concentration of agarose. Clin Exp Rheumatol 8: 371–378.
Von der Mark K (1986): Differentiation, modulation and dedifferentiation of chondrocytes. Rheumatology 10: 272–315.

APPENDIX: MATERIALS AND SUPPLIERS

Materials	Suppliers
Antibodies for c-*myc* oncoprotein	Oncogene Science, Inc
Antibodies for SV40 large-T antigen	Oncogene Science, Inc
Cell Tak	Collaborative Biomedical Products
Cloning rings	Belco
Collagenase	Worthington Biochemical Corp.
Fetal calf serum (FCS)	IBF-Biotechnics
Geneticin (G418 sulfate)	Gibco/BRL
Gentamicin	Gibco/BRL
Ham's F12	Gibco/BRL
Hyaluronidase (Type I-S)	Sigma
ITS+	Collaborative Biomedical Products
Metronidazole	Sigma
Plasmid Maxi kit	Oiagen
Porcine pancreatic trypsin	Biosys
Serum substitute Nutridoma-SP	Boehringer Mannheim
Silicone paste (grease), 70.428	Rhône-Poulenc

13

B-Lymphocytes

Bryan J. Bolton and Nigel K. Spurr

European Collection of Animal Cell Cultures, Centre for Applied Microbiology and Research, Porton Down, Salisbury, SP4 0JG, England (B.J.B.) and Human Genetic Resources, Imperial Cancer Research Fund, Clare Hall Laboratories, Blanche Lane, South Mimms, Potters Bar, Herts. EN6 3LD, England (N.K.S.)

Culture of Immortalized Cells, Edited by R. Ian Freshney and Mary G. Freshney.
ISBN 0-471-12134-7

1. INTRODUCTION

The agent used to immortalize human B-lymphocytes is the Epstein–Barr virus (EBV). This is a herpes virus infecting human lymphocytes in vitro causing infectious mononucleosis and is also aetiologically associated with two human tumors: Burkitt's Lymphoma and anaplastic nasosopharyngeal carcinoma [Epstein and Achong, 1986]. B-lymphocytes recently immortalized using EBV *in vitro* express about 10 of the 100 or so genes of EBV. There is no direct evidence that any particular gene is required for immortalization; however, there is circumstantial evidence that at least 3 of the 10 expressed genes may contribute to the immortalization process. For a more detailed genetic analysis of the immortalizing function of EBV, see Hammerschmidt and Sugden [1989].

B-Lymphoblastoid cell lines established from normal human blood, infected *in vitro,* are usually polyclonal in derivation. They have a lymphoblastoid morphology with protrusions which appear as "club feet" or "hand-mirrors". The cells are diploid when first transformed, but then become polyploid over longer-term culture. They express normal B-cell markers such as HLA Class I and II antigens and as a result are often used as controls in tissue typing. Another use for B-lymphoblastoid cell lines is as a permanent source of genomic DNA for human genetic research. This use has expanded dramatically over the last 5 years because of the many national human genome mapping initiatives.

B-lymphoblastoid lines are generally easy to propagate, as long as they are not overdiluted more than 1 : 3 on subculture and the medium is not allowed to become alkaline. In fact, they prefer to be grown at quite high densities (above 4×10^5 cells/ml) and at a relatively low pH.

2. PREPARATION OF INFECTIVE EBV

2.1. Preparation of EBV Containing Supernatant from the B95-8 Cell Line

Infective EBV for the immortalization of B-lymphocytes can be obtained from the supernatant of the marmoset cell line B95-8. This cell line is available from recognized culture collections such as the ECACC (85011419) and ATCC (CRL 1612). The advantage of obtaining the cell line from a culture collection is that it will be certified to be free of mycoplasma contamination. This is extremely important because, if the B95-8 cell line is contaminated, the mycoplasma will be passed onto all lymphoblastoid cell lines generated from it. It is considered good practice to check the B95-8 cell line at regular intervals for mycoplasma using various standard tests such as the Hoechst

33258 DNA stain, Microbiological culture [Mowles, 1990], Gen Probe kit (Lab Impex) ELISA kit (Boehringer Mannheim GmbH), and PCR (Stratagene). The merits of these various tests are discussed in detail elsewhere [Doyle and Bolton, 1994].

The protocol below outlines a temperature induced method used in both the authors' laboratories.

Protocol 2.1

Reagents and Materials

- B95-8 cells
- RPMI 1640 + 5% FBS
- RPMI 1640 + 2% FBS

Protocol

(a) Maintain a culture of the B95-8 cell line in RPMI1640 + 5% FBS at between 3–9 $\times$ 10^5 cells/ml by diluting the cells 1:2 - 1:4 every 3–4 days.

(b) When sufficient cells have been obtained (depending on the volume of supernatant required) dilute to 2 $\times$ 10^5 in RPMI 1640 + 2% FBS.

(c) Gas the flasks with 5% CO_2, screw the tops tight, and incubate at 37°C overnight, then at 33°C for between 2–3 weeks without any medium changes.

(d) On the day of harvest, stand the flasks on their ends to allow cells to settle out.

(e) Decant the supernatant into sterile tubes and centrifuge at 400g for 10 min to remove any cell debris.

(f) Filter the supernatant through a 0.45-μm filter, syringe, or bottle top type, depending on volume, to remove cell debris and large particles but not the virus.

(g) Aliquot the supernatant at required volumes and store at -80°C or below, preferably in vapor phase of liquid nitrogen. Freeze/thawing of the virus should be avoided as this will reduce the efficacy of EBV infection.

(h) Prior to use, thaw at 37°C and use immediately.

It is important to note that the B95-8 cell line, although predominantly lymphoblastoid in morphology, does contain some fibroblast-like cells, but there is no need to remove these cells when subculturing the cell line.

Some laboratories recommend the use of chemical inducing agents such as 12-0-tetradecanoyl phorbol-13-acetate (final concentration 20 ng/ml) or sodium-n-butyrate (3mM) [Walls and Crawford, 1987].

However, the authors' opinion is that they do not significantly increase the titre of the virus and may interfere with activation of the B-lymphocytes.

Prior to using the virus supernatant, it should be tested and compared to previous batches for its activity. This can be done by titration of the EBV supernatant using cord blood lymphocytes (see Section 2.2) or by simply testing its ability to transform control lymphocytes that are known to be susceptible to immortalization by EBV.

2.2. Titration of EBV Supernatant Using Cord Blood Lymphocytes

This method essentially provides a means of quantifying the virus, enabling comparison of different batches of virus supernatant.

Protocol 2.2

Materials and Reagents

- RPMI10: RPMI 1640 + 10% FBS
- Viral supernatant from B95-8 cells (see Protocol 2.1)
- 96-well plates

Protocol

(a) Separate the mononuclear cells from the cord blood as described in Section 3.2 and resuspend them in RPMI10 at, 2–3 $\times$ 10^6 cells/ml.

(b) Use immediately if possible or culture overnight in a flask at 37°C. Cells give satisfactory results if cultured overnight but not if kept longer. It is not advisable to pool donors.

(c) Remove adherent cells by incubating for 30 min on plastic. This step may be left out if the cells have been cultured overnight.

(d) Prepare serial 10-fold dilutions from neat to 1:10^6 of the virus supernatant to be tested in RPMI10.

(e) Aliquot the cord mononuclear cells into centrifuge tubes to give 2×10^6 cells per tube.

(f) Centrifuge at 150g for 10min.

(g) Pour off supernatant.

(h) Add 100 µl of the different virus dilutions to the tubes of cells. As a control, add 100 µl RPMI10 to a tube of cells without virus.

(i) Resuspend cells by tapping the base of each tube gently and incubate at 37°C for 1 h.

(j) Add 1 ml of RPMI10 to each tube and mix gently.

(k) Plate cells out onto a 96-well flat-bottomed microtitre plate (200 µl per well). Use two separate plates for each titration and two different batches of medium when feeding.

(l) Feed once a week by removing 100-μl medium without disturbing the cells and replacing with fresh medium.
(m) Incubate the plates at 37°C in 5% CO_2/95% air.

After 10–14 days examine the cultures using an inverted microscope. Positive wells will be seen as proliferating foci of B-lymphocytes, which, after about 28 days, are usually visible macroscopically as large clumps of cells. The negative control cultures and cultures failing to be immortalized should contain only dying cells and debris.

The efficiency of immortalization is defined as the negative log to the base 10 of the virus dilution, which induced 50% immortalization of the cultures. A titre of at least 10^{-3} preferably 10^{-4} is desirable for routine use.

3. SEPARATION OF LYMPHOCYTES FROM WHOLE PERIPHERAL BLOOD

There are many protocol variations for the collection of blood, such as type of anticoagulant and subsequent separation, depending on which laboratory you are in. The methods listed below are by no means exhaustive but are protocols that have been proved to work successfully in the authors' laboratories. The methods that have not been verified by the authors are indicated.

3.1. Blood Collection

Since 1990 over 12,000 blood samples have arrived at the ECACC for immortalization as part of the U.K. Human Genome Mapping Project. Initially, sodium heparin tubes were recommended as the anticoagulant of choice; however, over the years ECACC must have received blood in every type of anticoagulant and tube type on the market. These include lithium or sodium heparin, potassium EDTA, and citrated blood. None of these have been shown to be better than any other. The important criteria for obtaining blood in an optimal condition for establishing an immortalized B-cell line are as follows:

Temperature	Ambient (20–25°C) avoiding extremes of heat and cold
Anticoagulant	Li or Na Heparin, K-EDTA, ACD tubes*
Volume	Ideally 5–10 ml; however, volumes as low as 500 μl can be successfully used
Time delay	Blood should reach laboratories within 72 h to obtain optimal success rates*

* Preliminary results have shown these tubes to enable blood to be kept at ambient temperature for up to 2 weeks.

The protocol below describes the method used at ICRF for collecting blood, enabling it to be stored for up to one week at ambient temperature prior to immortalization.

Protocol 3.1

Materials and Reagents

- ***Tri-sodium citrate (TSC), 3.3% stock solution***

Tri-sodium citrate.$2H_2O$	33 g
Water (UPW) to	1 liter

Dispense into 40-ml aliquots, autoclave, and store at room temperature.

- ***2-mercaptoethanol (ME), 1M stock solution***

ME 14.4M (Biorad 161-0710)	1 ml
Water (UPW)	13.4 ml

- ***ME, 5×10^{-3}M final stock solution***

1M stock solution	0.5 ml
Water (UPW)	1 liter

Filter through a 0.2-μm nylon filter, dispense into 5-ml aliquots, and store at 4°C for up to 2 months.

- ***Anticoagulant***

RPMI/HEPES (Gibco)	200 ml
TSC, 3.3% stock solution	40 ml
ME 5x10-3M	2 ml

Add 25 ml of this mixture to sterile 50-ml red capped flat-bottomed tubes and seal cap with Parafilm. Label, date, and store for up to 2 months at 4°C.

Protocol

(a) Remove blood bottles from fridge several hours before use.

(b) Add 25 ml of blood to the anticoagulant tube immediately after virus screening and mix gently.

Blood samples will now be stable for up to a maximum of one week after collection.

3.2. Separation of the Blood

Mixed populations of T- and B-lymphocytes, referred to as *peripheral blood mononuclear cells*, may be separated from whole blood samples by density gradient centrifugation. The lymphocytes may then be stored in liquid nitrogen until needed for immortalization with EBV. It is essential that all steps are carried out under sterile conditions. Human blood samples must be handled in a Class 2 safety cabinet while wearing gloves (see Section 5 and Chapter No. 2).

The following protocol should be followed if the blood collection protocol described previously (Section 3.1) was used. If starting with blood collected in a standard manufacturer's blood tube, (e.g, Li/Na Heparin, ACD, K-EDTA) then skip steps (a)–(f) and go straight to step (g).

Protocol 3.2

Reagents and Materials

Sterile

- RPMI/HEPES
- $CaCl_2$, 1M
- Freezing medium: RPMI10 with 10% DMSO
- 250 ml conical flask with foil cap
- Glass beads, 4 mm, previously aliquoted in lots of 20, and autoclaved
- Lymphoprep, 14 ml in 50-ml centrifuge tubes
- 1-ml syringe

Nonsterile

- Acetic acid, 4%
- Dye exclusion viability stain: Nigrosine, trypan blue or naphthalene black
- Gyratory shaker

Protocol

(a) Pour contents of the red-capped blood tube into a 250-ml conical flask and rinse out tube with 4 ml RPMI/HEPES (at room temperature).
(b) Add 20 sterile glass beads to the conical flask.
(c) Replace foil on the flask and place flask on a gyratory shaker.
(d) Add 0.6 ml sterile 1M $CaCl_2$ through the foil using a 1-ml syringe and needle. Immediately start shaker, or the blood sample will clot (dispose of syringe and needle in accordance with local safety guidelines).
(e) Defibrinate blood for 15 min (minimum time) at approximately 320 revolutions on the shaker. (This step may be carried out by swirling the sample by hand.) If the sample has defibrinated, the glass beads should have been trapped in a clot of material that is clearly visible.
(f) Add 20 ml RPMI/HEPES to the flask; this step is essential; otherwise the gradient (see below) will not form correctly.
(g) Divide the defibrinated blood between two 50-ml tubes, each containing 14 ml of Lymphoprep, by overlaying very carefully with a 5-ml pipette attached to a pipette aid.
(h) Rinse out the flask carefully with a 5–6 ml RPMI/HEPES and add

to tubes, ensuring that they are evenly loaded. Underlaying samples may be preferred to overlaying.

(i) If starting with blood from a standard blood tube, dilute the blood 1:1 with RPMI (HEPES may or may not be included). Then, at a ratio of 1 volume lymphocytes to 2 volumes blood/RPMI mix, overlay or underlay as described in step (f).

(j) Spin tubes at 1800 rpm for 20 min in a Beckman GP centrifuge or 2000 rpm in Sorval RT6000 (~750-g), at 20–25°C, with no brake or refrigeration.

(k) Remove the cells at the interphase (seen as a faint haze) with a Pasteur pipette.

(l) Divide the cells into two Universals and dilute 1:1 with RPMI/HEPES.

(m) Count the cells from both tubes. There will be some residual red cells in the interphase, and these need to be eliminated from the total cell count.
    (i) Remove 100 μl of cells from the Universal
    (ii) Add a drop of 4% acetic acid (this lyses the red cells).
    (iii) Add 100 μl of nigrosine. Dead cells will take up stain, live cells will remain translucent and refractile.

(n) After counting, centrifuge cells at 500-g for 10 min at room temperature without the brake on.

(o) After centrifugation, aspirate the RPMI/HEPES from the cell pellet and freeze cells in 1-ml aliquots containing at least 3×10^6 cells in freezing medium, using rate-controlled freezing methods similar to those used for established cell lines.

3.3. Accuspin Blood Separation System

The Accuspin system is available from Sigma (Cat. A6929 and A7054) and consists of polypropylene tubes fitted with a high-density filter below which the separation medium is preloaded. Full instructions on their use are provided with each kit; however, the principle of the method is the same as that described in Section 3.2, except that it has several advantages listed as follows:

- No need to dilute the blood with RPMI.
- No need to layer the blood onto the Lymphoprep as the whole blood can be poured directly into the tube.
- Lymphocyte interphase can be poured off as the red cells and granulocytes remain below the filter.
- The method is much less labor-intensive, and the whole separation process can be carried out in about half the time of the traditional method.

The one disadvantage is that the tubes are relatively expensive. However, when staff time and the cost of the individual components of the traditional method are taken into account, the costs of both methods are comparable. The quality of the separations are comparable to the traditional method; however, if blood samples over 72 h old are used, the first centrifugation step (3.2 h) may need to be repeated.

4. METHODS OF IMMORTALIZATION

Although there are many variations of the methods used for immortalization of B-lymphocytes using EBV, most of them follow similar principles. However, each laboratory has its own variation, and it is suggested that you find a method that works in your laboratory and achieves the desired success rate. The method described below has a success rate of 95% so long as lymphocytes less than 72 h old are used (fresh or frozen). Outlined in this section are the methods used in the authors' laboratories. The advantages and disadvantages of some of the variations that can be used are also discussed.

4.1. Standard Immortalization Method

The protocol described below is essentially that of Walls and Crawford [1987] with modifications made after the experience of carrying out thousands of transformations in both of the authors' laboratories.

Protocol 4.1

Reagents and Materials (sterile)

- Medium: RPMI1640 + 10% FBS
- Mitogen medium: RPMI1640 with 20% FBS, 1% v/v phytohemagglutinin (PHA) (Gibco-BRL) penicillin (100 U/ml), streptomycin (100 μg/ml), neomycin (100 μg/ml)
- EBV: B95-8 supernatant containing EBV, virus titer 10^{-3}–10^{-4} (see Section 2.1)
- Centrifuge tube, 15 ml

The following protocol is for 3–6 × 10^6 cells, specifically, a typical yield of lymphocytes from a 10-ml blood sample.

Protocol

(a) (If using fresh lymphocytes, proceed to the next step). Thaw a vial of frozen lymphocytes (~3–6 × 10^6 cells) at 37°C and transfer to a 15-ml centrifuge tube. Add pre-warmed medium to 15-ml and mix. Centrifuge at 300g for 5 min.

(b) Discard supernatant and resuspend cell pellet in 1 ml of pre-warmed freshly thawed EBV.
(c) Incubate at 37°C for 1–1½ h with gentle agitation half way through incubation.
(d) Add medium to 5 ml and centrifuge at 150g for 10 min. Ensure that a pellet forms before discarding the supernatant.
(e) Resuspend cells in mitogen medium to give 1-2 × 10^6 cells/ml.
(f) Pipette 1 ml of cells into the equivalent number of wells of a 24-well flat-bottomed culture dish.
(g) Incubate at 37°C in 5% CO_2.
(h) Feed the cultures twice weekly by removing half the supernatant and replacing it with fresh medium (as above but with no PHA added) without disturbing the cells.
(i) After 1–2 weeks, foci of B cells should be visible under an inverted microscope and the initial proliferation of Tcells should regress (See Fig. 13.1)
(j) Once established, usually between 2 and 3 weeks, often indicated by the media turning yellow overnight and lots of large clumps of cells present, the cultures may be expanded into 25-cm^2 tissue culture flasks. At this stage the FBS content of the medium can be reduced to 10%. The flasks should be kept upright and the cultures split 1 : 3 every 2–4 days depending on their growth rate.

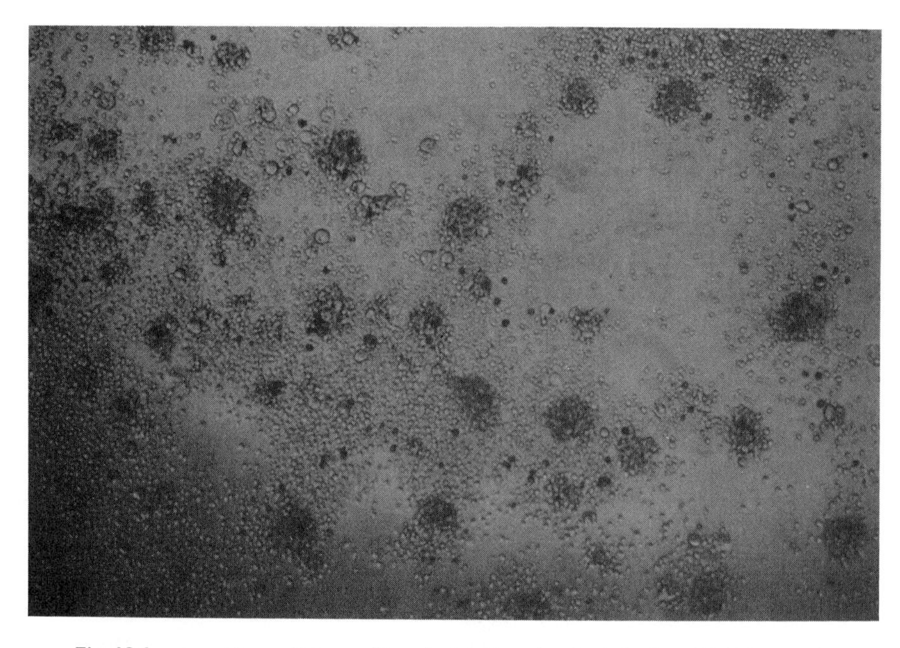

Fig. 13.1. A positive well from a 24 well plate 2 weeks after infection with EBV. Foci of immortalized B-lymphoblastoid cell lines are easily distinguishable.

Steps (b)–(d) may be omitted by resuspending the washed lymphocyte pellet directly into 1 ml of the medium in step (e) along with the addition of 200 μl of EBV supernatant. This is now the method of choice at ECACC.

4.1.1. Use of Cyclosporin A instead of PHA

Cyclosporin A is toxic to the T cells leaving the B cells, which are infected by EBV, to proliferate. This mode of action is different from that of PHA, which stimulates the proliferation of the T cells. While the benefits of adding cyclosporin A are obvious, the rationale for PHA is not so clear. It is thought that proliferation of the T-cells helps to condition the medium with growth factors.

Protocol 4.1.1

Reagents and Materials

(a) Dissolve 3 mg of cyclosporin A powder in 200 μl absolute ethyl alcohol.
(b) Add to this 60 μl Tween 20.
(c) Add 740 μl of serum-free RPMI1640 dropwise using a whirlimixer after every drop.
(d) Then add 2 ml RPMI10 again dropwise with whirlimixing after each drop.
(e) Make suitable size aliquots and store at -20°C. The stock solution should remain stable for several months.
(f) The cyclosporin A stock solution should be used at a 1:1000 dilution to give a final concentration of 1 μg/ml in the immortalization medium.

Protocol

Use this cyclosporin-containing medium for steps (e)–(g) in the protocol in Section 4.1.

4.1.2. Use of feeder cells

Feeder cells are not normally required as long as there are sufficient numbers of healthy lymphocytes to begin with, (i.e. $> 1 \times 10^6$/ml.) For lymphocytes separated from whole blood that has been in transit for over 72 h or for very low yields of lymphocytes, the use of a feeder cell layer is extremely beneficial.

At ECACC freshly prepared mouse peritoneal macrophages are used [McBride, 1993], whereas ICRF use irradiated or mitomycin C-treated normal human fibroblasts [Fox, 1993]. The feeder cells should be seeded at approximately 2×10^3 cells per well of a standard 24-well plate or equivalent.

4.1.3. Immortalization directly into flasks

This method has the advantage of being quicker and easier to set up but cannot be used easily for small numbers of lymphocytes (i.e, $<3 \times 10^6$ cells/ml) or if many samples are done at the same time.

Protocol 4.1.3

Reagents and Materials

- 15-ml centrifuge tube
- RPMI 1640 + 10% FBS
- Mitogen medium: RPMI1640 with 20% FBS, 1% V/V phytohaemaglutinin (PHA) (Gibco-BRL) penicillin (100 U/ml), streptomycin (100 μg/ml), neomycin (100 μg/ml)
- 25-cm^2 flasks
- EBV supernatant

Protocol

(a) Thaw a vial of frozen lymphocytes (at least 3×10^6 cells) at 37°C and transfer to a 15-ml centrifuge tube.
(b) Add prewarmed medium to 15-ml, mix gently, and centrifuge at 150g for 5 min.
(c) Resuspend pellet in 4 ml of mitogen medium [see Section 4.1, step (e)] and transfer to a 25-cm^2 flask.
(d) To the flask add 1 ml of freshly thawed EBV containing supernatant.
(e) Incubate at 37°C in a 5% CO_2 in air atmosphere.
(f) Leave the cultures for 5 days then feed twice weekly by removing 2 ml of medium and adding 2 ml of fresh medium (without PHA).
(g) After 2–3 weeks the medium should turn yellow overnight after feeding and large clumps of cells should be visible macroscopically.
(h) At this stage expand the cultures as described in Section 4.1, step (j).

4.2. Immortalization from Whole Blood

This method is particularly useful for the immortalization of small volumes of cryopreserved whole blood, especially when limited amounts of blood are available from the patient or when it is not feasible to separate the lymphocytes prior to cryopreservation because of the large number of subjects being collected.

4.2.1 Cryopreservation of whole blood

Blood is collected using the anticoagulant of choice (see Section 3.1). This method is essentially that of Penno et al. [1993].

Protocol 4.2.1

Reagents and Materials

- DMSO
- 2-ml cryovials

Protocol

(a) Add 1 ml of whole blood combined with 100 μl of DMSO to 2-ml cryovial and mix gently. Freeze the blood by standard rate-controlled methods used for established cell lines, that is, aiming for an approximate cooling rate of -1°C/min. If a programmable freezer is not available, then the vials may be placed in a expanded polystyrene box (such as an empty 15-ml polystyrene test-tube rack) in a -80°C freezer overnight.

(b) Store the frozen blood in vapor-phase nitrogen until required.

4.2.2. Preparation of lymphocytes from frozen blood and immortalization

This is essentially the same as for fresh blood apart from an extra step that includes removing the red cell debris from the lymphocyte pellet.

Protocol 4.2.2

Reagents and Materials

- *RPMI:* serum-free RPMI1640
- PBSA.
- *RPMI10:* RPMI1640 containing 10% FBS
- EBV supernatant
- 50mM Tris buffer, pH 9.5.
- Lymphoprep
- 15 ml sterile conical test tubes
- *Pretreated plates:*
  (a) Pretreat overnight a 24-well tissue culture plate with sterile 50mM Tris buffer, pH 9.5.
  (b) Add RPMI10 to the wells and incubate for 1 h at 37°C and then rinse with PBSA.

Protocol

(a) Thaw vials of frozen blood at 37°C immediately after removal from storage in vapor-phase nitrogen.

(b) Dilute samples by adding 1 ml RPMI to the cryovial and mix gently.

(c) Layer the blood/RPMI, mix gently onto 3 ml of Lymphoprep in a 15-ml sterile conical test tube, and centrifuge at 400g for 25 min.

(d) Recover the lymphocyte interphase and make up to 10 ml with RPMI. Centrifuge at 1500g for 10 min.
(e) Remove supernatant and resuspend cell pellet in 1 ml sterile PBSA.
(f) Add cells to the pretreated 24-well plate and incubate for 60 min at 37°C.
(g) Remove cells and place in a 15-ml conical test tube.
(h) Wash wells with 1 ml PBSA and collect any additional nonadherent cells.
(i) Make up to 10 ml with RPMI and centrifuge at 150g for 10 min.
(j) Resuspend pellet in EBV supernatant and follow procedures as in Section 4.1, preferably using feeder cells.

This method has been attempted by one of the authors (B.J.B.) with seven out of eight samples attempted being successful. Penno et al. (1993) report a success rate of 83% from a sample size of 115 frozen blood samples compared to approximate success rate of 95% for fresh or frozen lymphocytes.

5. SAFETY PRECAUTIONS

All stages of the process from handling of blood to cell line investigation and subsequent B-lymphoblastoid culture should be carried out to ACDP Containment Level 2 in a Class II Biosafety cabinet (see Chapter 2).

5.1. Handling Whole Blood

Before handling blood it is advisable to read the ACDP guidelines "HIV—the causative agent of AIDS and related conditions."

- Laboratory staff handling the blood should wear a gown and gloves, which should be removed before leaving the laboratory.
- All materials that have been in contact with blood are placed into a container that is closed inside the cabinet and autoclaved prior to disposal.
- It is advisable for all laboratory staff handling blood on a regular basis to be vaccinated against hepatitis B virus, particularly if the clinical history of the patient is not known.

5.2. Handling EBV

EBV and cell lines established using EBV should be handled at ACDP containment level 2.

REFERENCES

Doyle A and Bolton BJ (1994). Quality control of cell lines and the prevention, deletion and cure of contamination. In Davis JM (ed): "Basic Cell Culture - A Practical Approach." Oxford University Press, pp 243–270.

Epstein MA, Achong BG (eds) (1986): "The Epstein-Barr Virus." London: William Heineman.

Fox M (1993): Irradiated cells and chemical treatment of cells for feeder layer production. In Doyle A, Griffiths JB, Newell DG (eds): "Cell and Tissue Culture: Laboratory Procedures." Wiley, and Sons, Chichester, UK: module 2D:3.

Hammerschmidt W, Sugden B (1989): Genetic analysis of immortalizing functions of Epstein-Barr Virus in human B lymphocytes. Nature 340: 393–397.

McBride BW (1993): Production of peritoneal macrophage feeder layers. In Doyle A, Griffiths JB, Newell DG (eds): "Cell and Tissue Culture: Laboratory Procedures." Wiley, Chichester, UK: module 2D:1.

Mowles J (1990): Animal cell culture. In Pollard JW Walker. JM (eds): "Methods in Molecular Biology," Vol 5, Humana Press, Clifton, NJ: pp 65–74

Penno MB, Pedrotti-Krueger M Rat T (1993): Cryopreservation of whole blood and isolated lymphocytes for B-cell immortalization. J Tiss Cult Meth 15: 43–48.

Walls EV, Crawford D.H. (1987): Generation of human B lymphoblastoid cell lines using Epstein-Barr Virus. In Klaus GGB (ed): "Lymphocytes—a practical approach" Oxford University Press, pp 149–162.

APPENDIX: MATERIALS AND SUPPLIERS

Materials	Suppliers
Accuspin blood separation system	Sigma (Cat. A6929, A7054
Acid citrate dextran (ACD) tubes	Becton Dickinson
Anticoagulant tube	Becton Dickinson
B95-8	ECACC (85011419), ATCC (CRL 1612).
Cyclosporin A	Sandoz Pharma
Fetal bovine serum	Sigma
Filters, nylon	Millipore
Glass beads (4 mm, undrilled)	Fisons
Gyratory shaker	Edmund Buhler
Heparin, Li or Na	Sigma
HEPES	Sigma
K-EDTA	BDH/Merck
Lymphoprep	Nycomed
2-Mercaptoethanol, 14.4M	Biorad 161-0710
Microtitration plates	Nunc
Mycoplasma ELISA kit	Boehringer Mannheim GmbH
Mycoplasma kits	Gen Probe (Lab Impex)
Mycoplasma PCR kit	Stratagene
Naphthlene black	Sigma
Neomycin (100 μg/ml)	Life Technologies Gibco BRL
Nigrosine, 34059	Gurr/BDH
PBSA.	Life Technologies Gibco BRL
Penicillin (100 U/ml)/streptomycin 100 μg/ml	Life Technologies Gibco BRL
PHA	Life Technologies Gibco BRL
Phytohaemaglutinin (PHA)	Life Technologies Gibco BRL
Plates, 24-well, 96-well	Nunc.
RPMI1640	Life Technologies Gibco BRL
Tris	Sigma
Trypan blue	Sigma
Tubes, flat-bottomed	R and L Slaughter Ltd.
Tween 20	Sigma

14

Immortalization of Human Astrocytes

J. F. Burke, T. N. C. Price and L. V. Mayne

Biochemistry Group, School of Biological Sciences, University of Sussex, Falmer, Brighton, BN1 9QJ, England. (J.F.B.) and Trafford Centre for Medical Research, University of Sussex, Falmer, Brighton BN1 9RY, England (T.N.C.P., L.V.M.)

Culture of Immortalized Cells, Edited by R. Ian Freshney and Mary G. Freshney.
ISBN 0-471-12134-7

I. INTRODUCTION

Astrocytes are the most abundant cell type in the brain and, like neurons, are derived from the neuroepithelium of the primitive neural tube. Early studies suggested a passive role for astrocytes, providing physical support for neurons [Peters et al., 1976] but in the last decade, it has become evident that astrocytes play a key role in the physiology, development, and pathology of the nervous system [Kimelberg and Norenberg, 1989].

Astrocytes are the first glial cell type to differentiate, and their putative precursors, the radial glia, provide an important substrate for neuronal migration in the cerebral cortex and for axon growth in the corpus callosum [Hatten, 1990]. The embryonic development of neurons and glia progresses through a series of central nervous system (CNS) precursor cells, that demonstrate remarkable plasticity. The differentiated glial and neuronal cell types found in the brain result from a complex hierarchical progression of progenitor cells. In some regions, such as the retina, a single precursor cell can give rise to several different classes of neurons and glia [Turner and Cepko 1987] while in other regions of the CNS, different cell lineage relations exist [Price et al., 1987; Luskin et al., 1988].

Astrocytes possess a great variety of receptor systems, suggesting that they are sensitive to a broad spectrum of neurotransmitters (reviewed in Murphy and Pearce [1987]). *In vitro,* astrocytes respond to an extensive range of neurohormones, indicating that astrocytes, as well as neurones, may be targets for these peptides. The presence of neurotransmitter receptors on astrocytes *in vitro* implies that, far from being passive architectural elements, they may respond *in vivo* in a complex way to changing conditions in the brain.

Astrocytes are relatively easy to isolate from brain tissue and to grow in culture. Like all other human cell types, they have a limited proliferative potential *in vitro* and, like many other differentiated cell types, this *in vitro* lifespan is very short. In general, cultures will undergo only 10 population doublings (PD) and this severely limits their use for longer-term research. Other shortcomings are found with the use of primary astrocytes, including limitations in the number of cells that can be isolated by available procedures, even from fetal tissue and considerable variation in the quality of brain tissue available. There are major functional and biochemical differences between astrocytes from different regions of the brain and most procedures for purifying astrocytes are not able to separate these functionally different groups of astrocytes. Their limited lifespan prevents the further separation and isolation of subgroups of functionally distinct astrocytes.

Immortal cell lines have proved to be an invaluable tool for analysis of the biochemical, molecular, and physiological aspects of cell and tissue biology, and the development of cell lines derived from neurones and astrocytes is important for studies of neuronal communication and survival. Unfortunately, lifespan and proliferative potential are tightly controlled in human cells, preventing the spontaneous appearance of long-lived immortal cell lines. Similarly, human cells are refractory to the methods developed in rodent cells for generating immortalized cell lines. While a number of important rodent neuronal and glial cell lines have been established, virtually no human cell lines of nontumor origin have been reported. The use of rodent-derived cell lines (e.g., Radany et al., [1992]) has produced an enormous amount of information on many aspects of neuronal cell biology; however, it is not clear whether these observations can be applied directly to human cells either *in vitro* or *in vivo,* and it has, therefore, been imperative that human cell lines are developed as well.

Two major issues are relevant when considering the development of an astrocyte cell line: (1) the choice of immortalization procedure and (2) the effect such a procedure may have on the differentiated characteristics of the cell. Our understanding of the immortalization process has come largely from studies of fibroblasts and, to a lesser extent, from studies of a limited number of differentiated cell types. The most widely successful agent for immortalising human cells is SV40 T-antigen. Work from our laboratory [Price et al., 1994; Mayne, unpublished observations] validated the use of SV40 T-antigen as a generic approach for the immortalizaton of a wide range of human cell types. When expressed from a suitable promoter, SV40 T-antigen is able to extend the *in vitro* lifespan of both fibroblasts and more specialized cells, but is not able to directly immortalize. It does, however, promote genetic changes that are likely to result in the removal of the normal restrictions on cellular lifespan. These changes are relatively rare events, and the frequency of immortalization in SV40-transformed cultures is estimated at 1×10^{-7} [Huschtscha and Holliday, 1983; Shay et al., 1993].

Studies on human and rodent cells indicated that SV40 T-antigen may lead to a loss of differentiation functions and the subsequent appearance of a developmentally immature phenotype [Frederiksen et al., 1988; Wynford-Thomas et al., 1990]. Therefore, considerable interest has focused on developing conditionally immortal cell lines whereby the differentiated phenotype can be maintained under conditions where T-antigen expression is reduced. This approach has been used successfully in rodents and cell lines have been generated from the developing rat nervous system using a temperature-sensitive mutant of SV40 T-antigen [Frederiksen et al., 1988]. These cell lines display the characteristic cytochemical profile of glial progenitor cells under permissive growth conditions; however, at the nonpermissive temperatures, the cells usually differentiate into either glia or neurons.

The method described here for the development of human astrocyte cell lines relies on the use of SV40 T-antigen. Vectors expressing T-antigen are introduced by transfection into primary cultures of human astrocytes to generate an SV40-transformed, precrisis culture. These cultures have an extended *in vitro* lifespan and can provide substantial, valuable material for research. However, they are not immortal and will eventually enter crisis with many similarities to that seen in other cell types such as fibroblasts. If enough cells are cultured and carefully nurtured, rare immortal derivatives appear, which when expanded give rise to an immortal, permanent cell line. In our experience, these SV40-immortalized cells retain many of the key features of astrocytes.

2. ISOLATION AND PURIFICATION OF ASTROCYTES

Astrocytes may be isolated from fetal or adult brain. For work on human tissue, we have modified the method published by McCarthy and de Vellis [1980] originally developed for the culture of rat pup cerebral tissue. Before obtaining human tissue samples, it is necessary to obtain ethical permission from your local governing body. Work on fetal tissue in the United Kingdom is further regulated under guidelines issued by the Polkinghorne Committee [1989].

Brain tissue should be removed as soon as possible after death, under aseptic conditions, and transported to the laboratory in culture medium placed on ice. (Astrocytes will remain viable if left overnight at 4°C.) Care should be taken to remove vascular tissue and membranes. The tissue is then cut into small pieces, if necessary, and gently triturated to produce a single cell suspension. The cell viability in this crude cell suspension varies from 30 to 50%. All subsequent manipulations with this cell suspension are performed at 4°C. The crude cell suspension is washed several times in complete medium to remove any lipids present and is then passed through a 200-μm Nitex nylon membrane filter to remove any large particles that remain. A viable cell count is then obtained in the presence of 0.4% trypan blue, and the cells are plated at high density (2×10^5 cells/cm^2) onto standard tissue culture plates in complete medium. The cultures should be left undisturbed for 3 days at 37°C in a humidified incubator with 5% CO_2 and the medium replaced regularly after 4-5 days.

Neuronal cells do not survive this isolation procedure and after 7–9 days the plates are covered with a mixed glial population. Fibroblasts may contaminate these cultures, but the risk is reduced by careful removal of vascular tissue and membranes and by maintaining the cultures with inhibitors of fibroblast growth such as *cis*-4-hydroxy-L-proline or D-valine. On inspection, the mixed glial cultures are composed of a monolayer of astrocyte-like cells with oligodendrocytes resting on top. These oligodendrocytes can be efficiently removed by gently shaking the cultures on a rotary shaker overnight and removing the dislodged oligodendrocytes with the medium. The monolayer of astrocytes is subcultured by trypsinisation and replated as purified astrocytes at a density of 1×10^4 cells/cm^2.

2.1. Isolation of Mixed Glial Cell Cultures

Protocol 2.1

Reagents and Materials

Sterile medium

- Astrocyte culture medium [serum-supplemented medium (SSM)]:
- DMEM/F12, 50:50, with 1.2 g $NaHCO_3$/l, and 15mM HEPES
  100 U/ml penicillin, 100 μg/ml streptomycin
  110 μg/ml sodium pyruvate
  100nM phorbol 12,13-dibutyrate
  200 μg/ml *cis*-4-hydroxy-L-proline
  10% fetal calf serum
- The basal medium, DMEM/F12, can be replaced with MEM-D-valine.
  This medium selectively inhibits the growth of fibroblasts without affecting astrocyte proliferation.

Sterile equipment

- Petri dishes, tissue culture grade, 90 mm
- Plastic bulb pipettes, 2
- Centrifuge tubes, 2×50 ml
- Selection of cell culture flasks, dishes, multiwell plates
- Nitex (200 μm) monofilament nylon cloth (Tekmar Inc., USA)

Nonsterile equipment

- Hemocytometer counting chamber

Protocol

(a) Remove brain tissue (usually by pathologist). May be stored overnight at 4°C in serum-supplemented medium (SSM). Safety precautions:
   (i) Ship brain samples on ice by private courier service and handle as if infected on arrival at the laboratory.
   (ii) Work on the assumption that all human blood and tissue samples are a potential source of human pathogens.

(iii) Perform all manipulations in a Class II laminar flow cabinet in a room with restricted access.

(iv) Soak all glassware and materials in fresh disinfectant (Chloros) immediately after use before further processing.

(b) Transfer the brain tissue to a petri dish and select pieces for processing. These pieces should be the most intact and undissociated fragments in the sample.

(c) Remove any pieces of meninges and vascular tissue with forceps. Cut large pieces to smaller fragments, ~2 mm diameter.

(d) Clean the brain fragments by washing in medium.

(e) Place the tissue pieces in a 30–50 ml centrifuge tube and add 10 ml SSM.

(f) Invert several times to wash.

(g) Centrifuge for 2 min at 400g.

(h) Repeat this wash cycle until the medium is clear.

(i) After washing, dissociate the tissue by trituration. Add 10 ml SSM to the tissue pieces and gently triturate using a wide-bore plastic bulb pipette until the fragments have been dissociated.

(j) Remove lumps of tissue that have failed to dissociate by straining through a Nitex 200-μm filter.

(k) Estimate the number of viable cells in this crude cell suspension by performing viable cell counts using 0.4% trypan blue.

(l) The cells should be stored on ice until seeding.

The purity of the astrocyte preparation can be ascertained by immunohistochemical staining for the astrocyte specific marker, glial fibrillary acidic protein (GFAP). GFAP is a cytoskeletal protein found only in mature astrocytes. The isolation method described above routinely gives preparations that are 95–100% pure astrocytes as judged by staining for GFAP. Primary astrocyte cultures are relatively robust and can be routinely subcultured with trypsin and frozen in liquid nitrogen in a similar manner to fibroblasts [see Freshney, 1994, for general cell culture conditions and handling] see Appendix for astrocyte medium.

3. *IN VITRO* GROWTH OF PURIFIED HUMAN ASTROCYTES

At least two populations of astrocytes can be distinguished on the basis of their morphological and biochemical properties and these are demonstrated in Figure 14.1. Type 1 astrocytes are polygonal in shape with a more diffuse GFAP staining pattern while type 2 astrocytes are stellate in appearance and stain brightly for GFAP. The isolation

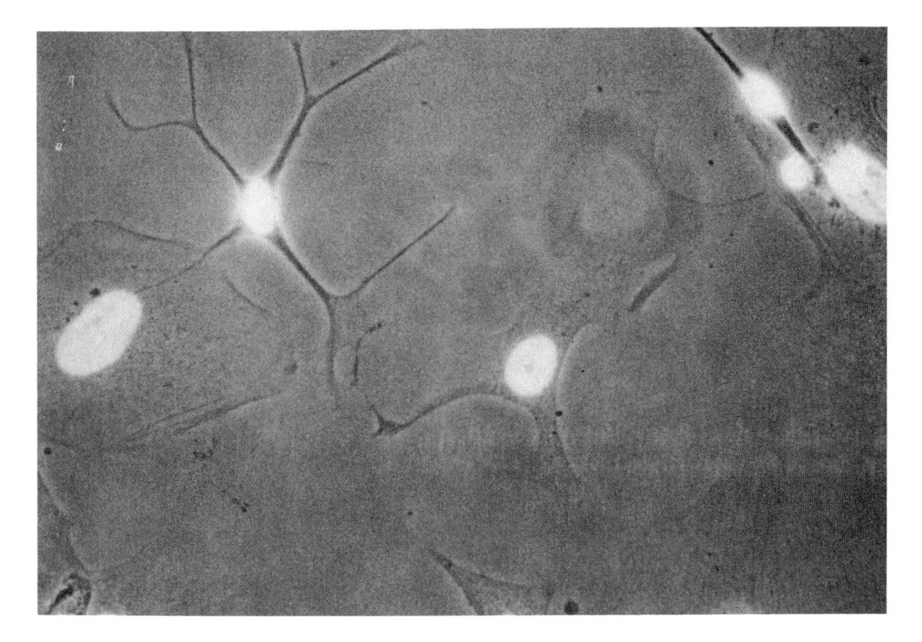

Fig. 14.1. Phase constrast picture of type 1 and type 2 astrocytes isolated by the method described in the text. (See upper left corner of photograph.)

method described above produces cultures containing both type 1 and type 2 astrocytes, but as the cells are passaged, fewer and fewer stellate cells are present. These astrocyte cultures will grow until confluence and can be passaged several times. After approximately 10 PD, the culture will no longer proliferate as it has come to the end of its natural lifespan. The cells are then characterized by a senescent appearance similar to that seen for fibroblasts, and though they no longer divide, they will remain viable for extended periods of time.

3.1. Preparation of Purified Astrocyte Culture

Protocol 3.1

Reagents and Materials

- SSM
- Petri dishes

Protocol

(a) Plate 2×10^5 viable cells/cm^2 in SSM in standard cell culture dishes.

(b) Incubate at 37°C in a humidified environment with 5% CO_2. The cultures should not be disturbed for at least 2 days postseeding to allow the cells to settle and adhere to the plastic.

(c) After 4–5 days postseeding, replace the medium and, thereafter, fresh medium should be added twice weekly.
(d) Maintain the cultures in SSM until a confluent monolayer of mixed glial cells is generated (about 7–9 days postseeding). Phase-dark, process bearing oligodendrocytes will be seen residing on top of the phase-light astrocyte monolayer.
(e) Remove nonastrocytes from these mixed glial cultures:
    (i) Change the medium and allow it to equilibrate with the cells for 2 h in the incubator.
    (ii) Remove the oligodendrocytes by placing the flask in a controlled environment, shaking incubator.
    (iii) Rotate the flasks gently at 200 rpm overnight. This treatment detaches the oligodendrocytes into the medium, leaving the astrocyte monolayer attached to the flask.
    (iv) Remove the oligodendrocytes with the culture medium and replace fresh medium into the flask.
(f) Trypsinize the astrocyte monolayer and replate at 1×10^4 cells/cm^2 in SSM. The ensuing culture will be highly enriched for astrocytes (>95%), as gauged by cytochemistry for the astrocyte marker, glial fibrillary acidic protein (GFAP) (see below).

4. TRANSFECTION OF DNA INTO ASTROCYTE CULTURES

Human astrocytes are as efficient as fibroblasts for DNA uptake and the same methods that work well for fibroblasts work equally successfully with astrocytes (for protocol, see [Mayne et al., Chapter 4, this volume]. The most commonly used methods for DNA transfection are electroporation, lipofection, and calcium phosphate precipitation. Each of these methods, under optimal conditions, produces similar transfection frequencies. The calcium phosphate precipitation method has been routinely used with success in our laboratory for transfecting human astrocytes as it is relatively easy to perform and inexpensive.

The calcium phosphate precipitation method [Graham and Van der Eb, 1973] requires the formation of a DNA precipitate in the presence of calcium and phosphate ions. This DNA precipitate is then added to the medium of the cultured cells where it is taken up by the cells. Not all cells within the culture are competent for uptake and healthy, well-cared-for cells are generally better recipients for gene transfer experiments. Details of the calcium phosphate precipitation method are presented in Mayne et al. [Chapter 4, this volume].

4.1. T-Antigen Expression Vectors

A range of T-antigen expression vectors are available, and the essential differences between these vectors are (1) the choice of dominant selectable marker gene, (2) the source of the promoter that drives T-antigen expression, and (3) alternative forms of the T-antigen itself. The presence of a dominant selectable marker gene allows direct selection for cells that have successfully taken up the vector during transfection. These genes confer resistance to a number of selective toxic agents. The most frequently used dominant selectable marker genes are *gpt, neo,* and *hyg B.* Constitutive and inducible promoters that function successfully in human fibroblasts are equally successful in promoting gene expression in astrocytes. In our experience, both the metallothionein (mMTI) and the native SV40 promoter efficiently drive gene expression in astrocytes. The identification and cloning of the GFAP promoter [Besnard et al., 1991] provides an astrocyte-specific promoter sequence if cell-specific expression is required. From our previous work [Price et al., 1994], it is clear that many of the deleterious effects of T-antigen on cell growth and morphology can be minimized by using expression constructs that produce low levels of T-antigen. To reduce the potential effects of T-antigen on the astrocyte-specific functions of your cells, we advise the use of low-level expression vectors such as p735.6 [Price et al., 1994]. T-antigen is expressed, in this vector, from a modified mouse metallothionein promoter. In the absence of any inducing agent, very low levels of T-antigen are produced, which are sufficient to maintain cell viability and make the cells immortalization competent, yet the cells retain normal growth properties and appearance.

4.2. Posttransfection Care of Cultures

During the weeks following transfection, the bulk of cells in the transfected cultures will die as a result of the presence of the selective agent. Medium changes should be frequent to remove any cell debris and dying cells. Once the bulk of cells has gone, there is no need for further medium changes. Between 4 and 6 weeks after transfection, resistant colonies will become apparent, and these can be subcultured individually, or the colonies can be combined together to form one culture. From our experience, selecting individual colonies or "bulking up" cultures does not have any obvious effect on the likelihood of obtaining an immortalized derivative later. Aliquots of the ensuing "precrisis" cultures should be frozen in liquid nitrogen at the earliest opportunity and further freezer stocks made at frequent and regular intervals.

In all cases, we have observed a greatly extended *in vitro* lifespan after transfection with vectors expressing SV40 T antigen. Despite the short *in vitro* life span of astrocytes, when compared with fibroblasts, the lifespan of transfected cultures of both cell types is very similar, averaging 55–60 PD. Despite the greatly extended lifespan, these cells are not yet immortal, and in almost every case, the proliferative rate and appearance of the culture will decline as the cells enter crisis. As for fibroblasts, the appearance of crisis can be quite variable and the length of time taken for immortalized derivatives to appear can vary from 3 to 24 months. As the cells degenerate in crisis, it is often difficult to know how to handle them. Large amounts of cell debris can be removed by trypsinizing the cells and replating back into the used flask. Debris sticking to the flask can be washed away with trypsin, while the cells have been removed. As the number of cells declines, parallel cultures can be combined.

The long periods of crisis often make the cultures very susceptible to contamination. If flasks are lost with contamination, then duplicate ampoules can be recovered from liquid nitrogen if ample freezer stocks have been made. In general, we do not freeze cells in crisis, but maintain large freezer stocks one or two splits before the onset of crisis. This permits enough cell divisions to build up a pool of cells at the threshold of crisis. The greater the number of flasks carried at this point, the greater the likelihood of obtaining a postcrisis immortal cell line.

Crisis can vary significantly between cultures. In some cases, there is only a limited reduction in the growth rate of the culture and only a few obviously degenerate cells. In other cases, the culture will deteriorate to the point where few, if any, viable cells are apparent. This variability presumably reflects the different timing of events responsible for removing restrictions on lifespan; the earlier this event occurs, the less severe and prolonged crisis may be. The emergence of an immortal cell line is marked by the appearance of robust healthy cells that are actively dividing. These cells form a new, self-renewing culture after subculturing. As SV40 T-antigen does extend *in vitro* lifespan, in almost all cases, and as crisis can be minimal, it may be difficult to know when you have really isolated a cell line with unlimited growth potential. For our working definition, any cell culture that has undergone more than 100 PD posttransfection with an SV40 T-antigen expression vector can be subcloned, with individual cells expanded into separate cultures, and is likely to be immortal. When crisis is marked and postcrisis derivatives are subcloned and expanded, one can be confident of isolating a cell line with unlimited growth potential.

It is essential that any cell line derived is checked against the original starting cell culture to ensure that it is a true immortal derivative and not a contaminant from another culture. DNA fingerprinting provides an easy and reliable method to confirm the line's origins. We routinely subclone (see Freshney [1994] for method) our cultures before further use and build up substantial freezer stocks of those clones that form our long-term stocks.

5. CHARACTERIZATION OF ASTROCYTE CELL LINES

To be of value, any immortal cell line must retain the essential features of the primary cells, and it is therefore important to examine these characteristics. Many differentiated cell types are not well characterized, particularly those from different developmental stages. Wherever possible, it can be useful to keep a small piece of tissue or a frozen stock of the original starting cells for subsequent comparisons between the new cell line and the original cells. The key features that we have examined in our astrocyte cell lines are cell surface markers, expression of GFAP and other cytoskeletal proteins, and neurotransmitter uptake.

Few cell types can be defined simply by the presence or absence of any single genetic or cell surface marker. It usually requires a number of factors in the right context. An important starting point is to ensure the purity of the initial cultures, to minimize any chance of contaminating cell types, such as fibroblasts, being selectively immortalized. High purity cultures are obtained by scrupulously removing the brain tissue away from membranes and vascular tissue, and by growing the cells in the presence of specific inhibitors of fibroblast growth.

The main cell types of the brain are astrocytes, oligodendrocytes, and neurons. The isolation procedures and growth conditions described here do not permit the growth of neurons. Although oligodendrocytes are removed during the isolation process, it remains possible that contaminating cells remain in the culture. Cytochemistry can be used to distinguish oligodendrocytes from astrocytes. Mature oligodendrocytes stain with antibodies specific for galactocerebroside C (galC) and do not stain with antibodies against GFAP. In contrast, mature astrocytes are $GFAP^{+}$, $GalC^{-}$. A more detailed description of the cytochemical profile of mature and immature astrocytes is given in Fedoroff et al. [1984].

Astrocytes are a heterogeneous population of cells and differ both structurally and functionally depending on the region of the brain they are derived from and the period of development when they are isolated. Thus, many cellular properties will be unique to astrocytes of various types. However, several properties are shared between astrocytes of many types. These include expression of glutamine synthetase and active uptake of neurotransmitters. Glutamine synthetase is a key enzyme in the metabolism of the neurotransmitters glutamate and γ-aminobutyric acid (GABA), and in the detoxification of ammonia [Norenberg, 1979]. In the brain, expression of glutamine synthetase is restricted to astrocytes. This enzyme can be easily detected by either immunocytochemistry or Western blotting using standard techniques.

Astrocytes, as well as neurons, actively take up GABA and glutamate. For both neurotransmitters, uptake is dependent on the presence of sodium ions. GABA uptake also shows a cell-specific response to inhibitors, with β-alanine inhibiting the process in glial cells and diaminobutyric acid (DABA) inhibiting the process in neurons. Expression of glutamine synthetase, in conjunction with neurotransmitter uptake, provides a valuable demonstration of astrocyte function in your cell lines.

5.1. Characterization of Astrocyte Cultures

Protocol 5.1

Reagents and Materials

Antibodies for histocytochemistry and Western blotting:

Primary antibodies

- GFAP — Mouse monoclonal from Serotec
- Glutamine synthetase — Mouse monoclonal from Affiniti

Secondary antibody

- Rabbit anti-mouse from Dako

Protocol

Standard immunoperoxidase or immunofluorescent technique [Freshney, 1994].

5.2. Astrocyte-Specific Neurotransmitter Uptake

5.2.1. γ-Aminobutyric acid (GABA)

The ability of astrocyte cultures to take up the neurotransmitter [^{3}H] γ-GABA can be measured using the method of Evrard et al. [1986].

Protocol 5.2.1.

Reagents and Materials

- 24-well plates
- PBSA+B
- β-Alanine, 1.0mM
- Diaminobutyric acid, 1.0mM (DABA)
- *GABA:* 30nM 4-amino-*n*-[2,3-^{3}H] butyric acid (94 Ci/mmol)
- 1% Triton-X 100 in 0.1M NaOH
- 1% Triton-X 100 in 0.1M HCl
- Liquiscint scintillation fluid
- BioRad Protein assay kit

Protocol

(a) Seed cells into 24-well plates and allow to grow until they are confluent. Each treatment should be done in triplicate, and you must ensure that the proper controls are included (as described below).

(b) Remove the medium, rinse the cells twice with PBSA+B by adding the buffer to the plates and aspirating off immediately, and drain well.

(c) Add 500 μl PBSA+B to cover the monolayer of cells. [*Note:* Neuronal cells and astrocytes are sensitive to different inhibitors of GABA uptake. Therefore, to demonstrate astrocyte-specific uptake, use the inhibitor 1mM β-alanine, which specifically inhibits glial uptake or 1mM diaminobutyric acid (DABA), which specifically inhibits neuronal uptake.]

(d) Add the inhibitors to parallel wells and incubate at 37°C for 15 min. Include control wells, without inhibitors.

(e) After pretreatment with the inhibitors, incubate the treated and untreated wells (without inhibitors) for a further 15 min in the presence of 30nM GABA.

(f) Remove the PBSA+B ± inhibitor and replace with 30nM GABA diluted in PBS A+B with or without the inhibitor, as appropriate. A mock treated well should be included that is not incubated with radiolabeled GABA.

(g) After incubation, remove the PBSA+B solution by aspiration.

(h) Wash any free GABA away by rinsing the cells 3 times with PBSA+B and aspirating off. Drain well and ensure that all the buffer is removed.

(i) Lyse the cells directly in the wells by adding 250 μl 1% Triton-X 100 in 0.1M NaOH, followed by 250 μl 1% Triton-X 100 in 0.1M HCl.

(j) Remove 450 μl of this total cell extract and add to 4 ml of Liquiscint scintillation fluid in a scintillation vial.

(k) Determine the [^{3}H] content by liquid scintillation counting.

(l) The remaining 50 μl of cell extract is used to determine the protein concentration of the samples by the BioRad Protein Assay kit (BioRad) following the manufacturer's instructions. The radioactive counts are then expressed as counts per minute (cpm) per milligram of protein.

Active uptake of GABA is very dependent on the presence of sodium. To demonstrate sodium dependence of GABA uptake, replace the sodium ions in the PBS buffer with lithium chloride and follow the steps described above.

The kinetics of neurotransmitter uptake may be calculated from the velocity of GABA uptake in the concentration range (25–500 μM). To measure the rate of GABA uptake, set up the experiment, as described above, using a range of GABA concentrations, but incubate for 40 min. The bulk of the GABA concentration is made up with unlabeled (cold) GABA (Sigma) and a constant amount (30nM) of radiolabeled GABA.

5.2.2. Glutamate

Measurement of glutamate uptake by astrocyte cultures follows a method similar to that described above for GABA, but without the inhibitors. Use 30nM [^{3}H]-glutamate (20–40 Ci/mmol, Amersham International) and incubate for 15 min at 37°C. Process as described above. Like GABA uptake, active uptake of glutamate requires sodium. Sodium-dependence of the reaction can be determined by incubating in PBSA+B where sodium chloride is replaced by lithium chloride. To measure the rate of glutamate uptake, incubate the cells with 30nM [^{3}H] glutamate in 25–1000μM total glutamate for 20 min at 37°C. Use unlabeled (cold) glutamate to make up to the concentration required.

ACKNOWLEDGMENTS

This work was initiated with support from the LINK Genetic Engineering Programme in collaboration with SERC/BBSRC and DTI, Celltech Research Ltd, Glaxo Group Research (Greenford) Ltd. and Unilever Research, Colworth and continued with further funding from the BBSRC, Glaxo Group Research (Greenford) Ltd. and Celltech Research Ltd. L.V.M. was a recipient of a Wellcome Trust Senior Research Fellowship (Basic Biomedical Sciences) and an R.M. Phillips Senior Research Fellowship into Ageing.

REFERENCES

Besnard F, Brenner M, Nakatani Y, Chao R, Purohit HJ, Freese E (1991): Multiple interacting sites regulate astrocyte-specific transcription of the human gene for GFAP. J Biol Chem 28:18877–18883.

Evrard C, Galiana E, Rouget P (1986): Establishment of "normal" nervous cell lines after transfer of polyoma virus and adenovirus into murine brain cells. EMBO J 5:3157–3162.

Fedoroff S, McAuley WAJ, Houle JD, Devon RM (1984). Astrocyte lineage. V. Similarity of astrocytes that form in the presence of db cAMP in cultures to reactive astrocytes in vivo. J Neuroscience Res 12:15–27.

Frederiksen K, Jat PS, Valtz N, McKay RDG (1988): Immortalisation of precursor cells from the mammalian CNS. Neuron 1:439–448.

Freshney RI (ed) (1994): "Culture of Animal Cells, Manual of Basic Technique," 3rd ed. Chichester, UK: Wiley.

Graham FL van der Eb AJ (1973): A new technique for the assay of infectivity of human adenovirus 5 DNA. Virology 52:456–467.

Hatten ME (1990): Riding the glial monorail: a common mechanism for glial-guided migration in different regions of the development brain. Trends Neurosci 13: 179–184.

Huschtscha LI Holliday R (1983): Limited and unlimited growth of SV40-transformed cells from human diploid MRC-5 fibroblasts. J Cell Sci 63:77–99.
Kimelberg HK Norenberg MD (1989): Astrocytes. Sci Am 260:44–52.
Luskin MB, Pearlman AL, and Sanes JR (1988): Cell lineage in the cerebral cortex of the mouse studied in vivo and in vitro with a recombinant retrovirus. Neuron 1:635–647.
Mayne et al (1996), this volume, Chapter 4.
McCarthy KD, de Vellis J (1980): Preparation of separate astroglial and oligodendrocyte cell cultures from rat cerebral tissue. J Cell Biol 85:890–902.
Murphy S, Pearce B (1987): Functional receptors for neurotransmitters on astroglial cells. Neuroscience 22:381–394.
Norenberg MD (1979). The distribution of glutamine synthetase in the rat central nervous system. J Histochem Cytochem 27:756–762.
Peters A, Palay SL, Webster H (1976): The neurons and the supporting cells. In "The Fine Structure of the Nervous System." Philadelphia: Saunders, pp 231–263.
Polkinghorne Committee, Report of (1989) Her Majesty's Stationery Office, London.
Price J, Turner D, Cepko C (1987): Lineage analysis in the vertebrate nervous system by retrovirus-mediated gene transfer. Proc Natl Acad Sci USA 84:156–160.
Price TNC, Moorwood K, James MR, Burke JF, and Mayne LV (1994): Cell cycle progression, morphology and contact inhibition are regulated by the amount of SV40 T antigen in immortal human cells. Oncogene 9:2897–2904.
Radany EH, Brenner M, Besnard F, Bigornia V, Bishop JM, Deschepper CF (1992): Directed establishment of rat brain cell lines with the phenotypic characteristics of type 1 astrocytes. Proc Natl Acad Sci USA 89:6467–6471.
Shay JW, van der Haegen BA, Ying Y, and Wright WE (1993): The frequency of immortalisation of human fibroblasts and mammary epithelial cells transfected with SV40 large T-antigen. Exp Cell Res 209:45–52.
Turner DL, Cepko C (1987): A common progenitor for neurons and glia persists in rat retina late in development. Nature 328:131–136.
Wynford-Thomas D, Bond JA, Wyllie FS, Burns JS, Williams ED, Jones T, Sheer D, Lemoine NR (1990): Conditional immortalisation of human thyroid epithelial cells: a tool for analysis of oncogene action. Mol Cell Biol 10:5365–5377.

APPENDIX: SOURCES OF MATERIALS

Materials	Suppliers
β–Alanine	Sigma
γ-aminobutyric acid (GABA)	Sigma
4-Amino-*n*-[2,3-^{3}H] butyric acid (94 Ci/mmol)GABA	Amersham International
Astrocyte culture medium	Life Technologies Gibco BRL
cis-4-Hydroxy-L-proline	Sigma
Diaminobutyric acid, DABA	Sigma
DMEM/F12	Life Technologies Gibco BRL
GFAP antibody	Serotec
Glutamine synthetase antibody	Affiniti
Liquiscint scintillation fluid	National Diagnostics
MEM-D-valine	Life Technologies Gibco BRL
Nitex monofilament nylon cloth, 200 μm	Tekmar, Inc., USA
PBSA+B	Oxoid
Penicillin	Life Technologies Gibco BRL
Phorbol 12,13-dibutyrate	Sigma
Plates, 24-well	Falcon
Protein assay kit	BioRad
Rabbit antimouse IgG	Dako
Sodium pyruvate	Life Technologies Gibco BRL
Streptomycin	Life Technologies Gibco BRL
Triton-X 100	BDH/Merck

15

Megakaryocyte Cell Lines From Transgenic Mice

Katya Ravid

Department of Biochemistry, Boston University School of Medicine, Boston, Massachusetts 02118

Culture of Immortalized Cells, Edited by R. Ian Freshney and Mary G. Freshney.
ISBN 0-471-12134-7

1. INTRODUCTION

Pluripotent bone marrow stem cells give rise to 2N megakaryoblasts that are subsequently converted to mature polyploid megakaryocytes. The mature megakaryocytes then fragment into small anucleate platelets that circulate within the blood [Hawiger et al., 1987]. Past studies of megakaryocyte biochemistry and differentiation have been hampered because these cells constitute only 0.05–0.1% of the nucleated cells of bone marrow. The currently available cell lines, all derived from patients with leukemia, exhibit multilineage properties and/or low-ploidy states [Martin and Papayannopoulou, 1982; Tabilio et al., 1983, 1984; Ogura et al., 1985; Greenberg et al., 1988; Sledge et al., 1986]. Therefore, the use of a molecular biological approach to generate immortal megakaryocytic cell lines that maintain lineage fidelity and become polyploid would constitute a significant advance in this field of research.

We attempted to generate hematopoietic cell lines that could provide models for lineage programming decisions. Retrovirus-mediated introduction of oncogenes into pluripotent mouse hematopoietic cells has been reported previously [Williams et al., 1984]. However, we used the platelet factor 4 (PF4) promoter linked to a temperature-sensitive SV40 large-T antigen to target oncogene expression in transgenic mice to a specific hematopoietic cell lineage. In prior studies, transient expression experiments as well as transgenic mice were used to define the tissue-specific elements of the PF4 gene [Doi et al., 1987] and the promoter region, which allows selective expression in megakaryocytes and platelets [Ravid et al., 1991a,b]. The use of the conditional oncogene allowed the large-T antigen to be maintained in an inactive state in transgenic mice, which have a body temperature higher than the permissive one [Kaplan et al., 1983]. This latter feature permitted most of the mice to exhibit normal hematopoietic cell function for at least 3 months and prevented the emergence of hematopoietic tumors during this period of time. However, the leaky nature of the conditional oncogene in conjunction with slight variations in body temperature caused death in the colony. Since the focus of our work was to generate immortalized megakaryocytes, we did not study potential changes in the hematopoietic systems of these mice with aging. The bone marrow cells obtained from the young transgenic mice were cultured at 34°C which yielded immortalized cell lines expressing megakaryocytic characteristics. The inactivation of the oncogene by a shift of temperature from 34°C to 39.5°C induced initiation of megakaryocyte maturation.

2. PREPARATION OF REAGENTS AND MEDIA

2.1. Reagents for Preparation of Transgenic Mice

2.1.1. Plasmid construction

The plasmid PF4SVtsA58 was constructed by employing pPF4GH [Ravid et al., 1991b], which contains the rat 1.1 kbp PF4 (platelet factor 4) promoter linked to the HGH (human growth hormone) gene. This construct has a unique EcoR I site at the 3' end of the HGH gene and a unique Ban II site 20 basepairs downstream of the transcriptional start. A unique KpN I site was introduced at the 5' end of the PF4 promoter in pPF4GH, via linkers, to produce $pPF4GH_{KB}$. This later plasmid was digested with Ban II and EcoR I to remove the HGH gene and a Bgl II/EcoR I SVtsA58 fragment [Frederiksen et al., 1988] was introduced, via Ban II/Bgl II linkers, to generate pPF4SVtsA58. The orientation of insertion of the SVtsA58 gene was confirmed by DNA sequencing. The PF4 promoter/SVtsA58 gene of 3.6 kb was obtained by cutting pPF4SVtsA58 with KpN I and EcoR I and purifying the appropriate fragment by agarose gel electrophoresis. This fragment was used to produce transgenic mice as described below.

2.1.2. Solutions for transgenic mice

- *Injection buffer:*

Tris buffer, pH 7.5	10 mM
EDTA	0.25 mM

  Sterilize through a 0.2-μm filter.
- *DNA used for microinjections to mice embryos:* 1 μg/ml in injection buffer.

2.2. Media and Reagents for Preparation of Bone Marrow Cells

2.2.1. IMDM

IMDM1: Iscove's modification of Dulbecco's medium (IMDM) is purchased as a sterile liquid solution, with L-glutamine (4 mM) to be added, and supplemented with penicillin (2001 U/ml), streptomycin (200 μg/ml), and horse serum (20%). The medium may be kept at 4°C for up to one week.

IMDM2: IMDM1 supplemented with the following hematopoietic growth factors:

Erythropoietin	1 U/ml
Interleukin (IL)-3	5 ng/ml
IL-6	5 ng/ml
GM-CSF (granulocyte–macrophage colony-stimulating factor)	5 ng/ml

Use the medium immediately on addition of the growth factors.

2.2.2. HBSS

A stock of 10-fold concentrate of HBSS is prepared to yield 10×HBSS as follows:

KCl	4.0g
KH_2PO_4	0.6g
$NaHCO_3$	3.5g
NaCl	80g
Na_2HPO_4	0.475g
Dextrose	10g
H_2O, pH7.05	1 l

Filter the solution through a 0.2-μm filter and store at 4°C.

2.2.3. Adenosine Theophylline (AT) Solution

A stock of 10-fold concentrate of AT is prepared to yield 10×AT solution as follows:

Adenosine	0.267 g
Theophylline	0.36 g
H_2O	100 ml

Filter the solution through a 0.2-μm filter, aliquot to 10 ml, and store frozen at −20°C.

2.2.4. Citrate solution, 129mM

Citric acid, trisodium dihydrate	3.8 g
H_2O	100 ml

Filter the solution through a 0.2-μm filter and store at 4°C.

2.2.5. CATCH Solution

Citrate adenosine theophylline (CATCH) solution is prepared as follows:

10×HBSS solution	10 ml
10×AT solution	10 ml
129mM sodium citrate	10 ml

Deoxyribonuclease I (DNAse type IV) 3000 U/ml	1.0 ml
Fetal bovine serum	5.0 ml
H_2O pH 7.8	65 ml

Filter the solution through a 0.2-μm filter and use immediately.

2.2.6. Lysis buffer

Tris base	2.06 g
NH_4Cl	7.5 g
H_2O	1 liter

Filter the solution through a 0.2-μm filter and store at 4°C.

2.2.7. Trypsin EDTA

Trypsin, 0.05% (w/v), EDTA, 0.53mM in 1×HBSS, Ca^{2+}-/Mg^{2+}-free.

2.2.8. Freezing solution

Fetal bovine serum	50% (v/v)
IMDM1	40% (v/v)
Dimethyl sulfoxide	10% (v/v)

3. GENERATION OF TRANSGENIC MICE AND ISOLATION OF IMMORTALIZED MEGAKARYOCYTES

3.1. Generation of Transgenic Mice

The construct employed to generate transgenic mice contains 1104 basepairs of the 5' upstream region of the rat PF4 gene linked to the tsA58 mutant SV40 large-T antigen, thus generating 3.6-kbp PF4SVtsA58 . The fragment PF4SVtsA58 (Fig. 15.1), diluted in injection buffer (see Section 2.1.2), was used for the microinjection of embryos to produce transgenic mice following a technology well detailed by Hogan et al. [1986]. Foster mice females were of ICR strain (Harland, Frederick, M.D.), and the microinjected eggs were of FVB

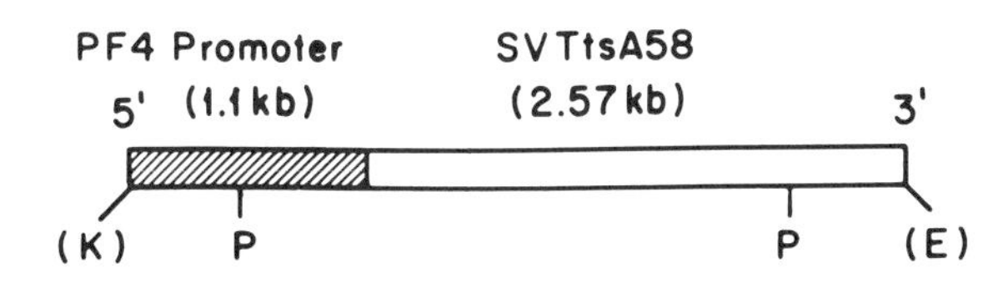

Fig. 15.1 The PF4SVtsA58 construct used for production of transgenic mice. The construct contains 1.1 kbp of the 5' upstream region of the rat PF4 promoter linked to 2.57 kbp of STtsA58 gene (K, KpN I cohesive ends; E, EcoR I cohesive ends; P, Pst I site.)

strain (Taconic Farms, Germantown, NY). Mice were screened for transgene integration by Southern blot analyses of tail DNA [Hogan et al., 1986], and transgene expression was detected by Northern blot analyses [Ravid et al., 1991a], using the large-T antigen gene as a probe (Fig. 15.2).

3.2. Isolation of Bone Marrow Cells

Protocol 3.2

Reagent and Materials

Sterile

- Citrate solution
- Lysis buffer

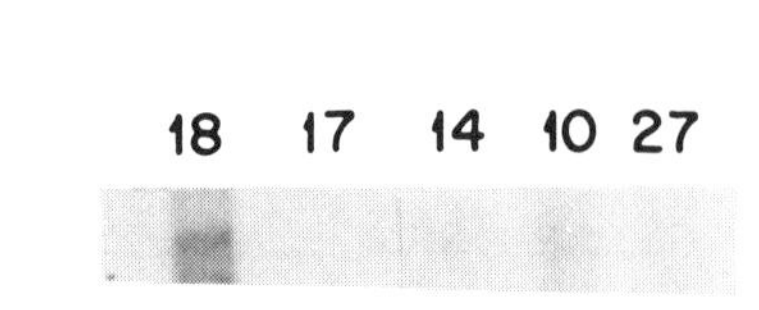

Fig. 15.2 Identification of transgenic mice.(A) Southern blot analyses of 10 μg of DNA isolated from different founders, digested with Pst I, and fractionated by 1.2% agarose gel electrophoresis served to determine transgene integration into the mouse genome. (B) Northern blot analysis of 20 μg of total RNA prepared from bone marrow cells of the offspring of different founders (18,17,14,10,27). The blots were probed with the T-antigen gene fragment and the bands shown correspond to the T-antigen message. As shown in the figure, only founders 18>10>27 expressed the transgene.

- CATCH solution
- 1-ml syringe
- 20G needles
- 18G needles
- 250-μm mesh nylon filter

Nonsterile

- Hemocytometer

(a) Anesthetize transgenic mice (>6 weeks old) with ether
(b) Bleed approximately 50% of their blood by direct cardiac puncture using a 1-ml syringe, with a 20G needle, and filled with 0.1 ml of citrate solution, reaching the heart through a small cut in the chest [Kuter et al., 1989].
(c) Remove femurs and tibias with dissecting scissors and place into cold CATCH solution (see Section 2.2.5).
(d) Flush marrow cells out of the bones using an 18G needle filled with CATCH solution (2 ml per femur and tibia).
(e) Disperse bone marrow cells in CATCH solution with a syringe.
(f) Filter through a 250-μm mesh nylon filter.
(g) Pellet the cells at 400g for 5 min.
(h) Resuspend in an equal volume of lysis buffer and incubate at 37°C for 10 min to lyse red blood cells.
(i) Pellet cells as above.
(j) Wash twice with CATCH solution.
(k) Resuspend in CATCH solution and determine cell number using a hemocytometer.

3.3. Clonal Isolation of Immortal Cells

Protocol 3.3

Reagents and Materials, All Sterile

- Six-well plates, 35-mm-diameter wells
- IMDM2 (see Section 2.2.1)
- Centrifuge tube, 5–15 ml
- Culture dishes, 100-mm diameter
- Flasks, 75 cm^2
- Cryotubes, 2 ml
- Reagents and materials for ring cloning (see Chapter 11 and Freshney [1994])
- Reagents and materials for cell freezing [Freshney, 1994]

Protocol

(a) Harvest bone marrow cells either from a normal or transgenic mouse expressing the transgene (Fig. 15.2B).

(b) Culture cells in 35-mm-diameter six-well plates at a density of 2×10^7cells/3 ml IMDM2.
(c) Incubate cells in 5% CO_2 at 34°C for >3 weeks.
(d) Feed the cells every 4 days as follows:
    (i) Collect 2.5 ml of the medium in a centrifuge tube and spin down the cells in suspension at 400*g* for 5 min.
    (ii) Remove 60% of the medium and replace it with fresh IMDM2.
    (iii) Add back the resuspended cells to the original plate.
(e) After 4–5 weeks of culture, the nonadhering and adhering cells originating from the control mouse will die; however, colonies of adhering cells derived from transgenic mice will fill the dishes. Clone these cells by the ring-cloning technique, expand to obtain stocks and subsequently freeze:
    (i) Seed 200–500 cells/10 ml IMDM2 in a 100-mm-diameter culture dish.
    (ii) Isolate individual colonies with a glass cloning ring (Freshney [1994]; see also Chapter 11), remove from the plate by trypsinization (0.1 ml, trypsin-EDTA solution) and grow in a 35-mm dish as above.
    (iii) When cells are semiconfluent, remove from dish with 1 ml trypsin EDTA and expand by culturing at a density of 10^6 cells/12 ml IMDM2 in a 75-cm^2 flask.
    (iv) Freeze stocks of cells (10^6 cells/ml of freezing solution) in 2-ml vials by packing in an insulated container of approximately 15-mm wall thickness, place directly at –70°C and transfer to liquid nitrogen after 24 h.

4. CHARACTERIZATION AND VALIDATION OF THE CELL LINES

4.1. Determination of Lineage Markers

The immortalized cell lines (MegT) were analyzed for the presence of lineage-specific markers with immunohistochemical and Northern blot analyses [Ravid et al., 1991a,b). MegT37 exhibited a strong specific staining with an antibody to rat PF4 and also possessed the megakaryocytic marker acetylcholine esterase [Jackson, 1973]. MegT37 cells also showed weak positive staining with an antibody to CD45 antigen, but were otherwise negative with antibodies to the macrophage-specific antigen Mac-1, T-helper-cell-specific antigen CD4, and mast-cell-specific antigen IgE [Spangrude et al., 1988]. Similar results were obtained with the other cell lines cloned from the transgenic mice. Northern blot analyses were utilized to detect expression of markers on the cell lines to which antibodies to rodent antigens are not available and to compare the levels of the T-antigen message in the different cell lines. A significant amount of the large-T-antigen message was noticed in all cell lines. The mRNA for von Willebrand Factor (vWF) was not detected in the cells, nor was the erythroid marker (α-globin [Nishioka and Leder, 1979], while the messages for the megakaryocytic markers GPIIb and the PF4 were detectable at either 34°C or 39.5°C. The PF4 message in all the cell lines cultured at the permissive conditions was similar to the level detected when the cells were cultured at 39.5°C, while the level of the GPIIb message increased by about twofold on shifting the cells to 39.5°C. The existence of the PF4 and GPIIb messages in the cell lines was also confirmed by amplifying reverse transcribed mRNA by the polymerase chain reaction (data not shown). Amplifying reverse transcribed mRNA by the polymerase chain reaction was used to exclude the presence of mRNA for the T-cell surface marker CD2 [Diamond et al., 1988] to which rodent cDNA was not available to us. The results of all lineage markers are summarized in Table 15.1.

To ascertain whether inactivation of large-T antigen would allow MegT37 to become polyploid, the cell line was analyzed for DNA content per cell by flow cytometry as outlined elsewhere [Ravid et al., 1993a, Kuter et al., 1989]. When grown at 34°C, the majority of MegT37 (95%) exhibited 2N or 4N nuclei, with very few cells possessing 8N nuclei (Fig. 15.3). When grown at 39.5°C, the frequency of MegT37 with 2N nuclei decreased to about 25%, and those with 8N nuclei increased to about about 30%, with some cells possessing 16N nuclei (Fig. 15.3). Phase-contrast photomicrographs of the cells grown at 34°C (Fig. 15.4a) or at 39.5°C (Fig. 15.4b) were taken prior to flow cytometry analyses. At 34°C, cells were spindle-shaped and adherent;

Table 15.1. Lineage Properties of MegT Cells[a]

Hematopoietic cells	Lineage markers	MegT cells
Megakaryocytes	PF4	+
	Acetylcholine esterase	+
	GPIIb	+
	vWF	−
Myelomonocytes	Mac1	−
T cells	CD4	−
	CD2	−
Mast cells	IgE	−
Erythrocytes	α-globin	−
Leukocytes	CD45	+ (weak)

[a]This table indicates the presence (+) or absence (−) of markers of different hematopoietic lineages in MegT cells. The distribution of the various markers was identical at 34°C and at 39.5°C. This table is reproduced from Ravid et al. [1993b].

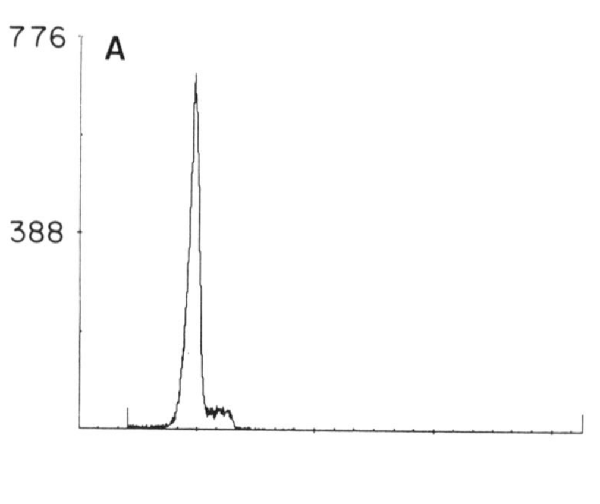

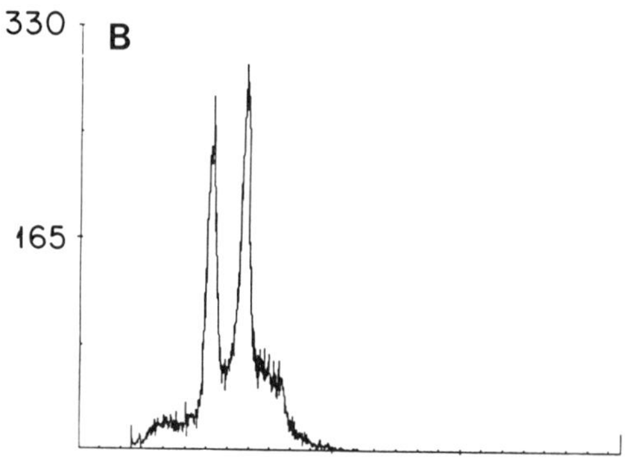

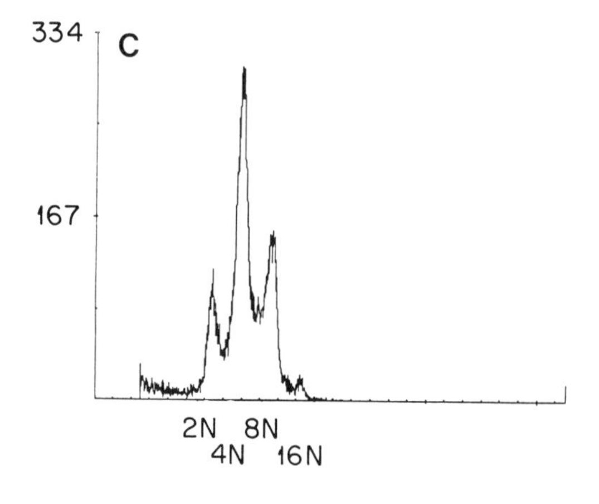

Fig. 15.3 The flow cytometric examination of MegT37. Flow cytometric analysis of normal mouse bone marrow cells (A), or MegT37 cells cultured for 4 days at 34°C (B), or at 39.5°C (C), respectively. The abscissa shows the DNA content, determined based on fluorescence due to Hoechst staining, and the ordinate reflects the number of cells at each DNA value (linear scale).

while at 39.5°C, a large fraction of the cells detached from the plate. All of these nonadhering cells appeared oval or round, and about 50% of them had a diameter larger than that of a 2N cell (>10–15 μm), large nuclei, and multiple nucleoli. The electron-microscopic analyses of MegT37 grown at 39.5°C confirmed the presence of a large multi-lobulated nucleus and a high nuclear-cytoplasmic ratio, which is characteristic of polyploid megakaryocytes. However, typical (α-granules

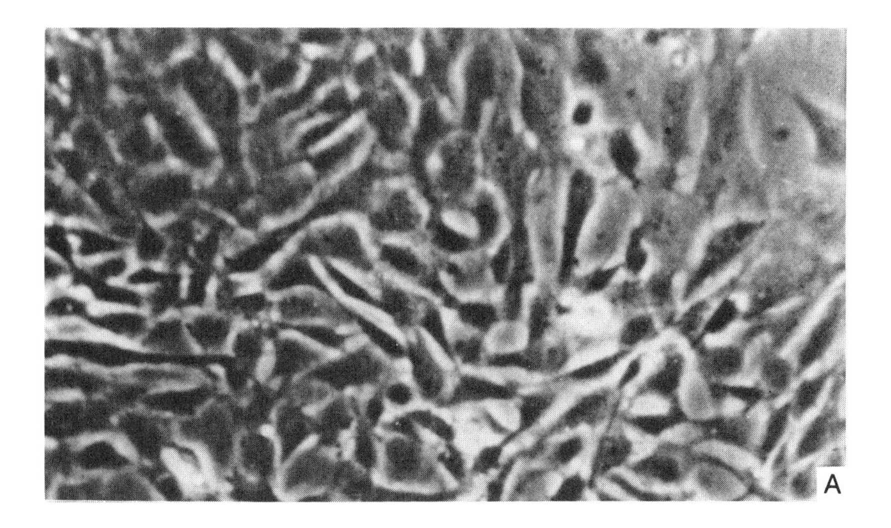

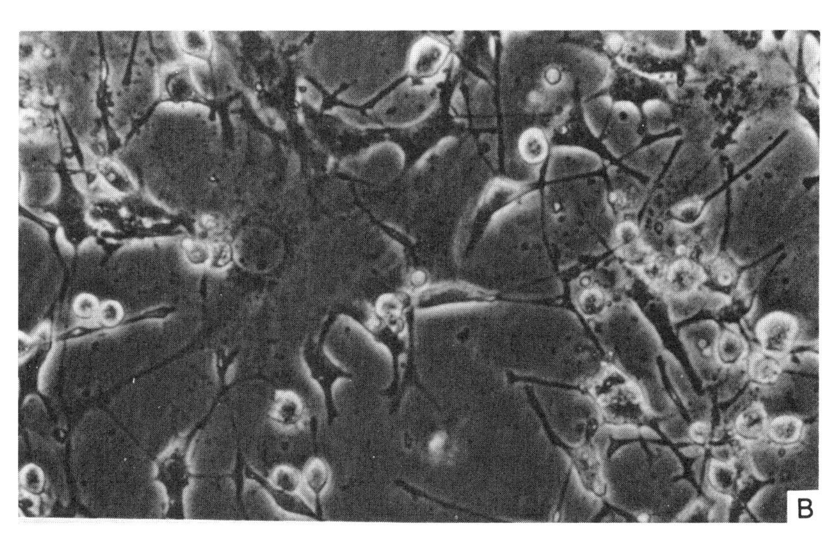

Fig. 15.4 The microscopic examination of MegT37.Phase-contrast photomicrographs of samples of cells grown at 34°C (A), or at 39.5°C (B). Original magnification ×400. This figure is reproduced from Ravid et al. 1993b.

were absent while lysosomes were readily apparent. The large nucleus and lysosomes were not present in cells grown at 34°C.

4.2. Conditions for Regulating Growth and Differentiation

The immortalized megakaryocytic cell lines (MegT37) were cultured in the presence of hemopoietic growth factors (IMDM2 medium, described in Section 2). Withdrawing these growth factors from medium of cells cultured at 34°C resulted in a decrease in the doubling time from 26 to 22 h. However, these growth factors did not have an effect on the extent of ploidization of cells cultured at 39.5°C. The profile of cellular ploidization did depend, however, on the cell concentration, with 2×10^5 cells/5ml medium in a 25-cm^2 culture flask proved to be optimal.

5. ADVICE ON SAFETY PROCEDURES

5.1. Handling Primary Bone Marrow Cells

Since bone marrow is harvested from transgenic mice expressing a viral oncogene in the hemopoietic system, it is advisable to perform all procedures with gloved hands and to avoid contact with needles used for collecting blood and bone marrow. This precaution is taken as no data is available on the extent to which viral DNA is released into the plasma on natural or injury-induced cell death.

5.2. Culturing Early-Passage Cell Lines

Since the cultured cells harvested from the transgenic mice contain the large-T antigen integrated in the genome there is no significant risk of human contamination via equipment or the culture medium. However, it is recommended to dispose of vessels used for aspirating the conditioned medium into disinfectant (see Chapter 4), as well as to use disposable pipettes. All dishes and pipettes used for culturing the cells are considered as biohazardous: collect separately into disinfectant or autoclave.

5.3. Immortalization Procedure

Our immortalization procedure involves microinjection of single-stranded DNA containing the large-T antigen into one cell embryos collected from mice. Although this oncogene is active only on integration into the genome, one should avoid contact with the needle used for microinjection. The whole procedure should be performed with gloved hands and the needles used for microinjecting large-T antigen-containing DNA should not be reused.

5.4. Handling Immortalized Cell Lines

The same precautions used for handling the early-passage cell lines (Section 5.1) are used for growing the immortalized cell lines.

5.5. Biosafety of Products Derived from Immortalized Cell Lines

All dishes used for handling protein and RNA prepared from the immortalized cell lines are disposable and discarded postautoclaving.

6. ALTERNATIVE IMMORTALIZATION METHODS

Several different oncogenes encode nuclear proteins that are able to immortalize somatic cells *in vitro*. In some cases, oncogene expression induces the formation of tumor cells that synthesize many of the major differentiation products of normal cells from which they are derived [Amsterdam et al., 1988; Garcia et al., 1986; Moura Neto et al., 1986; Muller and Wagner; 1984]. In other situations, oncogene-mediated cell immortalization inhibits expression of the terminally differentiated state [Beug et al., 1987; Cherington et al., 1986; Dmitrovsky et al., 1986; Falcone et al., 1985]. The availability of conditional oncogenes makes it potentially possible to establish cell lines that cycle at the permissive temperature, and differentiate under nonpermissive conditions in the presence of appropriate growth factors. Indeed, retroviral vectors driving the expression of conditional oncogenes such as SV40 large-T or *myc* have been employed *in vitro* to produce cell lines having the potential to differentiate [Frederiksen et al., 1988; Iujuidin et al., 1990; Eilers et al., 1989]. Viral and cellular nonconditional oncogenes have been also used *in vivo* to produce transgenic mice [Andres et al., 1987; Leder et al., 1986; Harris et al., 1988; Mahon et al., 1987; Reynolds et al., 1988; Bender and Pfeifer, 1987; Efrat et al., 1988]. In each case, tumors were noted in tissues that are targets for high levels of oncogene expression, which frequently caused the death of the transgenic mice. Our work represents the first attempt to immortalize megakaryocytes. The use of a targeted expression via a tissue specific promoter ensured specificity, while the application of the immortalization method in transgenic mice allowed the mouse biological system to perform the initial cellular cloning of cells with stable integration of the oncogene.

REFERENCES

Amsterdam A, Zauberman A, Meir G, Pinhasi-Kimhi O, Suh BS, Oren M (1988): Cotransfection of granulosa cells with simian virus 40 and Ha-*ras* oncogene generates stable lines capable of induced steroidogenesis. Proc Natl Acad Sci USA 85: 7582–7586.

Andres AC, Schonenberger CA, Groner B, Henninghausen L, LeMeur M, Gerlinger P (1987): Ha-*ras* oncogene expression directed by a milk protein gene promoter: tissue-specificity, hormonal regulation, and tumor induction in transgenic mice. Proc Natl Acad Sci USA 84: 1299–1303.

Bender W, Pfeifer M (1987): Oncogenes take wing. Cell 50: 519–520.

Beug H, Blundell PA, Graf T (1987): Reversibility of differentiation and proliferative capacity in avian myelomonocytic cells transformed by *ts*E26 leukemia virus. Genes Devel 1: 277–286.

Cherington V, Morgan B, Spiegelman BM, Roberts TM (1986): Recombinant retroviruses that transduce individual polyoma tumor antigens: effects on growth and differentiation. Proc Natl Acad Sci USA 83: 4307–4311.

Diamond DJ, Clayton LK, Sayre PH, Reinherz EL (1988): Exon-intron organization and sequence comparison of human and murine T11 (CD2) gene. Proc Natl Acad Sci USA 85: 1615–1619.

Dmitrovsky E, Kuehl WM, Hollis GF, Kirsch IR, Bender TP, Segal S (1986): Expression of a transfected human *c-myc* oncogene inhibits differentiation of a mouse erythroleukaemia cell line. Nature 322: 748–750.

Doi T, Greenberg SM, Rosenberg RD (1987): Structure of the rat platelet factor four gene: a marker for megakaryocyte differentiation. Mol Cell Biol 7: 898–904.
Efrat S, Linde S, Kofod H, Spector D, Delanney M, Grant S, Hanahan D, Baekkeskov S (1988): Beta-cell lines derived from transgenic mice expressing a hybrid insulin gene-oncogene. Proc Natl Acad Sci USA 85: 9037–9041.
Eilers M, Picard D, Yamamoto KR, Bishop JM (1989): Chimaeras of *myc* oncoprotein and steroid receptors cause hormone-dependent transformation of cells. Nature 340: 66–68.
Falcone G, Tato F, Alema S (1985): Distinctive effects of the viral oncogenes *myc, erb, fps,* and *src* on the differentiation program of quail myogenic cells. Proc Natl Acad Sci USA 82: 426–430.
Frederiksen K, Jat PS, Valtz N, Levy D, McKay, R (1988): Immortalization of precursor cells from the mammalian CNS. Neuron 1:439–448.
Freshney RI (1994): "Culture of Animal Cells, a Manual of Basic Technique." New York: Wiley-Liss, pp 169–171.
Garcia I, Sordat B, Rauccio-Farinon E, Dunand M, Kraehenbuhl J-P, Diggelmann H (1986): Establishment of two rabbit mammary epithelial cell lines with distinct oncogenic potential and differentiated phenotype after microinjection of transforming genes. Mol Cell Biol 6:1974–1982.
Greenberg SM, Rosenthal DS, Greeley TA, Tantravahi R, Handin RI (1988): Characterization of a new megakaryocytic cell line: The Dami cell. Blood 72: 1968.
Harris AW, Pinkert CA, Crawford M, Langden WY, Brinster RL, Adams JM (1988): The E mu-*myc* transgenic mouse. A model for high-incidence spontaneous lymphoma and leukemia of early B cells. J Exp Med 167:353–371.
Hawiger J, Steer M, Salzman E (1987): In Colman R, Hirsh J. Marder V, Salzman E (eds): "Thrombosis and Hemostasis: Basic Principles and Clinical Practice." Philadelphia: Lippincott, pp 710–730
Hogan B, Constantini F, Lacy E (1986): In Hogan B, Constantini F, Lacy E (eds): "Manipulating the Mouse Embryo: A Laboratory Manual." (Cold Spring Harbor Laboratories, pp 153–197.
Iujuidin S, Fuchs O, Nudel U, Yaffe D (1990): SV40 immortalizes myogenic cells: DNA synthesis and mitosis in differentiating myotubes. Differentiation 43: 192–203.
Jackson CW (1973): Cholinesterase as a possible marker for early cells of the megakaryocytic series. Blood 42: 413–421.
Kaplan HM, Brewer NR, Blair WH (1983): The mouse body temperature. "Biomedical Research," Vol III, "Normative Biology, Immunology, and Husbandry." Orlando, FL: Academic Press, pp 261–262.
Kuter DJ, Greenberg SM, Rosenberg RD (1989): Analysis of megakaryocyte ploidy in rat bone marrow cultures. Blood 74:1952–1962.
Leder A, Pattengale PK, Kuo A, Stewart TA, Leder P (1986): Consequences of widespread deregulation of the c-myc gene in transgenic mice: multiple neoplasms and normal development. Cell 45: 485–495.
Mahon KA, Chepelinsky AB, Khillan JS, Overbeek PA, Piatigorsky J, Westphal H (1987): Oncogenesis of the lens in transgenic mice. Science 235:1622–1628.
Martin P, Papayannopoulou T (1982): HEL cells: a new human erythroleukemia cell line with spontaneous and induced globin expression. Science 216: 1233
Moura Neto V, Mallat M, Chneiweiss H, Premont J, Gros F, Prochiantz A (1986): Two simian virus 40 (SV40)-transformed cell lines from the mouse striatum and mesencephalon presenting astrocytic characters. I. Immunological and pharmacological properties. Devel Brain Res 26: 11–22.
Muller R, Wagner EF (1984): Differentiation of F9 teratocarcinoma stem cells after transfer of c-*fos* proto-oncogenes. Nature 311: 438–442.
Nishioka Y, Leder P (1979): The complete sequence of a chromosomal mouse alpha-globin gene reveals elements conserved throughout vertebrate evolution. Cell 18: 875–882.
Ogura M, Morishima Y, Ohno R, Rato Y, Hirabayashi MN, Najura H, Saito H (1985): Establishment of a novel human megakaryoblastic leukemia cell line, MEG-01, with positive Philadelphia chromosome. Blood 66: 1384–1389.

Ravid K, Beeler DL, Rabin MS, Ruley HE, Rosenberg RD (1991a): Selective targeting of gene products with the megakaryocyte platelet factor 4 promoter. Proc Natl Acad USA 88: 1521–1525.
Ravid K, Doi T, Beeler DL, Kuter DJ, Rosenberg RD. (1991b): Transcriptional regulation of the rat platelet factor 4 gene: interaction between an enhancer/silencer domain and the GATA site. Mol Cell Biol 11: 6116–6127.
Ravid K, Kuter DJ, Beeler DL, Doi T, Rosenberg RD (1993a): Selection of an HEL-derived cell line expressing high levels of platelet factor 4. Blood 81: 2885–2890.
Ravid K, Li YC, Reyburn H, Rosenberg RD (1993b): Targeted expression of a conditional oncogene to the hemopoietic system of transgenic mice. J Cell Biol 123: 1545–1551.
Reynolds RK, Hoekzema GS, Vogel J, Hinrichs SH, Jay G (1988): Multiple endocrine neoplasia induced by the promiscuous expression of a viral oncogene. Proc Natl Acad Sci USA 85: 3135–3139.
Sledge GW, Glant M, Jansen J, Heerema NA, Roth BJ, Goheen M, Hoffman R (1986): Establishment in long term culture of megakaryocytic leukemia cells (EST-IU) from the marrow of a patient with leukemia and a mediastinal germ cell neoplasm. Cancer Res 46: 2155.
Spangrude GJ, Heimfeld S, Weissman IL (1988): Purification and characterization of mouse hemopoietic stem cells. Science 241: 58–62.
Tabilio A, Pelicci PG, Vinci G, Mannoni P, Civin CI, Vainchenker W, Testa U, Lipinski M, Rochart H, Breton-Gorius J (1983): Myeloid and megakaryocytic properties of K562 cell lines. Cancer Res 43: 4569–4574.
Tabilio A, Rosa JP, Testa U, Kieffer N, Nurden AT, Del Canzio MC, Breton-Gorius J, Vainchenker W (1984): Expression of platelet membrane glycoproteins and a-granule proteins by a human erythroleukemia cell line (HEL). EMBO J 3: 453.
Williams DA, Lemischka IR, Nathan DG, Mulligan RC (1984): Introduction of new genetic material into pluripotent haematopoietic stem cells of the mouse. Nature 310: 476–480.

APPENDIX: MATERIALS AND SUPPLIERS

Materials	Suppliers
[α-^{32}P]ATP	New England Nuclear Research Products
[α-^{35}S]ATP	New England Nuclear Research Products
Antibody against IgE	Nordic Immunological Lab
Antibody to large-T antigen	Oncogene Science
Deoxyribonuclease DNAse Type IV	Sigma
Diisopropylfluorophosphate	Sigma
Filter, nylon mesh	Tekmar
Filters, 0.2 and 0.45 μm	Millipore
Foster mice females, ICR strain	Harlan
HBSS, Ca^{2+}/Mg^{2+}-free	Gibco BRL, Life Technologies
Hematopoietic growth factors	R&D Systems
Hoechst dye	Sigma
Horse serum	Gibco BRL, Life Technologies
IMDM	Gibco BRL, Life Technologies
M-MLV reverse transcriptase	Gibco BRL, Life Technologies
Microinjected eggs, FVB strain	Taconic Farms
Monoclonal antibody against CD45R	Gibco BRL, Life Technologies
Monoclonal antibody against Mac1	Boehringer Mannheim
Monoclonal antibody L3T4 antibody against CD4	Becton Dickinson Immunocytometry System
Poly-D-lysine	Sigma
PCR reagents	Perkin-Elmer Cetus
Tissue culture media	Gibco BRL, Life Technologies
Trypsin EDTA	Gibco BRL, Life Technologies

16

Production and Growth of Conditionally Immortal Primary Glial Cell Cultures and Cell Lines

M. Noble and Susan C. Barnett

Huntsman Cancer Institute, Room 720, Biopolymers Building, University of Utah Health Sciences Center, Salt Lake City, Utah; (M.N.) and Departments of Neurology and Medical Oncology, University of Glasgow, Garscube Estate, Glasgow G61 1BD, Scotland (S.C.B.)

Culture of Immortalized Cells, Edited by R. Ian Freshney and Mary G. Freshney.
ISBN 0-471-12134-7

1. GROWTH FACTOR COOPERATION CAN PROMOTE EXTENDED PROLIFERATION OF PRECURSOR CELLS

There is increasing interest in the transplantation of well-defined populations of precursor cells as a means of repairing damaged tissue, and as a method of delivering therapeutic compounds to sites of injury or degeneration. For example, it has been possible to reconstitute a functional immune system by the transplantation of purified haematopoietic stem cells (e.g., Spangrude et al. [1988]), and transplanted skeletal myoblasts and keratinocytes have been demonstrated to participate in the formation of normal tissue in host animals [Morgan et al., 1987, 1994; Dhawan et al., 1991; Gussoni et al., 1992]. Cell transplantation in the central nervous system (CNS) has also been proposed as a means of correcting neuronal dysfunction in diseases associated with neuronal loss [Lindvall et al., 1990; Renfranz et al., 1991; Snyder et al., 1992]. Glial cell dysfunction might also be corrected by cell transplantation. For example, transplantation of oligodendrocyte precursor cells eventually might allow repair of demyelinating damage in the CNS (e.g., Groves et al., [1993a]).

1.1. Control of Division and Differentiation of O-2A^{perinatal} Progenitors

Our studies leading to the transplantation of glial precursors were based on a systematic analysis of the control of division and differentiation of oligodendrocyte-type-2 astrocyte progenitors isolated from optic nerves of perinatal rats (O-2A^{perinatal} progenitors). Such studies have revealed three pathways of differentiation controlled by at least two distinct mechanisms (for reviews, see, e.g., Raff [1989], Noble [1991], Noble et al. [1991, 1992], Richardson et al. [1990], Lillien and Raff [1990]). The *in vitro* generation of type 2 astrocytes requires the presence of appropriate inducing factors, the identity of which has been partly elucidated in studies by Raff and colleagues (for review, see Lillien and Raff [1990], but see also Mayer et al. [1994]). In contrast, promotion of the differentiation of O-2A^{perinatal} progenitors into oligodendrocytes does not require the presence of inductive signals, but instead appears to be regulated primarily by the limitation of cell division. For example, cessation of cell division, due to mitogen withdrawal, is associated with rapid differentiation of O-2A^{perinatal} progenitors into oligodendrocytes [Raff et al., 1983b; Noble and Murray, 1984]. Alternatively, if O-2A^{perinatal} progenitors are induced to divide by purified cortical (type 1) astrocytes [Noble and Murray, 1984], or by platelet-derived growth factor, PDGF, the O-2A progenitor mitogen secreted by type-1 astrocytes [Noble et al., 1988; Raff et al., 1988; Richardson et al., 1988], then most of these progenitors undergo a limited number of divisions before differentiating into oligodendrocytes [Raff et al., 1988; Temple and Raff, 1986]. In addition, O-2A^{perinatal} progenitors can also progress along a third differentiation pathway in which dividing O-2A^{perinatal} progenitors that do not differentiate into oligodendrocytes appear to differentiate into O-2A^{adult} progenitors [Wolswijk and Noble, 1989; Wolswijk et al., 1990, 1991; Wren et al., 1992], which express stem-cell-like properties more appropriate for the physiological requirements of the adult nervous system [Wolswijk and Noble, 1989; Wolswijk et al., 1991; Wren et al., 1992; Noble et al., 1992]. The cellular and molecular mechanisms that control differentiation along this third pathway are not yet known.

More recent studies on the response of O-2A^{perinatal} progenitors to stimulation with different mitogens have revealed that exposure of these cells to PDGF in the presence of basic fibroblast growth factor (bFGF) causes these cells to undergo continuous self-renewal in the absence of differentiation [Bögler et al., 1990]. For example, cultures prepared from optic nerves of 19-day-old rat embryos begin to generate oligodendrocytes after 2 days, when established in the presence of PDGF alone [Raff et al., 1988; Bögler et al., 1990], yet remained oligodendrocyte-free even after 10 days of growth in the presence of PDGF + bFGF [Bögler et al., 1990]. Further experimentation has demonstrated that O-2A^{perinatal} progenitors can be continually grown for many months *in vitro* so long as cells are continuously exposed to both of these mitogens (S.C.B. and M.N., unpublished observations).

O-2A^{perinatal} progenitors grown in this manner retain the ability to undergo differentiation into oligodendrocytes when removed from the presence of both mitogens. The inhibition of differentiation caused by cooperation between these two growth factors appears to be as effective as that induced by immortalizing genes, and has allowed the generation of large numbers of O-2A^{perinatal} progenitors without resorting to genetic manipulation.

1.2. Remyelination by Purified O-2A^{perinatal} Progenitors

To investigate the possibility that populations of purified O-2A^{perinatal} progenitor cells could be used to remyelinate demyelinating CNS lesions, a variety of preparations of purified O-2A^{perinatal} progenitors were transplanted into demyelinating lesions in the spinal cords of adult rats. These lesions were generated by injection of ethidium bromide into spinal cord white matter that had been exposed to 40 Gy of x-irradiation 3 days previously. This lesion model has been shown to consist of demyelinated axons residing in a glia-free environment that cannot be remyelinated by cells derived from the host animal [Blakemore and Crang, 1988, 1989; Crang et al., 1992].

In initial experiments, O-2A progenitor cells derived from optic nerves of 7-day-old rats were grown in culture in the presence of PDGF and bFGF. Cells were passaged after one week in culture, and again after a further two weeks. This procedure routinely yielded cultures containing >95% O-2A progenitor cells. A suspension of 6×10^4 cells in 1 μl was injected into spinal lesions of 10 syngeneic adult rats, with a series of nontransplanted lesioned animals and animals injected with medium alone serving as controls. Three weeks after transplantation the lesioned area of the spinal cord was examined by light and electron microscopy.

Out of the 10 lesioned animals receiving injections of O-2A progenitor cells, 8 displayed extensive remyelination by oligodendrocytes, in contrast to the complete lack of remyelination observed in control lesions. In the best experiments, up to 90% of the demyelinated axons were reinvested with myelin sheaths possessing ultrastructural features and periodicity characteristic of CNS myelin [Groves et al., 1993a].

We demonstrated that the remyelinating oligodendrocytes in our experimental lesions were unequivocally derived from the transplanted cells by expressing the bacterial β-galactosidase gene in O-2A progenitor cells whose purity had been enhanced to >99.5% by antibody-mediated cell capture [Barres et al., 1992; Mayer et al., 1993, 1994]. Cells infected with the BAG retrovirus [Price et al., 1987] produced remyelination after the same survival period, although to a lesser degree than that exhibited by uninfected cells. Nevertheless, oligodendrocytes expressing β-GAL clearly contributed to the repair of the lesion.

The above experiments demonstrate that highly purified populations of O-2A progenitor cells may be used to achieve extensive remyelination of demyelinated CNS lesions. These populations could be expanded *in vitro* for several weeks simply by the application of two cooperating growth factors, without the necessity for the introduction of immortalizing oncogenes, and could also be genetically modified to express a foreign protein in the lesion site. These studies thus extended earlier demonstrations of remyelination by transplantation of glial cells [Blakemore and Crang, 1988, 1989; Crang et al., 1992; Duncan et al., 1988; Franklin et al., 1991; Gansmuller et al., 1991; Gumpel et al., 1983; Rosenbluth et al., 1990] in utilizing purified precursor cells, by the use of cooperating growth factors to expand these populations and by the demonstration that these cells could be genetically modified and still retain function. These experiments also offered one of the first direct demonstrations that O-2A progenitor cells can give rise to myelinating oligodendrocytes *in vivo*.

The discovery that cooperation between growth factors can cause prolonged self-renewal of precursors revealed a previously unknown means of regulating self-renewal in a precursor population. Such cooperation may, however, represent a more general phenomenon, as indicated, for example, by the importance of growth factor cooperation in promoting the extended division *in vitro* of hematopoetic stem cells [Huang and Terstappen, 1992] and primordial germ cells [Matsui et al., 1992]. In addition, there may be alternative manipulations that allow the extension of the mitotic lifespan of O-2A progenitor cells (and, by extension, perhaps of other cell types). There is currently some controversy regarding the action of particular growth factors and hormones on differentiation of O-2A progenitor cells [Barres et al., 1994; Ibarrola, et al., manuscript in preparation], but it appears to us that growth of O-2A progenitor cells in a cocktail of PDGF, bFGF, neurotrophin-3, and forskolin, in the absence of thyroid hormones and ciliary neurotrophic factor, allows the most effective expansion of this precursor cell population. We are hopeful that continued study will reveal similar possibilities with many other types of precursor cells, and that it will eventually be possible to grow large quantities of any cell type of interest without resorting to the introduction of immortalizing genes. At such a point, the need for continuous cell lines will disappear.

2. GENERATION OF CONDITIONALLY IMMORTAL CELL LINES BY GENETIC MODIFICATION *IN VITRO*

As there are very few cell types for which growth factor combinations able to promote extended in vitro division have been identified, the use of cocktails of growth factors to produce large numbers of cells has limited application at the moment. In addition, preparation of cells from primary cultures has the disadvantage of requiring application of cell purification strategies, and generally does not yield the large numbers of cells that could be obtained through the growth of continuous cell lines. Thus, cell lines may offer a significant advantage when one wants to carry out molecular and biochemical studies requiring cell numbers sufficiently large for the isolation of protein, mRNA, DNA, and so on. In addition, expanded populations of primary cells are heterogeneous, in that even in purified populations there are many different clonal families of cells, each of which may have subtly different properties. Continuous cell lines, in contrast, offer the opportunity of working with populations all of which are members of a single clone.

The ever increasing importance of continuous cell lines in biological research has been associated with a pressure to develop ever more efficient means of producing such lines. The initial isolation of cell lines by growth of tumor samples *in vitro,* or as a result of spontaneous immortalization of primary cells during serial passage *in vitro,* was quickly supplanted by the use of a variety of transduction techniques that allowed insertion and expression of immortalizing genes in cells of interest during their *in vitro* growth. Thus, while earlier strategies of cell line production relied on chance, current approaches to this problem allow the researcher a slightly higher probability of obtaining continuous cell lines of interest.

Before discussing the targeted creation of continuous cell lines, it is important to point out one of the problems that can be associated with this general strategy. The introduction of immortalizing genes into cells can alter normal cellular physiology in two different ways. First, as the functional definition of an immortalizing gene is that it prevents cells from terminally differentiating, expression of characteristics of differentiated cells may be prevented in immortalized cell lines. In addition, expression of immortalizing genes can also alter the response of cells to exogenous signals, such as mitogens or regulators of differentiation. Such alterations can be so dramatic as to actually cause cells to respond to particular signals in a manner fundamentally different from that occurring in the normal counterpart of the immortalized cell. One example of this problem is offered by studies on cooperative interactions between the SV40 large T-antigen gene [an immortalizing oncogene, coding for the large T-antigen (LTAg)] and a mutationally activated *ras* gene (a transforming oncogene, the product of which is thought to be involved in the control of cell division). When expressed in Schwann cells harboring LTAg, *ras* converts slowly dividing cells dependent on exogenous mitogens to rapidly dividing cells capable of growing in mitogen-free conditions, an effect characteristic of a transforming oncogene. In contrast, when *ras* is expressed in Schwann cells that are not also expressing an immortalizing oncogene, then the constitutively activated *Ras* protein rapidly induces the diametrically opposite effect of proliferation arrest [Ridley et al., 1988]. As it is thought that the *ras* protein normally functions in nontransformed cells as a principal component of signal transduction systems (for review, see, e.g., Marshall [1991]), the ability of an immortalizing oncogene to alter the effects of *ras* activation raises the possibility that expression of immortalizing oncogenes can also alter the effects of other components of the signal transduction apparatus.

The potential alterations in cellular physiology associated with expression of immortalizing genes can be overcome, theoretically, through the use of conditional immortalizing genes, which allow the generation of cell lines in which the activity of the product of the experimentally introduced gene can be turned off by manipulation of the cellular environment. For example, the *ts*A58 mutant of LTAg [Tegtmeyer, 1975] encodes a thermolabile protein capable of immortalizing cells only at the permissive temperature of 33°C, and that has been successfully used in the generation of a variety of conditionally immortal cell lines, as discussed below. In addition, it has turned out that cell lines rendered immortal by overexpression of the c-*myc* gene are, in at least some cases, able to differentiate normally *in vivo.*

The construction of cell lines capable of differentiating has proved to be a powerful approach for studying the developmental biology of various neural cell types, including neurons and glia [Cepko, 1989; Frederiksen et al., 1988; Renfranz et al., 1991; Snyder et al., 1992; Barnett et al., 1994]. In these studies it has been possible to create cell lines that are able to participate in normal development after injection into the developing brain. The range of genes used to generate cell lines is large, and includes the SV40 large T gene (SV40LT) [Frederikson et al., 1988; Geller and Dubois-Dalcq, 1988; Peden, et al., 1989; Galiana et al., 1990; Allinquant et al., 1990; Almazan and McKay, 1992], polyoma virus large T and adenovirus 5 E1A [Evrard et al., 1988], *src* [Giotta et al., 1980; Trotter et al., 1989], and c- and v-*myc* [Snyder et al., 1992, 1995; Barnett and Crouch, 1995]. Recently, Louis and co-workers also have described a spontaneously generated cell line called CG-4 [Louis et al., 1992]; as this cell line arose spontaneously, its immortalizing mutation is unknown.

The capacity of neural cells to regain normal patterns of behavior after tsLTAg is inactivated at non-permissive temperatures (i.e., 37°–39.5°C) is indicated by both *in vitro* and *in vivo* experimentation. *In vitro,* in the studies on Schwann cells referred to earlier, the effects of expression of activated *ras* protein were identical in normal Schwann cells and in Schwann cells harboring *ts*LTAg and switched to nonpermissive temperatures [Ridley et al., 1988]. More dramatically, recent studies by McKay and colleagues [Renfranz et al., 1991; Cunningham et al., 1993] have demonstrated that a hippocampal cell line immortalized with *ts*LTAg *in vitro* can undergo neuronal differentiation and apparent integration into normal tissue when injected into the rodent CNS (where the normal body temperature is sufficiently high to cause *ts*LTAg to be rapidly degraded and thus inactivated). The capacity of c-*myc* expressing cells to integrate normally into CNS tissues is elegantly demonstrated by the recent studies of Snyder et al., [1992, 1995]. As discussed below, O-2A progenitor cell lines made with either tsA58 LTAg or *c-myc* are useful tools in the *in vitro* and *in vivo* study of this lineage.

The temperature sensitive tsA58 mutant gene of LTAg and the *c-myc* proto-oncogene also have been used to construct O-2A cell lines able to differentiate in a similar manner to O-2A progenitor cells from primary cultures [Barnett et al., 1994; Barnett and Crouch, 1995]. Retroviruses encoding a variety of other oncogenes (H-*ras,* c- and v-*mos, ts src;* Barnett and Crouch, [1995]) did not yield useful cell lines.

After cloning and expansion of cells, all cell lines were screened to assess their differentiation potential. This was carried out by plating the cells in the appropriate culture conditions for differentiation and immunolabeling cultures to identify progenitor, oligodendrocyte, and astrocyte phenotypes. For O-2A cell lines 4000 cells were plated on poly-L-lysine (PLL)-coated coverslips and incubated in DMEM containing defined additives but not mitogens (DMEM-BS) [Bottenstein and Sato, 1979], DMEM containing PDGF + bFGF [Bögler et al., 1990] or containing 10% fetal calf serum (DMEM-FCS), conditions that promote the generation of oligodendrocytes, progenitor cells or type-2 astrocytes, respectively. On day 3, 5 cells are immunolabeled with cell-type-specific markers and their differentiation potential established [Barnett et al., 1993]. The O-2A/LTAg and O-2A/c-*myc* were both able to differentiate in a similar manner to O-2A progenitor cells from primary cultures [Barnett et al., 1994; Barnett and Crouch, 1995].

The *in vitro* differentiation potential of the O-2A/LTAg cell line correlates well with its differentiation potential *in vivo.* Cell lines transplanted into white matter lesions depleted of host glial cells are able to remyelinate axons and also to generate astrocytes *in vivo* [Trotter et al., 1993; Barnett et al., 1994; Barnett and Crouch, 1995].

In addition, the O-2A/c-*myc* cell lines have been used for molecular studies. Using DNA mobility shift assays, we have been able to identify a potentially novel DNA-binding complex that appears to be differentially regulated during differentiation. This complex is present in progenitor cells but disappears in oligodendrocytes and type 2 astrocytes [Barnett et al., 1995].

3. CONDITIONALLY IMMORTAL CELL LINES DERIVED FROM H-2K^{b}tsA58 TRANSGENIC MICE

Despite the merits of using conditionally active oncogenes to immortalize cells, these approaches still share many of the problems associated with the introduction of constitutively active immortalizing genes into cells by *in vitro* means of gene insertion. Thus, the difficulty of targeting rare cells, the need to promote cell division to achieve integration and continued function of the immortalizing gene, and the need for extended growth of immortalized cells *in vitro* before enough cells exist for experimental use are all problems that are not affected by whether oncogene activity is conditional or constitutive.

An additional problem associated with the generation of cell lines by *in vitro* gene insertion is that every cell line produced by these procedures has the immortalizing oncogene integrated into a different site in the genome. For reasons presumably associated with these differing sites of integration, putatively identical cell lines can express markedly different levels of oncogene product and markedly different behaviors. This problem can even confound the use of conditional immortalizing genes. For example, introduction of *ts*A58 LTAg into rat embryo fibroblasts leads to the ready isolation of cell lines that can be grown indefinitely when maintained at 33°C (the permissive temperature for function of this gene), but that rapidly senesce when switched to 39.5°C (the nonpermissive temperature). However, the degree of conditionality expressed by different fibroblast lines varies over several orders of magnitude, with some lines showing only a modest reduction in growth at 39.5°C, while other cell lines derived from the same infected plate of fibroblasts exhibited an almost complete cessation of growth when switched to nonpermissive temperatures [Jat and Sharp, 1989]. As a consequence of this variability, it can be difficult to compare the properties of cell lines made in different laboratories, or even of lines made within a single laboratory.

3.1. Derivation of Cell lines by Targeting of Oncogene Expression in Transgenic Animals

One way to make certain that all cell lines produced in an experiment contain the same site of integration of the immortalizing gene is to create cell lines from transgenic animals. The possibility of making cell lines from such animals was recognized from the earliest studies in which immortalizing genes were inserted into the animal genome. For example, the first studies indicating that expression of oncogenes would disrupt normal development [Brinster et al., 1984] showed that mice expressing SV40 LTAg under the control of the metallothionein promotor developed tumors of the choroid plexus, and showed that cell lines could be isolated from the transformed tissue. Studies in which the 5′ regulatory sequences of the insulin gene were used to control the *in vivo* expression of SV40 LTAg showed that it was possible to use tissue-specific promoters to cause an immortalizing gene to be expressed precisely in the cell type of interest [Hanahan, 1985]. Further studies demonstrated that targeted oncogene expression allowed isolation of continuous cell lines from a variety of tissues in which the oncogene caused neoplastic development in vivo (e.g., Efrat et al., [1988], MacKay et al. [1988]). It has also been found that it was not necessary for full transformation to occur in order to allow generation of continuous cell lines from affected tissue. For example, nonmalignant hepatocyte cell lines (which become increasingly transformed with further growth in culture) can also be isolated from mice in which SV40 LTAg expression was driven by the mouse metallothionein promoter sequences [Paul et al., 1988a,b]. The use of targeted expression of immortalizing genes to create continuous cell lines of particular interest to neurobiologists has recently been demonstrated by studies of Hammang et al. [1990] and Mellon et al. [1990].

Despite the success of targeted transgene expression in the generation of neural cell lines, there are two major drawbacks to the use of this technique. First, in all cases reported thus far, cell lines have been isolated from tumor tissue. As tumor formation is associated with the acquisition of multiple genetic aberrations [Efrat et al., 1988; Hamilton and Vogelstein, 1988; James et al., 1988; Vogelstein et al., 1988; Thompson et al., 1989; Compere, 1989], it is likely that the resultant cell lines differ from their normal counterparts in a variety of important ways. Indeed, given that the acquisition of cooperating mutations is necessary to allow transformed growth to occur [Land et al., 1983a,b; Ruley, 1983], it is almost inconceivable that the resultant cell lines do not have multiple mutations. Furthermore, since expression of a single activated immortalizing gene seems to be sufficient to generate the phenotype of a benign tumour *in vivo* [Thompson et al., 1989], it is likely that these cells will have undergone a protracted period of abnormal growth *in vivo* prior to the point of having generated a tumor. The extent to which such cells can be expected to mimic their untransformed ancestors is unknown.

The second drawback associated with the approach of specifically targeting oncogene expression to individual cell types is the necessity of identifying cell-type-specific promoter elements for every cell type of interest. At present, there are relatively few cell types for which appropriate 5′ regulatory elements have been identified. In addition, as many of the cell-type-specific proteins identified thus far are expressed at relatively late stages of differentiation, it may be difficult to target oncogene expression to precursor populations by utilizing the promoter regions for such proteins. In specific regard to the generation of precursor cell lines, it may be more useful to use promoter regions associated with such developmentally regulated genes as the homeobox proteins.

3.2. The H-2K^{b}tsA58 Mouse, a Multipotential Source of Conditionally Immortal Cell Lines

An approach to cell line production that would overcome many of the difficulties described thus far would be to create a strain of transgenic animals in which expression of a conditionally active immortalizing gene was regulated in such a way that the gene was not functionally active in vivo (and thus did not perturb normal development) but could be turned on in cells isolated from any tissue of the body by simple manipulation of tissue culture conditions. Ideally, the oncogene utilized should be capable of immortalizing as wide a range of cell types as possible, and the regulatory regions used to control expression of this gene should also be capable of functioning in the widest possible range of cell types. Theoretically, transgenic animals of this generic class would allow the direct derivation of continuous cell lines from a wide variety of tissues simply by dissection and growth of cells in an appropriate *in vitro* milieu. In addition, each cell line generated would possess an identical integration site of the immortalizing gene. Such animals would also have two further distinct advantages. First, the presence of the immortalizing gene in the genome of the transgenic animal would not cause neoplastic development *in situ,* because functionally active protein would not be expressed *in vivo* at levels sufficient to perturb normal development. Second, expression of the functionally active immortalizing gene could be turned off again after desired continuous cell lines were obtained in tissue culture, thus allowing study of normal processes of differentiation either *in vitro* or, following cell transplantation, *in vivo.*

The first transgenic animals expressing the ideal characteristics described above are the H-2K^{b}tsA58 transgenic mice [Jat et al., 1991]. These mice harbor the tsA58 LTAg gene under the control of the H-2K^b Class I antigen promoter [Weiss et al., 1983; Kimura et al., 1986; Baldwin and Sharp, 1987]. The combination of existing studies on the effects of LTAg in tissue culture and in transgenic animals indicates clearly that this gene expresses its immortalizing function in a very wide variety of cell types [Bayley et al., 1988; Frederiksen et al., 1988; Williams et al., 1988; Burns et al., 1989; Lemoine et al., 1989; Renfranz et al., 1991; Santerre et al., 1981]. In addition, expression from the H-2K^b promoter can be induced, or enhanced, in almost all cell types by exposure of cells to interferons [Wallach et al., 1982; Basham and Merigan, 1983; Israel et al., 1986]. Thus, in cells or tissues (such as brain) that normally express little or no Class I antigens, exposure to interferons activates transcription from this promoter. Moreover, cells which constitutively express high levels of Class I antigens can still be induced to express even higher levels of expression by exposure to interferons.

In initial experiments on the H-$2K^b$tsA58 transgenic mice, 34 different transgenic founder animals were created [Jat et al., 1991]. Skin fibroblasts from normal and transgenic animals were placed in culture at 33°C, the permissive temperature for tsA58 LTAg, in the presence of γ-interferon (IFN-γ), which is known to increase expression from the H-$2K^b$ promoter. Skin fibroblasts derived from the transgenic mice all readily yielded proliferating cultures which could be continuously passaged when grown in permissive conditions. In contrast, fibroblasts stopped dividing within a limited number of passages, with a significant decline in even this limited passage number seen in cultures established from older animals.

The conditionality of growth observed in the fibroblasts derived from transgenic animals was correlated with the levels of tsA58 LTAg. In all cultures, the level of tsA58 LTAg was reduced by increasing temperature and/or by removal of IFN-γ. It is interesting that when the most conditional cultures were grown at 33°C in the absence of IFN-γ, a condition where these cells did not grow, low levels of LTAg could still be detected. This suggests that the level of LTAg produced in these conditions was below the threshold needed to support continued rapid cell growth or to allow single cell cloning.

In general, H-$2K^b$tsA58 transgenic mice appear to undergo normal development. However, these animals do routinely exhibit hyperplastic development of the thymus. This enlargement appears to be due to hyperplasia, rather than to malignant transformation, as thymic histology, T-cell repertoire, and T-cell clonality were all normal in the transgenic animals, and cells derived from enlarged thymuses did not generate tumors in syngeneic recipients. Despite their enlargement *in vivo,* thymuses of transgenic mice yielded conditionally immortal cultures containing cells of both epithelial and fibroblastic morphologies. Both morphological cell types could be readily cloned and exhibited optimal growth in fully permissive conditions and did not grow in nonpermissive conditions. Clones that exhibited epithelium-like morphologies expressed cytokeratin and had the ability to rosette T-lymphocytes. Thus, in our initial experiments we were able readily to derive conditionally immortal lines of epithelial cells, as well as of fibroblasts, from these mice.

One founder animal (H2ts6) survived to the age of 6 months and fathered multiple offspring which harbor a functional transgene. Sibling matings of transgenic offspring have generated homozygous animals that have been bred successfully through many generations. As discussed below, this transgenic mouse strain has already been found to be a ready source of novel cell lines.

3.3. Astrocyte Clones Derived from H-2K^{b}tsA58 Transgenic Mice Express Properties of Glial Scar Tissue

As part of our investigation of the use of H-2K^{b}tsA58 transgenic mice for neuroscience research, we generated clonal astrocyte cell lines that exhibit the properties of glial scar tissue. Such cell lines are of potential importance as it is felt that glial scarring within the central nervous system (CNS) may play an important role in inhibiting regrowth of axons [Reier et al., 1983; Barrett et al., 1984; Carlstedt, 1985; Smith et al., 1986; Carlstedt et al., 1987; Liuzzi and Lasek, 1987; Fawcett et al., 1989; Rudge et al., 1989; Rudge and Silver, 1990; Smith et al., 1990; Geisert and Stewart, 1991; McKeon et al., 1991] and perhaps also in inhibiting repair of extensive breakdown of myelin in the central nervous system (as occurs in multiple sclerosis; Ffrench-Constant et al. [1988]).

To generate clonal lines of conditionally immortal astrocytes, cortical astrocytes were purified from neonatal H-2K^{b}tsA58 transgenic mice by simple and well-established methods [Noble et al., 1984], with the only modificiation being that cultures were grown at 33°C in the presence of 25U/ml of IFN-γ. To generate clonal cell lines, cultures of purified astrocytes were infected with the BAG retrovirus, which expresses bacterial ß-galactosidase and the neomycin resistance gene [Price et al., 1987]. Subsequent selection in medium containing G418 allowed the ready generation of clonal colonies. Four of these clones were chosen for detailed analysis.

All four astrocyte lines chosen for further analysis expressed glial fibrillary acidic protein (GFAP, an astrocyte-specific cytoskeletal protein), albeit at lower levels than that seen in primary cultures of mouse cortical astrocytes. Characterization of expression of membrane and extracellular matrix molecules indicated that all of these clones expressed a phenotype much like that associated with glial scar tissue [Groves et al., 1993b]. All four lines expressed tenascin, laminin, and chondroitin sulphate proteoglycans, all of which are present in some glial populations during development and are particularly expressed in CNS lesions [Liesi et al., 1983; Liesi, 1985; McKeon et al., 1991; Laywell and Steindler, 1991; Laywell et al., 1992].

The extent of neurite outgrowth promoted by our astrocyte cell lines was consistent with the possibility that these cells expressed a phenotype functionally similar to glial scar tissue, in that monolayers of the four transgenic astrocyte lines were less effective than monolayers of primary cortical astrocytes at promoting outgrowth of cerebellar neurons. The mean total neurite length on all four astrocyte lines was less than 50% of that seen on primary astrocyte monolayers. Despite this dramatic failure to promote growth, the astrocyte cell lines did not cause the cerebellar neurons to clump together or fasciculate, as has been reported for neurons of the CNS growing on fibroblasts or meningeal cells derived from the CNS [Noble et al., 1984; Fallon, 1985]. Thus, although the extent of neurite outgrowth was markedly reduced on monolayers of our H-2K^{b}tsA58 astrocyte cell lines, the organization of neuronal cell bodies and processes on the surfaces of these astrocytes suggested that they still expressed glial, rather than non-glial, surface properties.

The four transgenic astrocyte cell lines were also less effective than primary astrocytes at supporting outgrowth of neurites from dorsal root ganglion neurons derived from 7 day old postnatal rats. However, there was no inhibition of the growth of dorsal root ganglion neurons derived from ganglia of 18-day-old embryos. In this respect also, the astrocyte cell lines appeared to behave like scar tissue, which is thought to be markedly less inhibitory for the growth of immature neurons as compared with mature neurons [Fawcett et al., 1989].

3.3.1. Inhibition of O-2A progenitor migration

As previously observed for purified cortical astrocytes [Noble et al., 1988; Richardson et al., 1988], all of the clonal astrocyte cell lines derived from H-2K^{b}tsA58 transgenic mice produced platelet-derived growth factor (PDGF) and promoted the division of O-2A progenitors *in vitro*. However, the astrocyte cell lines [Groves et al., 1993b] differed markedly from primary astrocytes in their support of O-2A progenitor migration. O-2A progenitors formed small, tight colonies on monolayers of all four astrocyte lines after seven days of growth, whereas on primary cortical astrocytes the O-2A progenitor cells were distributed much more evenly over the entire monolayer. The possibility that this failure to migrate represented a functional inhibition of migration was tested by preparing confrontation experiments. O-2A progenitors were either plated onto primary astrocytes and allowed to migrate onto the astrocyte cell lines, or were plated onto the astrocyte cell lines and allowed to migrate onto primary astrocytes. The interface between primary and transgenic astrocyte monolayers could be clearly seen, due to the lower level of GFAP expression in the transgenic astrocytes.

In experiments where O-2A progenitor cells growing on primary astrocytes were challenged with a monolayer of transgenic astrocyte lines, very few progenitors crossed the interface between the primary and transgenic astrocytes. Instead, the progenitor cells appeared to migrate to the astrocyte interface but no further, frequently aligning their processes along the interface. The apparent failure of O-2A progenitors to cross the interface from primary to transgenic astrocytes was not the result of a failure to cross an astrocyte interface *per se.* In the majority of cases where O-2A progenitors were plated onto transgenic astrocyte monolayers and challenged with monolayers of primary astrocytes, the progenitors were able to cross the astrocyte interface and migrate considerable distances over the primary astrocyte monolayer.

The availability of clonal cell lines that exhibit properties much like that of glial scar tissue offers us a simple *in vitro* system for analyzing the biochemical cues that hinder neurite outgrowth from mature neurons and for identifying factors that might inhibit migration of O-2A progenitors in demyelinated lesions.

3.4. Direct Derivation from H-2K^{b}tsA58 Transgenic Mice of Conditionally Immortal Myoblasts Able to Differentiate into Myotubes *in Vivo*

Transplantation of myogenic cells has been mooted as a treatment for genetic disease, in muscle and other tissues [Dhawan et al., 1991; Barr and Leiden, 1991], and also holds promise as a tool for investigating the regulation of expression of muscle genes *in vivo.* However, there are two major obstacles to full exploitation of these techniques: the finite mitotic capacity of primary myoblasts limits the cell numbers available from clones, while the tumorigenic tendency of established myogenic lines makes them difficult to use in vivo [Morgan et al., 1992].

To generate cell lines with the above characteristics, Drs. J. Morgan and T. Partridge, and their colleagues, isolated conditionally immortal myoblast cell lines from the limb muscles of 19 day-old embryonic, 10 day-old, and 4-week-old H-2K^{b}tsA58 transgenic mice [Morgan et al., 1994]. Cell growth was sufficiently robust to allow the generation of clones directly by growth at low density. Clones exhibiting a typical myogenic morphology grew readily in the permissive conditions, but not in semipermissive conditions (33°C, IFN-γ-). 14 out of 15 of the clones chosen, readily formed myotubes when grown at high density in nonpermissive conditions. Confluent cultures switched from permissive to nonpermissive conditions showed greatly reduced DNA synthesis and began to form myotubes within 24 h. The number of myotubes continued to increase for the next several days. Some fusion also occurred in permissive conditions when cultures were allowed to become very dense. In both cases, the resultant myotubes expressed dystrophin and muscle-specific myosin. Nonfused cells did not express muscle-specific myosin or dystrophin.

Cells from eight different clones were injected into the leg muscles of dystrophin-negative MDX nu/nu mice of the Gpi Is^a isotype in which either the right or both legs had been subjected to 18-Gy x-rays at 15–17 days of age. Irradiation inhibits the regeneration of *mdx* muscle, which eventually atrophies, thus presenting a good environment for the formation of new muscle from implanted myogenic cells [Partridge et al., 1989; Wakeford et al., 1991].

Seven out of eight clones injected *in vivo* formed new muscle histologically indistinguishable from that formed by injection of primary myoblasts [Morgan et al., 1990], but did not form tumors, in both irradiated and nonirradiated *mdx* muscles. No systematic difference in the ability to form muscle was seen between different clones, nor between irradiated verses nonirradiated legs.

Even after 19 passages (approximately 60 doublings) in culture, cells remained diploid and formed normal muscle in the absence of tumor formation on injection into irrradiated *mdx nu/nu* muscle, in contrast to the neoplastic behavior of established muscle cell lines [Wernig et al., 1991; Partridge et al., 1988; Morgan et al., 1992], no tumors were found up to 120 days post-implantation [Morgan et al., 1994].

One striking aspect of these experiments was that rare clones of H-2K^{b}tsA58-derived cells could be reisolated from injected muscle after several weeks of *in vivo* growth, injected into a second generation of hosts (where the injected cells formed new muscle), reisolated again, and injected to form normal muscle in still a third generation of hosts. Clones were re-isolated successfully from both irradiated and nonirradiated muscle. Only rare clones were obtained in these reisolation experiments, suggesting that myoblast cell lines derived from H-2K^{b}tsA58 mice are able to both form myotubes and enter the pool of slowly dividing, or quiescent, myogenic cells known to exist in normal muscle.

In contrast with the problems of senesence and neoplasia associated with the use of primary cells or spontaneously arising cell lines, respectively, clonal myoblast lines derived from H-2K^{b}tsA58 mice can be grown for extended periods *in vitro* without exhibiting any loss of conditionality or capability of undergoing normal differentiation *in vitro* or *in vivo*. Moreover, the ease of isolation of cell lines from H-2K^{b}tsA58 mice of different ages raises the possibility of comparing the properties of myoblast lines representing different developmental stages.

3.5. H-2K^{b}tsA58 Mice and the Generation of Mutant Cell Lines

One particularly exciting application of the H-2K^{b}tsA58 transgenic mice will be in the generation of cell lines from other strains of mice that are genetically aberrant, as a result of either being bred as a mutant strain of animals or a mutation introduced by choice (such as mice in which the function of a particular gene of interest has been disrupted). As cell lines can be generated from heterozygous H-2K^{b}tsA58 transgenic mice, it thus follows that mating of homozygote breeding stock with a homozygous mutant will yield animals expressing both properties of interest. If the mutation of interest is dominantly active, then direct generation of cell lines from F_1 litters would be appropriate, while recessive mutations would have to be rebred to homozygosity in order for the cell lines of interest to be derived. These strategies have already been applied to the study of myoblasts from *mdx* mice, which have a mutation in the dystrophin gene. H-2K^{b}tsA58 $\times$ *mdx* F_1 pups have proved to be a ready source of *mdx*-deficient skeletal myoblast lines, thus providing an important new tool for the study of dystrophin function [Morgan et al., 1994].

4. GENERAL PRINCIPLES GUIDING CELL LINE PRODUCTION FROM H-2K^{b}tsA58 TRANSGENIC MICE

A number of general principles emerge from our experiments to date. One point of central importance is that we think it most appropriate to utilize heterozygotes for the generation of cell lines in most (and possibly all) circumstances. The successful derivation of cell lines from heterozygotes offers a number of practical advantages, especially the fact that it is not necessary to maintain a homozygous breeding colony if one's intention is to use animals from only a limited series of dissections. Instead, male stud homozygotes can be mated with normal females, with cells of interest being derived from the heterozygous F_1 animals. It is easy to obtain a steady stream of heterozygous animals with a few homozygous stud males and a harem of normal females.

It does seem reasonable to consider whether cell types that have proved difficult to grow from heterozygous animals may be more readily obtained from homozygotes, due to the higher level of LTAg expression achieved per unit of interferon applied. Thus far, however, our tendency has been to expend more effort on determining appropriate growth conditions for each cell type of interest. Part of the reason for choosing this approach is that we are primarily interested in using cells derived from H-2K^{b}tsA58 transgenic animals as a tool for exploring the cellular biology of normal cells, and we prefer to express the minimal amount of LTAg needed to achieve immortalization.

A second useful observation we have made is that one can often (although not invariably) turn off tsLTAg function by shifting temperature up to only 37°C. This reduces the need for extra incubators from 2 (33°C and 39.5°C) to 1 (33°C), with regular 37°C incubators being used to turn off LTAg activity. For a large number of laboratories, this second advantage is particularly useful. More importantly, the lack of an absolute requirement for shift to 39.5°C means that transplantation *in vivo* has a higher chance of turning off functional expression of tsLTAg, thus allowing cells to differentiate normally *in situ.*

A further point to note is that we have not seen any great differences in effectiveness of IFN-γ from different manufacturers. Moreover, it should be possible to substitute any other inducer of Class I antigen expression (e.g., other interferons) to promote expression from the H-$2K^b$ promoter, in the event that IFN-γ is specifically detrimental to the growth of the cells of interest (a problem we have not encountered).

Although it appears likely that the H-$2K^b$tsA58 transgenic mice, and their generic relatives, will greatly facilitate cell line production, it is important to note that the probability of successfully generating a cell line of interest is still enhanced by knowledge about the biological properties of the cell of interest. Such a result is expected from previous studies on cell lines produced by retroviral infection of cultured cells. For example, even though expression of LTAg in Schwann cells simplifies the growth factor requirements of these cells, promotion of cell division in the immortalized Schwann cells still requires the presence of 2 out of 3 of the mitogenic stimuli used in the growth of primary Schwann cells [Ridley et al., 1988]. To date, our studies with cells derived from the H-$2K^b$tsA58 transgenic mice have indicated that generation of novel cell lines is indeed a straightforward matter for many tissues. However, when attempting to grow cells for which little is known about the control of division in the cell type of interest, it has also become clear that experimentation is required to determine suitable conditions for optimizing cell line production.

In other experiments being conducted at present, we and our colleagues have readily isolated cell lines of enteric glia, colonic epithelium [Whitehead et al., 1993], primitive kidney cells [Woolf et al., 1995], liver cells, pancreatic cells, and various cell types of the embryonic limb bud of H-$2K^b$tsA58 mice. Thus, the H-$2K^b$tsA58 transgenic mice appear to be useful in the manner we originally envisaged. Moreover, the success of our transplantation studies with myoblast lines derived from H-$2K^b$tsA58 mice suggests that these animals might provide a generally useful source of cell lines suitable for application in studies in which transplantation and incorporation into normal tissue is a desired goal.

4.1. Availability of H-$2K^b$tsA58 Mice

Charles River Laboratories has agreed to supply H-$2K^b$tsA58 transgenic mice to the international scientific community. Requests to purchase animals should be directed to Charles River Laboratories, 251 Ballardvale St.,Wilmington, MA 01887, USA.

5. PROTOCOLS

5.1. General Comments

To establish glial cell lines, there are several details that need to be taken into consideration initially:

1. The ideal method for transfering the appropriate gene into the cell
2. The appropriate gene
3. The culture conditions and growth factor requirements that maintain the cells in their non-differentiated and differentiated state.

There are several methods for the construction of cell lines that depend on the target cells. If the target cell is a robust cell that can be grown in serum-containing medium, then the most common method for the transfection of DNA is with calcium phosphate [Sambrook et al., 1989]. However, primary glial cells are very sensitive to this technique, and we have found that retroviral gene transfer or lipofectin-mediated gene transfer is much more gentle with a higher viability of resulting glial cells. As mentioned above, it has been possible to create O-2A progenitor cell lines only with the tsLTAg and c-*myc* genes, and so these would be genes of choice for O-2A cell lines. Finally, many of the growth factor requirements of the O-2A lineage cells are largely known [Bögler et al., 1990; Barres et al., 1992; Mayer et al., 1994] and in particular the application of the two growth factors—platelet-derived growth factor (PDGF) and basic fibroblast growth factor (bFGF)—allows selection for progenitor cells in our construction of cell lines since PDGF and bFGF promote self-renewal and prevent the cells from differentiating [Bögler et al., 1990]. It is important that the cells are actually dividing during the transfer of DNA as DNA is poorly integrated in nondividing cells. In addition, oncogene-containing cell lines do not necessarily lose their growth factor requirements; therefore, the resulting cell lines still require the same culture conditions as their primary cell counterpart. If the resulting cell lines do lose these requirements, it is likely that they will also lose their resemblence to the primary cell.

5.2. Preparation of Rat Optic Nerve Cultures

Protocol 5.2

Reagents and Materials

Sterile

- Collagenase: stock of 1.33% in Leibowitz, L-15, medium (~155 U/mg: Cambridge Bioscience).
- L-15 + 25 μg/ml gentamycin.
- DMEM serum-free modified medium (DMEM-BS; see

Bottenstein and Sato [1979]), DMEM containing 4.5 g/liter glucose and supplemented with 25 μg/ml gentamycin (Gibco), 0.0286% (v/v) BSA Pathocyte, 2mM glutamine, 0.5 mg/ml bovine pancreas insulin, 100 μg/ml human transferrin, 0.2μM progesterone, 0.10μM putrescine, 0.45μM-L-thyroxine, 0.224μM selenium, and 0.49μM 3,3′,5-triiodo-L-thyronine

- DMEM(SD): DMEM with soybean trypsin inhibitor (SBTI), 0.52 mg/ml, DNAse, 0.04 mg/ml, and bovine serum albumin, 3 mg/ml.
- DMEM-PF: 10 ng/ml PDGF + 10 ng/ml bFGF in DMEM-BS.
- Hanks' balanced salt solution, Ca^{2+}/Mg^{2+} -free.
- Growth factors: 10 ng/ml of each PDGF bFGF; dissolve stock in PBS + 0.1% BSA at 1 μg/ml and aliquot and store at −20°C.
- Trypsin, 0.25%
- Activated papain: The papain is activated by dissolving it in DMEM in a 37° waterbath, and adding (grain by grain) HCl-cysteine until the pH is acidic. Then neutralize the solution with sodium hydroxide, using the pH indicator in the DMEM to indicate the correct pH. Filter this mixture directly onto the cells, but make sure to first wash the filter with 5–10 ml of DMEM or UPW. Many filters contain a small amount of detergent as a wetting agent, and this will kill the O-2A progenitor cells.
- In DMEM (SD).
- Poly-L-lysine (>100,000 MW) (PLL), stock: 4 mg/ml, dilute 1:300 in sterile ultrapure water (UPW) and use at a final concentration of 13.3 μg/ml.
- Neomycin (G418 Geneticin), 1 mg/ml in DMEM with 10% FCS.
- Hygromycin, 0.75 mg/ml in DMEM with 10% FCS.
- Plastic syringe, 1 ml, and 21G, 23G, and 25G needles.
- Scalpel
- Tissue culture plastics
- 25-cm^2 flasks
- 6-well plates
- Petri dish
- Bijou bottles or equivalent 5-ml container
- Sprague-Dawley rats, 7 days old
- Antibodies for purification of O-2A progenitor cells: Ran-2, GalC, and A2B5 or O4

This is a modification of the method of Wysocki and Sato [1978], as adapted to glial cell purification by Barres et al. [1992] and modified again by Mayer et al. [1993, 1994].

To purify O-2A progenitor cells, 3 antibodies are needed. These are the Ran-2 antibody [Bartlett et al., 1980], which binds to type 1 astrocytes and meningeal cells, antigalactocerebroside antibody, GalC [Ranscht et al., 1982], which binds to oligodendrocytes, and either the A2B5 antibody [Eisenbarth et al., 1979] or O4 antibody [Sommer and Schachner, 1981]. The first two antibodies are used for negative selection, and one of the last two is used for positive selection. The advantage of the A2B5 antibody is that it labels O-2A progenitor cells early in their development, while the disadvantage is that it is an O-2A lineage specific marker only in the optic nerve. The O4 antibody is completely O-2A lineage specific for all regions of the CNS, with the exception of the olfactory bulb (where it labels olfactory nerve ensheathing cells; see Barnett et al. [1993]), but it begins to label O-2A progenitor cells only at a stage slightly later than that labeled by the A2B5 antibody.

Preparation of panning dishes

1. Incubate 100 mm tissue culture plastic Petri dishes overnight with anti-IgG (H + L) for second antibodies of any of the IgG classes, or anti-IgM for antibodies of the IgM class. Ran-2 and anti-GalC are both IgGs; A2B5 and 04 are both IgMs. The anti-Ig antibodies are prepared in 50mM Tris, pH 9.5, at a concentration of 1 mg/ml, and 10 ml is added to each plate. Solutions can be re-used up to 5 times before they are depleted significantly.
2. Prior to use of the panning dishes, wash them 3 times with DMEM. Then incubate them with 10 ml of the cell-type-specific antibody for at least one hour at 37°C in the incubator. If the cell-type-specific monoclonal antibody has not been prepared in serum-free chemically defined medium, it is necessary to add bovine serum albumin as a carrier protein. It is critical in this protocol that the cell-type-specific antibodies not be prepared in the presence of 10% serum, and that the hybridoma supernatants have been harvested in a serum-free medium.
3. Immediately prior to adding cells, wash again 3 times with DMEM.

Protocol

(a) Stages of this procedure can be carried out while the panning plates are washing.
(b) Dissect optic nerves, or other regions of the CNS from which O-2A progenitor cells will be purified.

(c) Remove meningeal tissues as much as possible by stripping them away.
(d) Mince tissue into small pieces with a scalpel.
(e) Incubate in 1.3% collagenase for 30 min at 37°C.
(f) Centrifuge for 5 min at 1000 rpm and then for 1 min at 3000 rpm.
(g) Resuspend in 30 U/ml of activated papain.
(h) Leave the tissue chunks to incubate for 1 h at 37°C in the papain solution.
(i) Add to this cell suspension 0.04 mg/ml of bovine pancreatic DNAse, also made up in DMEM.
(j) Leave for 5 min.
(k) Centrifuge cells as described above.
(l) Resuspend in 1 ml DMEM-BS and triturate through fine needles. *This is a critical stage in this preparation.*
    (i) Start with 19G needles, and take the tissue through gently once or twice, being careful not to introduce air bubbles into the fluid.
    (ii) Pass 3 or 4 times through a 23G needle
    (iii) Pass 6–10 passes through a 27G needle. This must be done with extreme gentleness, to avoid great reduction in your cell yield. A good way of judging as to whether you have been sufficiently gentle is to examine a drop of your cell suspension under the microscope. If you can readily find cells with interesting looking processes attached to them, then you've done a good job. If cells all look round, you have either been too rough or have dissected the spleen by mistake!
(m) Make the suspension of cells up to 5 ml and add to the first panning dish, which is Ran-2 for negative selection.
(n) Leave for 20 min at 37°C and gently shake the plate from time to time.
(o) Collect the supernatant and add to the second negative selection plate (anti-GalC). In addition, wash the Ran-2 plate with 1 ml of DMEM-BS and add these cells to the second dish.
(p) Incubate for 20 min, gently shaking the plate from time to time.
(q) Remove the supernatant and add to the third A2B5 or 04 dish. This time don't wash the plate, as the oligodendrocytes come off too easily.
(r) After 20 min, again with gentle shaking of the plate from time to time, discard the supernatant, as the cells have now stuck to the plate.
(s) Wash with 2 ml of DMEM-BS and discard this wash.

(t) Add 1 ml of DMEM-BS and gently scrape off the cell with a cell scraper. Do not use trypsin to remove the cells, as this leads to a decrease in cell viability.

(u) Check whether all cells have been removed from the plate, transfer the cell suspension to a bijou bottle and count the cells.

Notes:

1. Do not spin the cells down, or you will lose them all.
2. The yield that can be expected is 2×10^5 pure O-2A progenitor cells from an initial 2×10^6 mixed cells from optic nerves dissected from 20 7day-old rat pups. If you isolate cells from optic nerves of younger animals, the yield of O-2A progenitor cells will be reduced because they will be proportionally less well represented in the optic nerve. A higher cell number can be obtained by using 04 antibody to positively select cells from the entire brain (with the olfactory bulbs removed). Again, the total number of cells obtained is higher if 7-day-old rat pups are used, but useful numbers of cells can be obtained from younger animals.

5.2.1. Growth of O-2A progenitor cells

The manner in which cells are grown depends on the experimental questions to be asked. As most of our experimentation is conducted in such a manner as to observe clearly the behavior of single cells and their progeny, most of our methods are specialized for low-density culture. This is the most demanding way in which to grow cells, and these techniques work perfectly well with higher-density cultures.

There are three general methods to grow progenitor cells as purified populations. One method allows expansion of progenitor populations, the second allows analysis of the behavior of clones undergoing both expansion and differentiation, and the third generates pure populations of oligodendrocytes or type 2 astrocytes.

In all cases, barring the generation of type 2 astrocytes, cells are grown in chemically defined [DMEM-BS (see Section 5.2)] medium on poly-L-lysine-coated glass coverslips, slide flasks, tissue culture flasks, and so on [PLL; Sigma; M_r 175,000 (relative molecular mass) 13 μg/ml].

Expansion of pure populations of O-2A progenitor cells To expand pure populations of O-2A progenitors, two different growth conditions can be utilized. Both of these conditions promote the continuous division of O-2A progenitor cells while preventing the generation of oligodendrocytes.

Protocol 5.2.1

Expose cells to DMEM-PF. In general, it is best to feed cultures by only replacing half of the medium at any feeding; this replacement of medium should be carried out every other day.

This method, originally described in Bögler et al. [1990], has been used to generate expanded populations of O-2A progenitor cells that are capable of forming oligodendrocytes and remyelinating demyelinated axons in experimental lesions *in vivo* [Groves et al., 1993a]. It is striking, however, that cells grown in this way change with time in culture, and eventually become unable to respond to PDGF when applied in the absence of bFGF. Our interpretation of these results is that the biological clock that appears to limit the mitotic lifespan of O-2A progenitor cells dividing in response to PDGF continues to measure elapsed time even when differentiation is blocked by the presence of both mitogens [Bögler and Noble, 1994].

A second way to expand O-2A progenitor populations is to grow them in the presence of PDGF and neurotrophin-3 (NT-3).

Protocol 5.2.2

(a) Add 10 ng/ml PDGF + 5 ng/ml of neurotrophin-3 (NT-3) in chemically defined medium containing no thyroid hormone.
(b) Repeat steps (b) and (c) as in Protocol 5.2.

It appears at present that cells grown in this manner are able to maintain a rapid rate of division for a longer period than cells grown in the presence of PDGF + bFGF. Little is known, however, about the ability of these cells to function normally *in vivo* following transplantation, for example.

5.2.3. Induction of differentiation

To study differentation in a manner that may be related to events *in vivo,* O-2A progenitors can be grown on monolayers of purified cortical astrocytes, in the presence of astrocyte-conditioned medium [Noble and Murray, 1984], or in medium supplemented with 10 ng/ml of PDGF (a potent O-2A progenitor mitogen secreted by purified cortical astrocytes; see Noble et al. [1988], Raff et al. [1988], Richardson et al. [1988]. Embryonic O-2A progenitor cells grown in this manner generate oligodendrocytes with a timing similar to that occuring *in vivo* [Raff et al., 1985, Raff et al., 1988].

5.2.4. Pure oligodendrocyte cultures

To generate pure cultures of oligodendrocytes, it is only necessary to deprive O-2A progenitor cells of mitogen while they are growing in chemically defined medium. They will then differentiate into oligodendrocytes within 3 days. It is important to note, however, that if they have been grown in the presence of bFGF, it is necessary to lightly trypsinize the cells to remove the bFGF, which binds to the extracellular matrix [Bögler and Noble, 1994].

5.2.5. Pure type 2 astrocyte cultures

To generate pure cultures of type 2 astrocytes it is necessary to take freshly isolated and purified O-2A progenitor cells and expose them to 10% fetal calf serum [Raff et al., 1983a] or other fetal sera. For reasons we do not yet understand, cultures grown in PDGF + bFGF for extended periods generate many oligodendrocytes when this switch is made, thus precluding the use of cultures expanded in this manner for the generation of large quantities of type 2 astrocytes.

5.3. Retroviral Infection and Lipofectin

For either of these methods it is necessary to have the appropriate gene in a vector that has the appropriate LTRs (long terminal repeats). We chose to subclone the c-*myc* and tsLTAg gene into the BamH1 or SnaB1 cloning site of the pBabe vector using standard recombinant DNA techniques [Barnett et al., 1994; Barnett and Crouch, 1995]. The pBabe vector contains the gene for hygromycin resistance (*hygr*) or neomycin resistance (*neor*) [Morgenstern and Land, 1990]. This vector can then be used to construct retrovirus or be transfected into cells using the lipofectin reagent (BRL) or equivalent reagent.

5.3.1. Retroviral construction

Protocol 5.3.1

Reagents and Materials (Sterile)

- DNA for transfection
- Retroviral packaging cell line, ψ2
- DMEM-10FCS: DMEM with 10% FCS
- DMEM : 10FCS containing 1 mg/ml neomycin or 0.75 mg/ml of hygromycin
- Materials for picking clones (see Clarke, Chapter 7, this volume)
- 3T3 cells

Protocol

(a) For the construction of retrovirus, transfect the retroviral packaging cell line, ψ2 cells with 10 μg of DNA using the calcium phosphate method [Sambrook et al., 1989].

(b) After 48 h, passage the ψ2 transfected cells growing in DMEM containing 10% FCS 1:10.

(c) Select in either 1 mg/ml neomycin or 0.75 mg/ml of hygromycin depending on the vector used.

(d) Pick at least 10 resistant clones.

(e) Expand in culture.

(f) Determine the virus titer on 3T3 cells.
To determine the titer:
   (i) Plate 8×10^5 3T3 cells in a 100-mm dish.
   (ii) Infect the cells the next day with a known concentration of virus.
   (iii) Split the 3T3 cells 1:20 directly into selection medium (1 mg/ml for *neo* and 150 μg/ml for *hygro*).
   (iv) Ten days later, without a change of medium, count the resistant colonies.
   Optimum values are 10^4–10^6 colonies per milliliter of virus. The concentration of antibiotic to select cells is cell related and it may be necessary to titrate the antibiotic with your particular stock of cells.

(g) To collect retrovirus for infection of glial cells:
   (i) Grow ψ2 cells with a high titer to about 80% confluence.
   (ii) Place in a minimum volume of DMEM-FCS.
   (iii) Leave overnight.
   (iv) The next day, collect the supernatant from these cells.
   (v) Filter through a 0.45-μm filter.
   (vi) Collect filtrate.
   (vii) Aliquot.
   (viii) Freeze on dry ice directly.
   (ix) Store at −20°C.

5.3.2. Lipofectin mediated gene transfer

The optimum concentrations of DNA and Lipofectin reagent together with the time that the reagent is left on the cells should be calculated for each cell type. Briefly, equal volumes of diluted DNA and Lipofectin reagent (50 μl in water) are mixed and left at room temperature for 15 min (total 100 μl). This mixture is then dripped over the cells and left an optimum time for the cells. For mitogen expanded O-2A progenitor cells it was found that 10–20 μg of DNA together with 40–50 μg of Lipofectin reagent left 5 h was optimum. Cells were washed gently once in DMEM-BS and refed with DMEM-PF. Three days later cells were selected with 30 μg/ml of neo or hygro for 10 days, and maintained in DMEM-PF until colonies were big enough to pick. Since O-2A progenitor cells do not strongly adhere to culture dishes, it is possible to pick colonies by gently scraping the colony with the tip of a small pipette while drawing up medium. The colonies can be expanded by continual feeding with DMEM-PF and checked for clonality by Southern blot analysis using the Tag insert as probe [Barnett et al., 1994].

5.4. Generation of Astrocyte Cell Lines from H-2K^{b}tsA58 Transgenic Mice

Enriched astrocyte cultures can be prepared by a modification of the methods of Noble et al. [1984].

Protocol 5.4

Materials and Reagents

Sterile

- EDTA solution (200 μg/ml solution of EDTA in Ca^{2+}/Mg^{2+}-free DMEM) containing 8500 U/ml trypsin
- DMEM-SD (see Section 5.2)
- DMEM-FCS with 0.4% BSA
- AraC: 2×10^{-5}M cytosine arabinoside
- SBTI-DNAse (see Protocol 5.2, Materials & Reagents)
- Syringe and 21G and 25G needles.
- 75-cm^2 flasks

Nonsterile equipment

- Rotary platform (capable of 100 rpm)

Protocol

(a) Remove cortices from perinatal mice.
(b) Dissect free of meninges.
(c) Chop finely.
(d) Incubate at 37°C for 30 min in EDTA solution.
(e) Add SBTI-DNAse at a ratio of 4 ml for every 7 ml of cortical cell suspension.
(f) Incubate for a further 5 min.
(g) Centrifuge at 200g for 5 min.
(h) Resuspend the tissue in DMEM-FCS with 0.4% BSA.
(i) Dissociate by repeated trituration through 21G and 25G needles.
(j) Centrifuge the dissociated cells through a cushion of DMEM-FCS containing 4% BSA at 200g for 5 min.

(k) Resuspend the pellet in DMEM-FCS
(l) Seed the cells into PLL-coated flasks at a density of 10^7 cells/75-cm^2 flask.
(m) Grow the cultures for 6 days.
(n) Remove the cells from the top of the monolayer by shaking overnight at 37°C on a rotary platform (100 rpm).
(o) After 24 h, pulse the cells with 2×10^{-5} M AraC for 4 days.

This procedure routinely produced astrocyte cultures of >95% purity as assessed by staining with a polyclonal antiserum against GFAP.

5.4.1. Generation of astrocyte cell lines

Cortical astrocytes can be purified from neonatal H-$2K^b$tsA58 transgenic mice as described above, except that the cultures should be prepared at 33°C in the presence of 25 U/ml of interferon (IFN)-γ.

Protocol 5.4.1

Reagents and Materials

- BAG retroviral supernatant [Price et al., 1987]
- DMEM with 10% FCS
- Polybrene: 10 mg/ml; Sigma
- DMEM-FCS containing 25 U/ml of IFN-γ
- DMEM-FCS with 25 μg/ml G418 (Geneticin)
- 1.1-kbp probe for the *neo* gene
- 96-well trays
- 6-well trays
- 25- and 75-cm^2 tissue culture flasks
- 25- and 75-cm^2 tissue culture flasks
- 33°C incubator

Protocol

(a) Passage the cultures at a ratio of 1 : 4 and refeed.
(b) The following day, incubate cultures with 1 ml of BAG retroviral supernatant, 1 ml of DMEM-FCS, and 2 ml of Polybrene for 2 h at 37°C.
(c) Wash the cultures with DMEM-FCS
(d) Feed with DMEM-FCS containing 25 U/ml of IFN-γ.
(e) Shift back down to 33°C.
(f) Passage the cells at a ratio of 1:4 once more the following day and feed with the same medium.
(g) After four days supplement the culture medium with 25 μg/ml G418 and feed the cultures every 3 days for 2 weeks, after which time drug-resistant colonies may be clearly visible.

(h) Passage about 50 colonies using autoclaved cloning rings (see Clarke, this volume, Chapter 7) and culture in 96-well trays.
(i) Expand the clones progressively to 24- and 6-well trays, and then to 25- and 75-cm^2 tissue culture flasks.
(j) The passage at which the lines can be plated into a 25-cm^2 flask may be designated as passage 1.
(k) Choose ten lines for Southern blot analysis using a 1.1-kbp probe for the *neo* gene to determine whether each line contains a single retroviral integration site.

REFERENCES

Allinquant B, D'Urso D, Almazan G, Colman DR (1990): Transfection of transformed *shiverer* mouse glial cell lines. Devel Neurosci 12: 340–348.

Almazan G, McKay R (1992): An oligodendrocyte precursor cell line from rat optic nerve. Brain Res 579: 234–245.

Baldwin AS Jr, Sharp PA (1987): Binding of a nuclear factor to a regulatory sequence in the promoter of the mouse H-2K^b class I major histocompatibility gene. Mol Cell Biol 7: 305–313.

Barnett SC, Hutchins A-M, Noble M (1993): Purification of olfactory nerve ensheathing cells from the olfactory bulb. Devel Biol 155: 337–350.

Barnett SC, Franklin RJM, Blakemore WF (1994): *In vitro* and *in vivo* analysis of a rat bipotential O-2A progenitor cell line containing the temperature sensitive mutant gene of the SV40 large T antigen. J Neurosci 5: 1247–1260.

Barnett SC, Crouch, DH (1995): The effect of oncogenes on the growth, differentiation and transformation of oligodendrocyte-type-2 astrocytes progenitor cells. (1995) Cell Growth Differ 6: 69–84.

Barnett SC, Rosario M, Doyle AC, Kilbey A, Lovatt A, and Gillespie DAF (1995): Differential regulation of AP-1 and novel TRE-specific DNA binding complexes during differentiation of oligodendrocyte-type-2-astrocyte (O-2A) progenitor cells. Development 121: 3969–3977.

Bar E, Leiden JM (1991): Systemic delivery of recombinant proteins by genetically modified myoblasts. Science 254: 1507–1509.

Barres BA, Hart IK, Coles HSR, Burne JF, Voyvodic JT, Richardson WD, Raff MC (1992): Cell death and control of cell survival in the oligodendrocyte lineage. Cell 70: 31–46.

Barres BA, Lazar MA, Raff MC (1994): A novel role for thyroid hormone, glucocorticoids, and retinoic acid in timing oligodendrocyte development. Development 120: 1097–1108.

Barrett CP, Donati EJ, Guth L (1984): Differences between adult and neonatal rats in their astroglial response to spinal injury. Exp Neurol 84: 374–385.

Bartlett PF, Noble MD, Pruss RM, Raff MC, Rattray S, Williams CA (1980). Rat neural antigen-2 (Ran-2): a cell surface antigen on astrocytes, ependymal cells, Muller cells and leptomeningeal cells. Brain Res 204: 339–351.

Basham TY, Merigan TC (1983): Recombinant interferon-gamma increases HLA-DR synthesis and expression. J Immunol 130: 1492–1494.

Bayley SA, Stones AJ, Smith CG (1988): Immortalization of rat keratinocytes by transfection with polyomavirus large T gene. Exp Cell Res 17: 232–236.

Blakemore WF, Crang AJ (1988): Extensive oligodendrocyte remyelination following injection of cultured central nervous system cells into demyelinating lesions in the adult central nervous system. Devel Neurosci 19: 1–11.

Blakemore WF, Crang AJ (1989): The relationship between type-1 astrocytes, Schwann cells and oligodendrocytes following transplantation of glial cell cultures into demyelinating lesions in the adult rat spinal cord. J Neurocytol 18: 519–528.

Bögler O, Wren D, Barnett SC, Land H, Noble M (1990): Cooperation between two growth factors promotes extended self-renewal and inhibits differentiation of oligodendrocyte-type-2 astrocyte (O-2A) progenitor cells. Proc Natl Acad Sci USA 87: 6368–6372.
Bögler O, Noble MD (1994): Measurement of time in oligodendrocyte-type-2 astrocyte (O-2A) progenitors is a cellular process distinct from differentiation or division. Devel Biol 162: 525–538.
Bottenstein JE, Sato GH (1979): Growth of a neuroblastoma cell line in serum-free. Proc Natl Acad Sci USA 76: 514–517.
Brinster RL, Chen HY, Messing A, van Dyke T, Levine AJ, Palmiter RD (1984): Transgenic mice harboring SV40 T-antigen genes develop characteristic brain tumors. Cell 37: 367–379.
Burns JS, Lemoine L, Lemoine NR, Williams ED, Wynford-Thomas D (1989): Thyroid epithelial cell transformation by a retroviral vector expressing SV40 large T. Br J Cancer 59: 755–760.
Carlstedt T (1985): Dorsal root innervation of spinal cord neurons after dorsal root implantation into the spinal cord of adult rats. Neurosci Lett 55: 343–348.
Carlstedt T, Dalsgaard C-J, Molander C (1987): Regrowth of lesioned dorsal root nerve fibres into the spinal cord of neonatal rats. Neurosci Lett 74: 14–18.
Cepko CL (1989): Immortalization of neural cells via retrovirus-mediated oncogene transduction. Annu Rev Neurosci 12: 47–65.
Compere SJ (1989): The *ras* and *myc* oncogenes cooperate in tumor induction in many tissues when introduced into midgestation mouse embryos by retroviral vectors. Proc Natl Acad Sci USA 86: 2224–2228.
Crang AJ, Franklin JRM, Blakemore WF, Noble M, Barnett SC, Groves, A, Trotter J, Schachner M (1992): The differentiation of glial cell progenitor populations following transplantation into non-repairing central nervous system glial lesions to adult animals. J Neuroimmunol 40: 243–254.
Cunningham MG, Nikkhah G, McKay RDG (1993): Grafting immortalized hippocampal cells into the brain of the adult and newborn rat. Neuroprotocols 3: 260–272.
Dhawan J, Pan LC, Pavlath GK, Travis MA, Lanctot AM, and Blau HM (1991): Systemic delivery of human growth hormone by injection of genetically engineered myoblasts. Science 254: 1509–1511.
Duncan ID, Hammang JP, Jackson KF, Wood PM, Bunge RP, Langford L (1988): Transplantation of oligodendrocytes and Schwann cells into the spinal cord of the myelin-deficient rat. J Neurocytol 17: 351–360.
Efrat S, Linde S, Kofod H, Spector D, Delannoy M, Grant S, Hanahan D, Baekkeskov S (1988): Beta-cell lines derived from transgenic mice expressing a hybrid insulin gene-oncogene. Proc Natl Acad Sci USA 85: 9037–9041.
Eisenbarth GS, Walsh FS, Nirenberg M, (1979): Monoclonal antibody to a plasma membrane antigen of neurons. Proc Natl Acad Sci USA 76: 4913–4917.
Evrard C, Galiana E, Rouget P (1988): Immortalization of bipotential glial progenitors and generation of permanent "blue" cell lines. J Neurosci Res 21: 80–87.
Fallon JR (1985): Preferential outgrowth of central nervous system neurites on astrocytes and Schwann cells as compared with nonglial cells in vitro. J Cell Biol 100: 198–207.
Fawcett JW, Housden E, Smith-Thomas L, Meyer RL (1989): The growth of axons in three-dimensional astrocyte cultures. Devel Biol 135: 449–458.
Franklin RJM, Crang AJ, Blakemore WF (1991): Transplanted type-1 astrocytes facilitate repair of bipotential and plastic glio-neuronal precursor cells. J Neurocytol 20: 420–430.
Frederiksen K, Jat P, Valtz N, Levy D, McKay R (1988): Immortalization of precursor cells from the mammalian CNS. Neuron 1: 439–448.
French-Constant C, Miller RH, Burne JF, Raff MC (1988): Evidence that that migratory oligodendrocyte-type-2 astrocyte (O-2A) progenitor cells are kept out of the rat retina by a barrier at the eye-end of the optic nerve. J Neurocytol 17: 13–25.

Galiana E, Borde I, Marin P, Cuzin F, Gros F, Rouget P, Evrard C (1990): Establishment of permanent astroglial cell lines, able to differentiate *in vitro,* from transgenic mice carrying the polyoma virus large T antigen: An alternate approach to brain cell immortalization. J Neurosci Res 26: 269–277.
Gansmuller A, Clerin E, Kruger F, Gumpel M, Lachapelle F (1991): Tracing transplanted oligodendrocytes during migration and maturation in the shiverer mouse brain. Glia 4: 580–589.
Geisert EE, Stewart AM (1991): Changing interactions between astrocytes and neurons during CNS maturation. Devel Biol 143: 335–345.
Geller HM, Dubois-Dalcq M (1988): Antigenic and functional characterization of a rat central nervous system-derived cell line immortalized by a retroviral vector. J Cell Biol 107: 1977–1986.
Giotta GJ, Heitzmann J, Cohen M (1980): Properties of the temperature-sensitive Rous sarcoma virus transformed cerebellar cell lines. Brain Res 202: 445–458.
Groves AK, Barnett SC, Franklin RJM, Crang AJ, Mayer M, Blakemore WF, Noble M (1993a): Repair of demyelinated lesions by transplantation of purified O-2A progenitor cells. Nature 362: 453–455.
Groves AK, Entwistle A, Jat PS, Noble M (1993b): The characterisation of astrocyte cell lines that display properties of glial scar tissue. Dev Biol 159: 87–104.
Gumpel M, Baumann N, Raoul M, Jacque C (1983): Survival and differentiation of oligodendrocytes from neural tissue transplantated into newborn mouse-brain. Neurosci Lett 37: 307–311.
Gussoni E, Pavlath GK, Lanctot AM, Sharma KR, Miller RG, Steinman L, Blau HM (1992): Normal dystrophin transcripts in Duchenne muscular dystrophy patients after myoblast transformation. Nature 356: 435–438.
Hamilton SR, Vogelstein B (1988): Point mutations in human neoplasia. J Pathol 154: 205–206.
Hammang JP, Baetge EE, Behringer RR, Brinster RL, Palmiter RD, Messing A (1990): Immortalized rat neurons derived from SV40 TAg induced tumors in transgenic mice. Neuron 4: 775–782.
Hanahan D (1985): Heritable formation of pancreatic ß-cell tumors in transgenic mice expressing recombinant insulin/simian virus 40 oncogenes. Nature 315: 115–122.
Huang S, Terstappen LW (1992): Formation of haemopoietic microenvironment and haemopoietic stem cells from simple human bone marrow stem cells. Nature 360: 745–749.
Ibarrola N, Mayer M, Rodriguez-Pena A, Noble M (in preparation).
Israel A, Kimura A, Fournier A, Fellous M, Kourilsky P (1986): Interferon response sequence potentiates activity of an enhancer in the promoter of a mouse *H-2* gene. Nature 322: 743–746.
James CD, Carlbom E, Dumanski JP, Hansen M, Nordenskjold M, Collins VP, Cavenee WK (1988): Clonal genomic alterations in glioma malignancy stages. Cancer Res 48: 5546–5551.
Jat PS, Sharp PA (1989): Cell lines established by a temperature-sensitive simian virus 40 large-T-antigen are growth restricted at the nonpermissive temperature. Mol Cell Biol 9: 1672–1681.
Jat PS, Noble MD, Ataliotis P, Tanaka Y, Yannoutsos N, Larssen L, Kioussis D (1991): Direct derivation of conditionally immortal cell lines from an H-2K^b-tsA58 transgenic mouse. Proc Natl Acad Sci USA 88: 5096–5100.
Kimura A, Israel A, Le Bail O, Kourilsky P (1986): Detailed analysis of the mouse H-2K^b promoter: enhancer-like sequences and their role in the regulation of Class I gene expression. Cell 44: 261–272.
Land H, Parada LF, Weinberg RA (1983a): Cellular oncogenes and multistep carcinogenesis. Science 222: 771–778.
Land H, Parada LF, Weinberg RA (1983b): Tumorigenic conversion of primary embryo fibroblasts requires at least two cooperating oncogenes. Nature 304: 596–602.
Laywell E, Steindler D (1991): Boundaries and wounds, glia and glycoconjugates: Cellular and molecular analyses of developmental partitions and adult brain lesions. Ann NY Acad Sci 633: 122–141.

Laywell E, Dörries U, Bartsch U4, Faissner A, Schachner M, Steindler D (1992): Enhanced expression of the developmentally regulated extracellular matrix molecule tenascin following adult brain injury. Proc Natl Acad Sci USA 89: 2634–2638.

Lemoine NR, Mayall ES, Jones T, Sheer D, McDermid S, Kendall-Taylor P, Wynford-Thomas D (1989): Characterisation of human thyroid epithelial cells immortalised *in vitro* by simian virus 40 DNA transfection. Br J Cancer 60: 897–903.

Liesi P, Dahl D, Vaheri A (1983): Laminin is produced by early rat astrocytes in primary culture. J Cell Biol 96: 920–924.

Liesi P (1985): Laminin-immunoreactive glia distinguish regenerative adult CNS systems from non-regenerative ones. EMBO J 4: 2505–2511.

Lillien LE, Raff MC (1990): Differentiation signals in the CNS-type-2-astrocyte development *in vitro* as a model. Neuron 5: 111–119.

Lindvall O, Brondin P, Widner H, Rehncrona S, Gustavii B, Frackowiak R, Leenders KL, Sawle G, Rothwell JC, Marsden D, Björkland A (1990): Science 247: 574–577.

Liuzzi FJ, Lasek RJ (1987): Astrocytes block axonal regeneration in mammals by activating the physiological stop pathway. Science 237: 642–645.

Louis JC, Magal E, Muir D, Manthorpe M, Varon S (1992): CG-4, a new bipotential glial cell line from rat brain, is capable of differentiating *in vitro* into either mature oligodendrocytes or type-2 astrocytes. J Neurosci Res 31: 193–204.

MacKay K, Striker LJ, Elliot S, Pinkert CA, Brinster RL, Striker GE (1988): Glomerular epithelial mesangial, and endothelial cell lines from transgenic mice. Kidney Int 33: 677–684.

Marshall CJ (1991): How does p21*ras* transform cells? Trends Genet 7: 91–95.

Matsui Y, Zsebo K, Hogan BLM (1992): Derivation of pluripotential embryonic stem cells from murine primordial gene cells in culture. Cell 70: 841–847.

Mayer M, Bögler O, Noble M (1993): The inhibition of oligodendrocytic differentiation of O-2A progenitors caused by basic fibroblast growth factor is overridden by astrocytes. Glia 8: 12–19.

Mayer M, Bhakoo K, Noble M (1994): Ciliary neurotrophic factor and leukemia inhibitory factor promote the generation, survival and maturation of oligodendrocytes *in vitro.* Development 120: 143–153.

McKeon RJ, Schreiber RC, Rudge JS, Silver J (1991): Reduction of neurite outgrowth in a model of glial scarring following CNS injury is correlated with the expression of inhibitory molecules on reactive astrocytes. J Neurosci 11: 3398–3411.

Mellon PL, Windle JJ, Goldsmith PC, Padula CA, Roberts JL, Weiner RI (1990): Immortalization of hypothalamic GnRH neurons by genetically targeted tumorigenesis. Neuron 5: 1–10.

Morgan JR, Barrandon Y, Green H, Mulligan RC (1987): Expression of an exogeneous growth hormone gene by transplantable human epithelial cells. Science 237: 1476–1479.

Morgan JE, Hoffman EP, Partridge TA (1990): Normal myogenic cells from newborn mice restore normal histology to degenerating muscles of the mdx mouse. J Cell Biol 111: 2437–2449.

Morgan JE, Moore SE, Walsh FS, Partridge TA (1992): Formation of skeletal muscle *in vivo* from the mouse C2 cell line. J Cell Sci 102: 779–787.

Morgan J, Beauchamp JR, Peckham M, Ataliotis P, Jat P, Noble M, Farmer K, Partridge T (1994): Myogenic cells derived from H-2K^{b}tsA58 transgenic mice are conditionally immortal *in vitro* but differentiate normally *in vivo.* Devel Biol 162: 486–498.

Morgenstern JP, Land H (1990): Advanced mammalian gene transfer: high titre retroviral vectors with multiple drug selection markers and a complementary helper-free packaging cell line. Nucleic Acid Res 18: 3587–3596.

Noble MD, Murray K (1984): Purified astrocytes promote the *in vitro* division of a bipotential glial progenitor cell. EMBO J 3: 2243–2247.

Noble MD, Fok-Seang J, Cohen J (1984): Glia are a unique substrate for the *in vitro* growth of central nervous system neurons. J Neurosci 4: 1892–1903.

Noble MD, Murray K, Stroobant P, Waterfield MD, Riddle P (1988): Platelet-derived growth factor promotes division and inhibits premature differentiation of the oligodendrocyte/type-2 astrocyte progenitor cell. Nature 333: 560–562.
Noble M (1991): Points of controversy in the O-2A lineage: clocks and type-2 astrocytes. Glia 4: 157–164.
Noble M, Ataliotis P, Barnett SC, Bevan K, Bögler O, Groves A, Jat P, Wolswijk G, Wren D (1991): Development, differentiation and neoplasia in glial cells of the central nervous system. Ann NY Acad Sci 633: 35–47.
Noble M, Wren D, Wolswijk G (1992): The O-2A^{adult} progenitor cell: a glial stem cell of the adult central nervous system. Sem Cell Biol 3: 413–422.
Partridge TA, Morgan JE, Moore SE, Walsh FS (1988): Myogenesis *in vivo* from the mouse C2 muscle cell-line. J Cell Biochem Suppl12C: 331.
Partridge TA, Morgan JE, Coulton GR, Hoffman EP, Kunkel LM (1989): Conversion of mdx myofibres from dystrophic-negative to -positive by injection of normal myoblasts. Nature 337: 176–178.
Paul D, Höhne M, Hoffmann B (1988a): Immortalized differentiated hepatocyte lines derived from transgenic mice harboring SV40 T-antigen genes. Exp Cell Res 175: 354–365.
Paul D, Höhne M, Pinkert C, Piasecki A, Ummelmann E, Brinster RL (1988b): Immortalisation and malignant transformation of hepatocytes by transforming genes of polyoma virus and of SV40 virus *in vitro* and *in vivo.* Klin Wochenstr 66: Suppl 11, 134–139.
Peden KWC, Charles C, Sanders L, Tennekoon G (1989): Isolation of rat Schwann cell lines: use of SV40 T antigen gene regulated by synthetic metallothionine promoters. Exp Cell Res 185: 60–72.
Price J, Turner D, Cepko CL (1987): Lineage analysis in the vertebrate nervous system by retrovirus-mediated gene transfer. Proc Natl Acad Sci USA 84: 156–160.
Raff MC, Miller RH, Noble M (1983a): A glial progenitor cell that develops *in vitro* into an astrocyte or an oligodendrocyte depending on the culture medium. Nature 303: 390–396.
Raff MC, Abney ER, Cohen J, Lindsey R, Noble M (1983b): Two types of astrocytes in cultures of developing rat white matter: differences in morphology, surface gangliosides and growth characteristics. J Neurosci 3: 1289–1300.
Raff MC, Abney ER, Fok-Seang J (1985): Reconstitution of a developmental clock *in vitro:* a critical role for astrocytes in the timing of oligodendrocyte differentiation. Cell 42: 61–69.
Raff MC, Lillien LE, Richardson WD, Burne JF, Noble MD (1988): Platelet-derived growth factor from astrocytes drives the clock that times oligodendrocyte development in culture. Nature 333: 562–565.
Raff MC (1989): Glial cell diversification in the rat optic nerve. Science 243: 1450–1455.
Ranscht B, Clapshaw PA, Price J, Noble MD, Seifert W (1982): Development of oligodendrocytes and schwann cells studied with a monoclonal antibody against galactocerebroside. Proc Natl Acad Sci USA 79: 2709–2713.
Reier PJ, Stensaas LJ, Guth L (1983): The astrocytic scar as an impediment to regeneration in the central nervous system. In Kao CC, Bunge RP, Reier PJ, (eds): "Spinal Cord Reconstruction." New York: Raven Press, pp 163–195.
Renfranz PJ, Cunningham MG, McKay RDG (1991): Region-specific differentiation of the hippocampal stem cell line HiB5 upon implantation into the developing brain. Cell 66: 713–729.
Richardson WD, Pringle N, Mosley M, Westermark B, Dubois-Dalcq M (1988): A role for platelet-derived growth factor in normal gliogenesis in the central nervous system. Cell 53: 309–319.
Richardson WD, Raff M, Noble M (1990): The oligodendrocyte-type-2 astrocyte cell lineage. Semin Neurosci 2: 445–454.

Ridley A, Paterson H, Noble M, Land H (1988): Ras-mediated cell cycle arrest is altered by nuclear oncogenes to induce Schwann cell transformation. EMBO J 7: 1635–1645.
Rosenbluth J, Hasegawa M, Shirasaki N, Rosen CL, Lui Z (1990): Myelin formation following transplantation of normal fetal glial into myelin-deficient rat spinal cord. J Neurocytol 19: 718–730.
Rudge JS, Smith GM, Silver J (1989): An *in vitro* model of wound healing in the CNS: Analysis of cell reaction and interaction at different ages. Exp Neurol 103: 1–16.
Rudge JS, Silver J (1990): Inhibition of neurite outgrowth on astroglial scars *in vitro.* J Neurosci 10: 3594–3603.
Ruley HE (1983): Adenovirus early region 1A enables viral and cellular transforming genes to transform primary cells in culture. Nature 304: 602–606.
Sambrook J, Fritisch EF, Maniatis T (1989): "Molecular Cloning: A Laboratory Manual", 2nd ed. Cold Spring Harbour Laboratory Press.
Santerre RF, Cook RA, Crisel RM, Sharp JD, Schmidt RJ, Williams DC, Wilson CP (1981): Insulin synthesis in a clonal cell line of simian virus 40-transformed hamster pancreatic beta cells. Proc Natl Acad Sci USA. 78: 4339–4343.
Smith GM, Miller RH, Silver J (1986): Changing role of forebrain astrocytes during development, regenerative failure and induced regeneration upon transplantation. J Comp Neurol 251: 23–43.
Smith GM, Rutishauser U, Silver J, Miller RH (1990): Maturation of astrocytes *in vitro* alters the extent and molecular basis of neurite outgrowth. Devel Biol 138: 377–390.
Snyder EY, Deitcher DL, Walsh C, Arnold-Aldea S, Hartwieg EA, Cepko C (1992): Multipotent neural cell lines can engraft and participate in development of mouse cerebellum. Cell 68: 33–51.
Snyder EY, Taylor RM, and Wolfe JH (1995): Neural progenitor cells engraftment correct lysomal storage throughout the MPS VII mouse brain. Nature 374: 367–371.
Sommer I, Schachner M (1981): Monoclonal antibodies (O1 and O4) to oligodendrocyte cell surfaces: and immunocytological study in the central nervous system. Devel Biol 83: 311–327.
Spangrude GJ, Heimfeld S, Weissman IL (1988): Purification and characterization of mouse haemopoietic stem cells. Science 241: 58–62.
Tegtmeyer P (1975): Function of simian virus 40 gene A in transforming infection. J Virol 15: 613–618.
Temple S, Raff MC (1986): Clonal analysis of oligodendrocyte development in culture: Evidence for a developmental clock that counts cell divisions. Cell 44: 773–779.
Thompson TC, Southgate J, Kitchener G, Land H (1989): Multistage carcinogenesis induced by *ras* and *myc* oncogenes in a reconstituted organ. Cell 56: 917–930.
Trotter J, Boulter CA, Sontheimer H, Schachner M, Wagner EF (1989): Expression of *v-src* arrests murine glial cell differentiation. Oncogene 4: 457–464.
Trotter J, Crang AJ, Schachner M, Blakemore WF (1993): Lines of glial precursor cells immortalised with a temperature-sensitive oncogene give rise to astrocytes and oligodendrocytes following transplantation into demyelinated lesions in the central nervous system. Glia 9: 25–40.
Vogelstein B, Fearon ER, Hamilton SR, Kern SE, Preisinger AC, Leppert M, Nakamura Y, White R, Smits AM, Bos JL (1988): Genetic alterations during colorectal cancer development. N Engl J Med 319: 525–532.
Wakeford S, Watt DJ, Partridge TA (1991): X-irradiation improves mdx mouse muscle as a model of muscle fibre loss in DMD. Muscle and Nerve 14: 42–49.
Wallach D, Fellous M, Revel M (1982): Preferential effect of g interferon on the synthesis of HLA antigens and their mRNAs in human cells. Nature 299: 833–836.
Weiss EH, Mellor A, Golden L, Fahrner K, Simpson E, Hurst J, Flavell RA (1983): The structure of the mutant H-2 gene suggests that the generation of polymorphism in H-2 genes may occur by gene conversion-like events. Nature 301: 671–674.

Wernig A, Irintchev A, Hartling A, Stephan G, Zimmerman K, Starizinski-Powitz A (1991): Formation of new muscle fibres and tumours after injection of cultured myogenic cells. J Neurocytol 20: 982.
Whitehead RH, VanEeden PE, Noble MD, Ataliotis P, Jat PS (1993): The establishment of conditionally immortalised epithelial cell lines from both colon and small intestine of H-$2K^b$tsA58 transgenic mice. Proc Natl Acad Sci USA 90: 587–591.
Williams DA, Rosenblatt MF, Beier DR, Cone RD (1988): Generation of murine stromal cell lines supporting hematopoietic stem cell proliferation by use of recombinant retrovirus vectors encoding simian virus 40 large T antigen. Mol Cell Biol 8: 3864–3871.
Wolswijk G, Noble M (1989): Identification of an adult-specific glial progenitor cell. Development 105: 387–400.
Wolswijk G, Riddle P, Noble M (1990): Co-existence of perinatal and adult forms of a glial progenitor cell during development of the rat optic nerve. Development 109: 691–698.
Wolswijk G, Riddle PN, Noble M (1991): Platelet-derived growth factor is mitogenic for O-$2A^{adult}$ progenitor cells. Glia 4: 495–503.
Woolf AS, Kolatsi-Joannou M, Hardman P, Andermarcher E, Moorby C, Fine LG, Jat PS, Noble MD, Gherardi E (1995): Roles of hepatocyte growth factor/scatter factor and the met receptor in the early development of the metanephros. J Cell Biol 128: 171–184.
Wren D, Wolswijk G, Noble M (1992): *In vitro* analysis of origin and maintenance of O-$2A^{adult}$ progenitor cells. J Cell Biol 116: 167–176.
Wysocki LJ, Sato VL (1978): "Panning" for lymphocytes: a method for cell selection. Proc Natl Acad Sci USA 75: 2844–2848.

APPENDIX: MATERIALS AND SUPPLIERS

Materials	Suppliers
Anti-IgG (H+L)	Southern Biotechnique
Anti-IgM	
Antibodies: A2B5	AATC
Ran-2	
Gal-C	
O4	
Ara-C (see Cytosine arabinoside)	
bFGF	Precision
Bovine pancreas insulin	Sigma
BSA0 Pathocyte	ICN Chemicals
Ca^{2+}, Mg^{2+} -free DMEM	(In house formulation)
Collagenase	Cambridge Bioscience
Cysteine-HCl	Sigma
Cytosine arabinoside	Sigma
DMEM	Gibco BRL, Life Technologies
DNAse, bovine pancreas	Sigma
EDTA, FDS-grade	Sigma
G418, Geneticin	Gibco BRL, Life Technologies
Gentamycin	ICN Flow Laboratories or Gibco BRL, Life Technologies
Glutamine	Sigma
HBS	Gibco BRL, Life Technologies
Human transferrin	Sigma
Hygromycin	Sigma
Interferon-γ	Peprotech
L-Thyroxine	Sigma

Appendix *(Continued)*

Materials	Suppliers
Leibowitz medium (L15)	Flow Laboratories
Lipofectin	Gibco BRL, Life Technologies
Neurotrophin (NT3)	Peprotech
Papain, activated	Sigma
PDGF	R and D Systems
Poly-L-lysine (PLL)	Sigma, M_r 175,000
Polybrene	Sigma
Progesterone	Sigma
SeleniumT	Sigma
Soyabean trypsin inhibitor	Sigma
Triiodo-L-thyronine	Sigma
Trypsin, 0.25%	Sigma
Trypsin, purified	Sigma

List of Suppliers

Advanced Magnetics Inc
61 Mooney Street
Cambridge, MA 02138
(800)343-1346; (617)499-1433; fax: (617)497-6927

Affiniti Research Products Ltds.
Mamhead Castle, Mamhead,
Exeter, EX6 8HD, England
01626 891010; fax 01626 891090

Ajax Chemicals
9 Short St.
Auburn, NSW 2144, Australia
(02)648-5222; fax (02) 647-1794

Aldrich Chemical Co.
The Old Brickyard, New Road
Gillingham, Dorset SP8 4JL, England
0800 717 181, 01747 822 211; fax 0800 37 85 38, 01747 823 779.

Amersham International
Amersham Place, Little Chalfont, Amersham, Bucks., HP7 9NA, England.
01494 544 000; Fax: 01494 542266
P.O. Box 22400, Cleveland, OH 44122.
(800) 321-9322; (216) 765-5000; Fax: (216) 464-5075, (800) 535-0898

Arnolds Veterinary Products Ltd.
Cartmel Drive, Harlescott
Shrewsbury, Shropshire, England
01142-231632; fax 01142- 52111

ATCC, American Type Culture Collection
12301 Parlawn Drive
Rockville, MD 20852-1776
800 638 6597; fax (301) 231-5826

Azlon Products Ltd
Powerscroft Road
Sidcup
Kent, DA14 5EF
England
0181 300 8886; fax: 0181 300 2247

Baker Company, The
P.O. Drawer; E. Sanford, ME 04073
(207)324-8773; (800)992-2537; fax (207) 324-3869

BDH/Merck
Broom Road, Poole
Dorset, BH12 4NN, England
01202 745 520; fax 01202 738 299.

B-D Labware
Two Oak Park
Bedford MA 01730
(800)343 2035, 617 275 0004; fax 617 275 0043.

Becton Dickinson
2 Bridgewater Lane
Lincoln Park, NJ 07035
(201) 628-1144

Bellco Biotechnology
340 Edrudo Road
Vineland, NJ 08360
(609)691-1075; fax (609) 691-3247

Bibby Sterilin. Ltd.
Tilling Drive
Stone, Staffordshire ST150SA, England
(0785) 81-2121; fax (0785) 81-3748

Biofluids Inc.
1146 Taft Street
Rockville, MD 20850
(301)424-4140; fax (301) 424-3619

Bio-Rad Laboratories
1414 Harbour Way, South Richmond, CA 94804.
(510) 741 1000
Mayland Avenue, Hemel Hempstead, Herts., HP2 7TD, England.
0442 232552; Fax: 0442 259118

Biosys S.A., 21 quai du clos des Roses
60200 Compiègne, France
33 44 86 22 75

Boehringer Mannheim AG
Industriestasse, 6343 Rotkreuz, Switzerland
+41 42 65 42 42; fax +41 42 64 54 52

Boehringer Mannheim Australia
31 Victoria Avenue
Castle Hill, NSW 2154, Australia
(02) 899-7999; fax (02)634-2949

Boehringer Mannheim Corporation
9115 Hague Road, P.O. Box 50414
Indianapolis, IN 46250-0414
(800) 262 1640; fax (317) 576 2754

Boehringer Mannheim GmbH Biochemica
Sandhoferstr. 116
D-6800 Mannheim 31, Germany
0621 7590; fax 0621 759 4044

Boehringer Mannheim UK
Bell Lane, Lewes
East Sussex, BN 71LG, England
44 0273 480 444; fax 44 0273 480 266

British Drug House, Merck Pty. Ltd., 207 Colchester Road
Kilsyth, VIC 3137, Australia
(03)728 5855; fax (03) 728 1351

Calbiochem-Novabiochem AG, Division Europe
P.O. Box 5334, Cysatstasse 23a
6000 Lucerne 5, Switzerland
+41 41 51 16 51; fax +41 41 51 45 64

Cambridge Bioscience
42 Devonshire Road
Cambridge CB1 2BL
England
01223 316 855

Cascade Biochem Ltd.
Whiteknights, Reading, Berkshire, England
01734 31 20 19.

CellMark Diagnostics
20271 Goldenrod Lane
Germantown, MD 20874
(301)428-4980; fax (301)428-4877

Celtrix Laboratories
Palo Alto, CA 95054
(408) 988 2500

Centaur Services Ltd.
Centaur House, Torbay Road
Castle Carey, Somerset, BA7 7EM, England
0963 350005; fax 0963 351161

Chance Propper Ltd.
Spon Lane South
Smethwick, Warley, West Midlands, B66 1NZ, England
0121 553 5551; fax 0121 525 0139

Clonetics Corporation
9620 Chesapeake Drive
San Diego, CA 92123
(619)541-2726; fax (619)541-0823

Collaborative Biomedical Products, Becton Dickinson Labware
Two Oak Park, Bedford, MA 01730
(617) 275-0004

Collagen Corporation
2500 Faber Place, Palo Alto, CA 94303-3334
(800)227-8933

Corning, Corning Glass Works
Corning, NY 14831
(607)737-1640

Dako Ltd.
16 Manor Courtyard, Hughenden Avenue
High Wycombe, Bucks. HP13 5RE, England
01494 45 20 16

Difco Laboratories
P.O. Box 331058, Detroit, MI 48232.
(313) 462-8500; (800) 521-0851; Fax: (313) 462-8517
P.O. Box 14B, Central Avenue, West Molesley, Kent KT8 2SE, England.
0181 979 9951; Fax: 0181 979 2506

Dow Corning;
see **Fisons** for U.K.

Dow Corning
Box 0994
Midland, MI 48686
(517) 496-4000

DuPont Company, Biotechnology Systems
Barley Mill Plaza, P-24, Wilmington, DE 19898.
(800) 551-2121; Fax: (617) 542-8468

Edmund Buhler GmbH & Co
Rottenburger Strasse 3
P.O. Box 12 24
D-7454 Bodelshausen
Germany
07471 7070; fax: 07471 70788

Fisher Scientific
711 Forbes Avenue, Pittsburg, PA 15219
(412) 562 8300; Fax: (412) 562-5344

Fisons Scientific Equipment
Bishop's Meadow Road
Loughborough, Leics., England
01509 231 166; fax 01509 231 893

Gelman Sciences
600 South Wagner Road; Ann Arbor, MI 48016
(313)665-0651; Fax (313)761-1208

Gemini Bioproducts, Inc.
5115 Douglas Fir Rd., Ste. M
Calabasas, CA 91302-1441
USA
818-591-3530

Gibco BRL, Life Technologies Inc.
8400 Helgerman Court, P.O. Box 6009
Gaithersburg, MD 20884-9980
(301) 840-8000

Gibco BRL, Life Technologies Ltd.
PO Box 35, Trident House, Renfrew Road
Paisley PA3 4EF, Renfrewshire, Scotland
44 041 814 6100; fax 44 041 887 1167

Gibco BRL, Life Technologies Inc.
8451 Helgerman Court
Gaithersburg, MD 20877
(301)840-4027; fax (301)258-8238

Gibco BRL, Life Technologies
Uferstrasse 90
4019 Basal, Switzerland
+41 61 65 03 00; fax +41 61 65 03 30

Hartmann, Dr. Daniel, Unité de Pathologie Cellulaire
CNRS URA 1459, Institut Pasteur de Lyon, Avenue Tony Garnier
69365 Lyon Cedex 07, France
33 72 72 25 14; fax 33 78 61 13 45

Hays Chemical Distribution General Chemicals Division
215 Tunnel Avenue, East Greenwich, London SE10 0QE, England
0181 858 8631/1172

Harlan Sprague Dawley, Inc
P.O. Box 29176, Indianapolis, IN 46229
USA
(317) 894-7521; fax: (317) 894-4473

Heat Systems-Ultrasonics, Inc.
1938 New Highway
Farmingdale, NY 11735
(800) 645-9846, (516) 694-9555; fax (516) 694-9412

Howard, Dr. Bruce
Laboratory of Molecular Growth Regulation, Building 6, Room 416, NICHHD
National Institutes of Health, Bethesda, MD 20892

IBF-Biotechnics
35 Avenue Jean Jaures
92390 Villeneuve la Garenne, France
33 1 47 98 83 53; fax 33 1 47 92 26 55

ICN Biomedicals Ltd.
Unit 18, Thame Park Business Centre
Wenman Road, Thame, Oxfordshire OX9 3XA
Eagle House, Periguine Business Park, High Wycombe, Bucks., HP13 7DL, England
01494 443 826; fax 01494 473162

ICN Immunobiologicals
3300 Hyland Avenue
Costa Mesa, CA 92626
(714)545-0113; fax 714 641 7275

Imperial Laboratories (Europe) Ltd.
West Portway, Andover, Hants SP10 3LF, England
01264 333 311

Jackson Immunoresearch Laboratories
872 W. Baltimore Pike, P.O.Box 9
Westgrove, PA 19390
(215)869-4024, (800) 367-5296; fax (215) 869-0171

Johnson & Johnson, Medical, Ltd.
Coronation Rd., Ascot, Berks., SL5 9EY, England
01344 871 000; Fax: 01344 872 599

Jencons (Scientific) Ltd.
Cherrycourt Way Industrial Estate, Stanbridge Road
Leighton Buzzard, Beds. LU7 8UA, England
01525 372010; fax 01525 379 547

Lab Safety Supply
Domestic: P.O. Box 1368; Janesville, WI 53547-1368
(800)356-0783; Fax (800)543-9910
International: (608)754-7160; Fax (608)754-3937

Merck Ltd., BDH Laboratory Supplies
Hunter Boulevard, Magua Park, Lutterworth, Leics. LE17 4XN, England
01455 558 600

Microgon Inc.
23022 Lacadena, Suite 100, Laguna Hills, CA 92653.
(714) 581 3880; Fax: (714) 855-6120

Millipore Corp.
80 Ashby Rd., Bedford, MA 01730.
(617) 275-9200; Fax: (617) 275-5550
Millipore Ltd.
The Boulevard, Blackmoor Lane, Watford, Herts. WD1 8YW, England.
01923 816375, Fax: 01923 816 128

Nalge Company
P.O. Box 20365; Rochester, NY 14602-0365
(716)586-8800; Fax (716)586-8800

National Diagnostics, Inc
305 Patton Drive
Atlanta GA 30336
USA
(404) 699-2121; (800) 526-3867; fax: (404) 699-2077

New England Nuclear
See DuPont.

Nordic Immunological Laboratories
Drawer 2517
Capistrano Beach, CA 92624-0517
USA
(800) 554-6655; (714) 498-4467; fax: (714) 361-0138

Nycomed Pharma. AS
PO Box 5019, Majorstua N-0301, Oslo, Norway.
+47 22 96 36 36

Oncogene Science Inc.
160 Charles Lindbergh Boulevard., Uniondale, NY 11553
(516)222-0023; fax (516)222-0114
80 Rogers Street
Cambridge, MA. 02142
(617) 492-7289; (800) 662-2616; fax (617) 492-3967; (800) 828-4871

Oster Corporation, USA
Cartmel Drive, Harlescott, Shrewsbury, Shropshire, England
01142-231632; fax 01142-52111

Oxoid UK, see **Unipath.**

Oxoid USA Inc.
P.O. Box 691, Ogdensburg, NY 13669, USA.
(800) 567-8378

Paragon Biotechnology Inc.
Hopkins-Bayview Alpha Center, 5220 Eastern Avenue
Baltimore, MD (410) 550-2919.

Paragon Razor Co.
Nursey Works, Little London Road, Sheffield, S8 0UJ, England
01142-551063; Fax 01142-586738

Peprotech
10 Fitzgeorge Avenue
London W14 0SN
0171 603 8288

Perkin-Elmer Cetus
761 Main Avenue
Norwalk CT 06859
USA
(800) 762-4000; (203) 762-1000; fax: (203) 762-6000

Pharmacia LKB
S-751 82, Uppsala, Sweden

Pharmacia Biotech Ltd.
Davy Avenue, P.O. Box 100
Knowhill, Milton Keynes, MK5 8PB, England
01908 661 101; fax 01908 690 001

Promega
Delta House, Enterprise Road, Chilworth Research Centre
Southampton SO16 7NS, England
0800 181 037

Propper; see **Chance Propper Ltd.**

Qiagen GmbH
Max-Volmer Strasse 4
4010 Hilden, Germany
(2103)892-230; fax (2103)892-222

Qiagen Inc.
9600 De Soto Avenue
Chatsworth, CA 91311
(800)426-8157; fax (800)718-2056

Qiagen Ltd.
Unit 1, Tilligbourne Court, Dorking Business Park
Dorking, Surrey, RH4 1HJ, England
44 306 740444; fax 44 306 875885.

R&D Systems Europe
4-10 The Quadrant, Barton Lane
Abingdon, Oxon. OX14 3Y5, England
01235 529449; fax 01235 533 420

Research Genetics
2130 Memorial Parkway South
Huntsville, AL 35801
(800) 533--4363; fax: (205) 536-9016

Rhône-Poulenc
Usines Silicones, BP 22
69191 Saint Fons Cedex, France
33 72 73 74 75.

Rocket of London
Imperial Way
Watford, WD2 4XX, England
0923 239791; fax 0923 230212

Sandoz Pharma: Dr. D. Roma/Dr. E. Rissi
Pharma Division
Preclinical Research
Sandoz, Ltd
CH-4002, Bask
Switzerland 061 241111

Schuco International
Woodhouse Road
London N12 0NE, England
0181 368 1642; fax 0181 361 3761

Sigma Chemical Co.
3050 Spruce Street
St Louis, MI 63103
(314)771 5750, (800)325-3010; fax (314)771-5757; (800)325 5052

Sigma Chemical Co. Ltd.
Fancy Road, Poole
Dorset BH17 7NH, England
44 0202 733114; fax 44 0800 378785

Sigma Chemie
Postfach 260
9470 Buchs SG, Switzerland
+41 81 755 2721; fax +41 81 756 7420

Sigma-Aldrich Pty. Ltd.
Unit 2, 10 Anella Avenue,
Castle Hill, NSW 2154, Australia
(02)899-9977 fax (02)899-9742

R&L Slaughter
Units 11 & 12
Upminster Trading Park
Warley Street
Upminster RM14 3PJ
England
01708 227 140; fax: 01708 228 728

Southern Biotechnology Associates, Inc.
P.O. Box 26221
Birmingham, AL 35226
(205)945-1774; fax (205)945-8768

Southern Syringe Services Ltd.
New Universal House, 303 Chase Road
Southgate, London N14 6JB, England
0181 822 1971

Stratagene
11099 North Torrey Pines Road
La Jolla, CA 92037
(619) 535-5462; fax: (619) 535 5430
Cambridge Innovation Centra
Cambridge Science Park
Milton Road
Cambridge, CB4 45F
England

Taconic Farms
273 Hover Avenue
Germantown, NY 12526
USA
(518) 537-6208; fax: (518) 537-7287

Tekmar Co.
P.O. Box 429576, Cincinnati, OH 45242-9576
(513) 247-7000; (800) 543-4461; Fax: (513) 247-7050

TCS Biologicals, Ltd
Botolph Claydon, Buckingham MK18 2LR, England.
0129 671 4071/4072; Fax: 0129 671 4806

Trace Biosciences
5/11 Packard Avenue
Castle Hill, NSW 2154, Australia
(2)899-1056; fax (2)899-1260

Unipath
Wade Road, Basingstoke, Hants. RG24 0PW, England.
01256 841 144; fax: 01256 463 388

United States Biochemical Corporation
PO Box 22400
Cleveland, Ohio 44122
(216)765-5000; fax (216)464-5075

Upstate Biotechnology Inc.
199 Saranac Avenue, Lake Placid, New York 12946
Fax (617)890-7738; (800)233-3991

Vector Laboratories, Inc.
30 Ingold Road
Burlinghame, CA 94010
(415) 697-3600; fax: (415) 697-0339
16 Wulfrie Laboratories, Bretton
Peterborough, PE3 8RF
England
01733 265 530

Ventrex Laboratories Inc.
217 Read Street
Portland, ME 04103
Fax (207)775-4734

VWR Scientific
Domestic: (800)932-5000
International: (908)757-4045; Fax (908)757-0313

Watson Marlow
Falmouth, Cornwall, TR11 4RU, England
0326 373461; fax 0326373461

Worthington Biochemical Corporation
Hall Mill Road, Freehold, NJ 07728
(908)462-3838; (800)445-9603; fax (908)308-4453; (800)368-3108

U.K. agent: **Lorne Laboratories Ltd.**
7 Tavistock Estate, Ruscombe Business Park
Twyford, Reading RG10 9NJ, England
01734 342 400; fax 01734 342 788

Index